Neuroscience for Clinicians

C. Alexander Simpkins
Annellen M. Simpkins

Neuroscience for Clinicians

Evidence, Models, and Practice

 Springer

C. Alexander Simpkins
Private Practice
San Diego, CA
USA

Annellen M. Simpkins
Private Practice
San Diego, CA
USA

ISBN 978-1-4614-4841-9   ISBN 978-1-4614-4842-6   (eBook)
DOI 10.1007/978-1-4614-4842-6
Springer New York Heidelberg Dordrecht London

Library of Congress Control Number: 2012944960

Printed on acid-free paper

Springer is part of Springer Science+Business Media (www.springer.com)

Contents

Introduction

Psychotherapy has traditionally addressed the mind, with its thoughts and feelings. Some therapies work with the body as well, recognizing body health can enhance mental health. But recent neuroscience research has revealed a new dimension to add to your treatments: the brain and its wider extension into the nervous system. We invite you to *think brain*, to view your clients through the exciting new lens of neuroscience that will focus your therapeutic sensitivities on a vast new reservoir of discoveries and potentials. As you learn how to initiate brain change for the better, you can add new dimensions to your treatments.

The Brain–Mind–Body–Environment Network

How does the brain relate to psychological experiencing? Consider the mind–brain–body–environment interaction as a network. Some of the activity is localized in a specific area, but its effects ripple through a multidimensional system in multiple ways.

How does this happen? Imagine a net, stretching all around the room where you are sitting now as you read this book. Now visualize a crystal-clear jewel glittering at every juncture in the net. Each jewel reflects all the other jewels and this reflecting extends infinitely out into the world in every direction. This beautiful image comes from the ancient Buddhist *Flower Garland Sutra, Avatamsaka Sutra*. The mind, brain, body, and environment function in a network, much like Indra's all encompassing, ever interacting net.

You no longer have to choose whether a problem is psychological, biological, or environmental, because in a sense, every problem includes them all to some degree in an interacting network. Psychological disorders can be approached from all of these perspectives, depending on the particular person, situation, and problem. As the metaphor of Indra's Net suggests, the different aspects interacting together bring about the phenomena. Treatment can begin at any jewel in the network and work from there. As you integrate this fuller perspective into your treatments, you add new dimensions to your work.

Therapy occurs in multiple dimensions that include the brain along with the mind, body, and social environment. Rather than simply reframing in two-dimensions as we are accustomed to doing, we invite you to think more broadly, to envision a multi-dimensional network.

Neuroplasticity

The brain can change the mind, but the mind and experience can also bring literal structural changes to the brain. Recent brain research reveals that the brain is continually changing. Neuroplasticity takes place all through life. This is extremely hopeful for therapy. You will discover where there is malleability in the mind–brain–body–environment system and how to use it therapeutically. You can make your interventions at most of these different levels: sensory, motor, cognitive, and emotional, altering the brain's networks in positive ways, and Neuroscience for Clinicians will show you how to do so.

This book will give you some of the latest models, methods, techniques, and findings from neuroscience. As a trained, sensitive therapist, you are in a unique position to use knowledge about the brain, and seamlessly apply your therapeutic sensitivities for facilitating changes. Neuroscience researchers do not always have the clinical experience that you do, and so they may not see the immediate application for treatment that you will be able to develop, once you understand the workings of the nervous system.

How to Use This Book

This book's purpose is three-fold. First, we provide a clear, and we hope illuminative guide to the brain, that anyone, even someone with no previous neuroscience training, can understand. The information we have chosen to include will be clinically relevant, and will help you know what to look for if you are interested in digging deeper. Second, we offer techniques and methods with specific exercises to follow that allow you to integrate the new brain science into your treatments. And third, we provide second-order learning, so that you will learn how to learn about the brain in terms of treatments for applying in therapy. We guide in developing links between brain science and therapy so that you will continue to integrate the new research and discoveries of the future into your work. It is our sincere hope that you will build on the suggestions, methods, and exercises we offer, to make neuroscience a helpful part of your therapeutic approach. Please feel free to adapt these insights and exercises, and evolve your own.

Every chapter provides information and research findings clearly stated along with sections on how to apply these findings to clinical practice. We offer exercises and techniques that are immediately useful. Please feel free to vary the exercises to fit your own style and the unique individuality of your clients.

Organization of the Book

Part I: Preliminaries begins with the fascinating history of neuroscience with many of the key innovators and their pivotal discoveries. Chapter 2 introduces the language of neuroscience so that everyone can begin with a shared lexicon for clarity and better understanding. Chapter 3 will stimulate thought about the relationship between brain and mind. There are no simple answers, and we invite you to think deeply as your perspective will stimulate ideas for your treatments.

Part II: Neuroscience Methods describes some of the methods used in neuroscience that help to better understand the brain. Chapter 4 shows how neuroscientists have made some of their discoveries about how the brain works at its best by studying what functions a brain damaged individual lacks. Chapter 5 explains what neuroimaging is really telling us. The brain pictures you see are inferences, not direct photographs. And so, the computational modeling methods in Chap. 6 show you how the neuroimaging data is being interpreted.

Part III takes you from the simple units of the nervous system, the neuron, and neurotransmitters (Chap. 7), to structures (Chap. 8), pathways (Chap. 9), and neural networks (Chap. 10). With these understandings, you will have a fundamental knowledge of the key systems to work with in therapy.

Part IV deals with the different ways the brain changes through time. Brain change is more pervasive than previously believed, and you can harness these capacities in your treatments. Chapter 11 gives a glimpse into how the brain has evolved over millennia. Human brains share much in common with the lower animals, and so you learn about the brain by studying its evolution. Chapter 12 deals with developmental change, which traces the development of the brain from conception and then provides pivotal developmental points, for working therapeutically, including early attachment, adolescence, and old age. But therapists know that clients wish for change at various periods of their lives, and so Chap. 13 on neuroplasticity and neurogenesis at any time, offers ways the brain can change moment-by-moment. Here we find some of the most exciting discoveries that offer promise for the powerful effects that psychotherapy can have on mind–brain change.

Clinicians want to implement their understandings of the brain. Part V shows how to enhance some of the primary brain functions to assist you in your treatments. Chapter 14 outlines some general principles to guide you in incorporating the brain into your work. Then, Chap. 15 shows how to view mental functions through the lens of brain science and apply it to add power to your techniques when working with attention, Chap. 16, emotions, Chap. 17, memory, and Chap. 18, through the interpersonal perspective.

Part VI gives specific ways to shift the imbalances of the nervous system that are found in common mental disorders. We now have an ever-growing body of research data about how psychological problems alter the brain in characteristic patterns.

And we know that many of these patterns are reversible by psychotherapeutic methods. Chapter 19 describes how the typically used forms of psychotherapy alter the brain in different ways. Chapter 20 shows how anxiety changes the brain, followed by some methods you can use to restore healthy balance. Chapter 21 covers problems with moods, and Chap. 22 deals with substance abuse. Your therapeutic interventions can alter the brain for healthier functioning, and these chapters open the way.

Part I
Preliminaries

What makes people do, think, and feel what they do? Humanity has asked these questions for millennia. Neuroscience believes many of the answers are to be found in the brain. Studying the brain will unravel the mysteries of human behavior, cognition, and emotion. The ever-growing body of neuroscience research sheds light on some of humanities most daunting problems such as mental illness adding evidence to our psychological theories. And with this new information, we will be able to facilitate positive potentials for learning, creativity, and health.

The quest is not an easy one. In the past, the brain has been an impenetrable black box. Past generations have struggled with how to investigate the living brain, and many ingenious methods were created. Chapter 1 traces the history of the discovery process, beginning with the ancients and traced through to modern times.

Chapter 2 provides the tools you need to navigate through the new brain science. Often knowledge of terminology is taken for granted. And so, you may have found the use of complex lexicons obscures rather than reveals. This chapter will give you what you need to be able to navigate like a pro, to understand modern studies, and communicate clearly with others about neuroscience.

Before you move forward in using the new brain science in your practice, it is important to consider some of the questions that it raises about the relationship between mind and brain. Is the mind simply the brain? Or is it a function of the brain? Or perhaps mind and brain are two separate things? Answers to these questions effect how you will utilize the findings from neuroscience in your own therapy. Chapter 3 discusses these thought—provoking issues and invites you into the ongoing discussion.

Chapter 1
The Birth of a New Science

We arise from our past, and can be enriched by engaging with it. Neuroscience is a relatively new area of study, and yet interest in the brain stretches back through the ages. Human beings have always been curious to understand the mind and how it relates to the body and the brain. Many contemporary neuroscience issues have ancient roots, as people continue to engage in the same questions. Diverse cultures and peoples contributed to what we know today. We offer you some of the highlights, with the understanding that this fascinating field is continuing to be explored and expanded. We hope the story of early discoveries will deepen and broaden your understanding of brain structure and how it is involved in mental functioning.

From Divine Reign to Brain

In ancient times, people believed that their behaviors, emotions, and cognitions were controlled by divine beings. But through centuries of objective observations, theories began to change as some people recognized that the brain seemed to be involved in regulating thoughts, feelings, and actions, rather than just divine will. They came to understand the structures of the human brain and learned how it relates to the mind and body.

Ancient Egypt

The first known written reference to the brain is found in an Egyptian papyrus known as the Edwin Smith Surgical papyrus, dated around 1600 B.C., devoted to treatment for trauma from battle. It provides objective observations of the brain's

C. A. Simpkins and A. M. Simpkins, *Neuroscience for Clinicians,*
DOI: 10.1007/978-1-4614-4842-6_1,
© Springer Science+Business Media New York 2013

Fig. 1.1 The Edwin Smith
Papyrus

anatomy, including the cranial surfaces and the protective coverings of the central nervous system, the meninges. The papyrus also described the flow and pressure of blood and how blood moves through the brain (Fig. 1.1).

Later in Egypt, around the fourth to third centuries B.C., the rulers encouraged dissections of convicted criminals. By careful observation, their scientists recognized that the nervous system is interconnected through the whole body, with sensory nerves that are ascending to the brain and motor nerves that are descending from the brain. They also saw cavities running through the skull and developed a ventricular localization model. The cavities were thought to be the site for mental and spiritual processes. This theory would be developed further in Roman times and accepted through the Middle Ages in Europe.

Ancient China

The Chinese conceived of the mind and brain as flowing patterns of interaction. The Chinese ideogram for the word *brain* exemplifies this Chinese theory. The word for brain is an ideogram with three sections. Each of these sections depends on the others for the meaning. The three arrows pointing to the left signify fire or stream when they are combined with the other elements. This part of the ideogram corresponds to flow of thought or stream of consciousness. The slender, ladder-like character when used in combination means meat, flesh, or body tissue. The square with an x within means intelligence. When paired with the symbols for sun and moon this symbol can indicate intelligence that shines like the sun and the moon. Together, we see a view of the qualities of mind and brain, structure and

Fig. 1.2 Brainideogram
(Credit line: Halfhill 2008)

function working together as the source for intelligence, ever flowing, both material substance (body) and consciousness (Fig. 1.2).

Ancient Chinese medicine is not antithetical to the modern idea of embodied cognition. The flow of chi, the energy that comes from the movement of opposites Yin and Yang in the universe and the five elements, are also embodied in the brain-body system. The brain does not function outside of these forces but is part of it. The functions of the brain are distributed among the five organs: heart, lung, liver, spleen, and kidney. The famous Yin-Yang symbol expresses this interaction of opposing forces always at play between them. Thus, any disorder of the brain, mind, or body, is not characterized in just one organ but is always understood as a dynamic functional system, and not a fixed disorder or isolated disease. Our modern models of the brain are coming to recognize how interconnected and interactive the different brain structures are. We can also represent the on–off signals between neurons as Yin-Yang opposites. We develop the parallels in our recent book, *The Dao of Neuroscience* (Simpkins and Simpkins 2010).

Ancient Greece and Rome

Much of the greatness of Greek thought rests on the philosophical wisdom of eminent thinkers such as Socrates (470–399) who initiated active discussions to probe deeper into the nature of the world and of human beings. Through careful observation, exacting dialogues, and clear distinctions, a scientific method began to evolve (Fig. 1.3).

Fig. 1.3 Socrates

Plato (420–347)

Plato offered one of the early models of mind that has filtered into prominent contemporary brain theories. In Book Four of Plato's *Republic*, (Plato 1997) Socrates investigated the nature of the soul. He developed a view of the soul as having three parts: appetitive, spirited, and rational. The appetitive part related to our lower desires such as hunger, thirst, and sex. The spirited was the emotional part. The rational part was the seat of thinking that makes judgments and decisions. In Plato's paradigm, reason should rule over passions and appetites to make a well-ordered soul. Freud's ego, superego, and id model may well spring from Plato's model (Fig. 1.4).

The brain has been conceptualized similarly, as an evolutionary triune model. In the 1960s, the American physician and neuroscientist, Paul D. MacLean (1913–2007), proposed that in the course of evolution, three distinct brains emerged, one after the other, that now exist together in the human brain (MacLean 1967). First was the reptilian brain that includes the upper brainstem, cerebellum, and thalamus, governing instincts and fundamental vital functions such as heart rate, breathing, body temperature, and balance, much like Plato's appetitive soul. Developing out of the brainstem in the first mammals came the limbic brain, known as the paleo-mammalian complex, which includes the hippocampus and amygdala. These structures regulate emotion, similar to the function of Plato's spirited soul. Finally,

Fig. 1.4 Plato

the neomammalian structure known as the neocortex is parallel to Plato's rational soul. It appeared most prominently in primates and developed fully in the human brain.

Aristotle (384–322 B.C.)

The great philosopher and scientist Aristotle found the seat of mental and emotional functioning in the heart. The brain, which was cold and bloodless, was merely a radiator to cool down the passions of the heart. History has criticized Aristotle's idea, and of course, everyone knows that it is false. But newer findings suggest a hint of truth in this view. The contemporary polyvagal theory (Porges 2011) has uncovered a heart-mind, that functions through the 10th cranial nerve, known as the vagus nerve. Brain, heart, and breath function together as part of cognition and emotion involved in interpersonal relationships, trust, and intimacy (Fig. 1.5).

In addition, every cognitive therapist knows that rational, clear thinking can serve to cool the passions of a disturbed client. Cognitive therapies such as CBT and REBT have been shown to activate the frontal lobes of the cortex, confirming Aristotle's idea that the brain can cool the passions. For example, during reappraisal tasks, one of the key methods used in cognitive therapies, activity in specific areas of the anterior cingulate and frontal lobes increases while activity in the amygdala, correlating with emotions, decreases (Banks et al. 2007).

Fig. 1.5 Aristotle

Today we have a more integrative perspective on how many systems of the brain function with the nervous system, and other body systems such as the heart and the breath to correlate with our thinking–feeling–sensing experience.

Hippocrates (460–380 B.C.)

Hippocrates is often considered the father of Western medicine. He was influenced by Socrates, which may account for his clear thinking approach. He believed that intelligence comes from the brain. He held the opinion that the brain exercises the greatest power over us, as the seat of our intelligent functioning (Adams 1886). He was one of the first to state that thoughts, sensations, emotions, and behaviors are guided through the brain. He saw the brain, mind, and body as a unity and developed treatments that addressed the whole person rather than just its parts. Illness was an imbalance of four humors, or fluids: blood, black bile, yellow bile, and phlegm. Therapy helped to restore this balance (Fig. 1.6).

Hippocrates observed that injury to one side of the brain correlated with paralysis on the opposite side of the body. He also clearly described the symptoms of epilepsy in his book on the topic titled, *On the Sacred Disease* (Adams 1886). Epilepsy is one of the brain disorders widely addressed by modern neuroscientists. During his time, people believed epilepsy was an act of God, but Hippocrates looked at it scientifically, recognizing that the brain was malfunctioning. "It is thus with regard to the disease called Sacred: it appears to me to be nowise more divine nor more sacred than other diseases, but has a natural cause like other affections" (Zanchin 1992, p. 94).

Fig. 1.6 Hippocrates

Galen (130–201 A. D.)

Galen was a Roman physician who had one of the longest lasting influences on the development of modern medicine and our understanding of the nervous system. He did many anatomical experiments that helped to identify major brain structures. During his time, it was illegal to perform autopsies. And so, Galen did what many modern neuroscientists do today, he dissected animals. He also studied brain-damaged individuals, mainly the Roman gladiators who received many types of brain injuries from their battles. By observing the effects of their injuries, he was able to learn about how the brain might be involved in functioning (Fig. 1.7).

Galen developed a theory based on an earlier model that mental and spiritual functions and processes were localized in the ventricles, running through cavities or cells in the skull where we now know blood flows. This theory became known as the Cell Doctrine. Galen proposed that the brain was a large clot of phlegm. A pumping system pushed "psychic pneuma" through the nerves. His theory about the structure of the brain partakes of the contemporary hydraulic aqueduct system that ran through the Roman cities. All functioning of the brain and other body systems depended on a balance of these fluids, which he referred to as psychic gases and humors and gave rise to mental functioning. He assigned each of the mental functions described by Aristotle (memory, attention, imagination, reason, and sensing) a specific location in the ventricles. Cognitive functions occurred from front to back in the ventricles. For example, the senses, connected to common sense, were located in the first ventricle.

Fig. 1.7 Galen

Change Comes from Evidence and Discovery

Galen's models of brain structure and function were accepted for more than 1200 years. Eventually this theory proved to be incorrect, but it would need strong empirical evidence to dismantle the long years of tradition. Talented individuals boldly questioned the status quo to open a new era for neuroscience.

Andreas Vesalius (1514–1564)

Andreas Vesalius had a unique pair of skills: He was both a talented artist and a skilled surgeon. He combined his unusual artistic talent with his ability to make careful observations of brain structures when he performed surgery. Vesalius performed dissections in front of groups of medical students who watched from surrounding circular balconies. He found no holes that corresponded to Galen's theory, and so began to question its correctness. His careful observations led him to state that Galen was wrong about the relationship between brain fluids and functions. Vesalius produced exquisitely detailed drawings of human anatomy, including the brain, in his famous book, *On the Fabric of the Human Body* (*De Humani corporis fabrica*) (Vesalius 1998–2009). To this day, Vesalius is often called the founder of modern human anatomy. Although he had some errors in the convolutions of the brain, his pictures have a detailed realism that helped to advance our understanding of the brain and body (Fig. 1.8).

Fig. 1.8 Vesalius Drawing

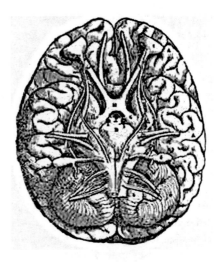

Fig. 1.9 Descartes

Rene Descartes (1596–1650)

During this period, people were searching for a single source for the mind. The Church firmly held that God directed the animal spirits. But a mechanistic vision of the human body was gaining in popularity. The famous philosopher, Rene Descartes, is well known for having separated the mind from the body, where the mind was spirit and the body a mechanical device. Reflexes responded along the nerves like cables, which ran through the body to the brain (Fig. 1.9).

Descartes explained that the mind and brain intersected in a single structure: the pineal gland, a small endocrine gland located in the center of the brain. He picked

Fig. 1.10 Descartes' pineal
gland diagram

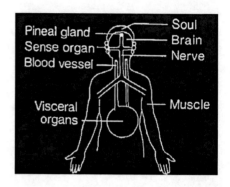

the pineal gland as the logical place for the mind, since it was the only single structure, not doubled in the two hemispheres. Thus, the non-material spirit expresses itself in the mechanistic, material body.

The two separate dimensions: mind and body, can influence each other both ways—mind over body or body over mind. The animal spirits flow from the pineal gland, stimulating automatic reflex actions. When we do a voluntary action, mind influences the animal spirits. Although Descartes was wrong about the pineal gland and the separation of mind from body, he did introduce the idea about how the mind, body, and brain are engaged in an interacting system. Pavlov also credits him with being the first to conceive of reflexes, for which Pavlov developed his famous theory of mechanism (Fig. 1.10).

The New Electricity in the Eighteenth Century

The Enlightenment brought a turn toward intellectual freedom and empirical verification. Great scientific progress was made during this exciting period. The refreshing new world view brought great scientific progress and deep cultural enrichment. One of the new scientific discoveries, electricity, illuminated the understanding of how the brain and nervous system worked.

Luigi Galvani (1729–1798)

Luigi Galvani was an Italian physician who made one of the early discoveries about the electrical properties of the nervous system. Like many great discoveries, it happened by accident. Galvani and his assistant were performing experiments with static electricity on frog skin. The assistant accidentally touched his metal scalpel to an exposed nerve on a dead frog's leg. Much to their amazement, sparks flew and the leg kicked as if it were alive. Galvani suddenly realized that electrical energy was the stimulus driving the movement of muscles. He named this force *animal electricity* and believed a fluid that moved through the nerves carried it.

Fig. 1.11 Galvani
experiment

Galvani's Experiment

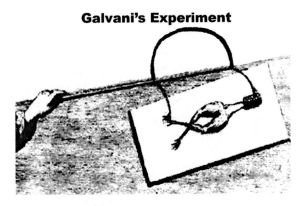

Another contemporary of the time, Alessandro Volta (1745–1842), disagreed with Galvani who believed the force could not be separated from the muscle. In order to prove his colleague wrong, Volta invented the first battery, which generated electricity independently of a body. So, in a sense we can thank neuroscience for paving the way for our electrical age (Fig. 1.11).

Johannes Peter Muller (1801–1858)

Johannes Muller was a German physiologist who made important contributions by bringing together many different disciplines for the study of physiology. His book, *Handbuch der Physiologie des Menschen, Elements of Physiology*, (Muller 2003) introduced his groundbreaking principles of the interaction between sensing and movement. He believed that the different kinds of perceptions we have, do not depend on the stimuli so much as on the sensory pathway over which the information is carried through the nervous system. Known as the law of specific nerve energies, this theory explained how all our senses respond to electrical stimulation, but that they react differently. One kind of nerve perceives light, another sound, and another smell. He claimed that sensation is not the state of the external stimuli but rather the state of our nervous system as it is excited by something external.

Roger Sperry would develop Mueller's idea even further and win a Nobel Prize for it in 1961. He showed that what determined the specificity of response in perceptual systems depended on the nerve's location in the brain. In amphibians, the optic nerve on the right side of the brain responds completely to stimuli to the left eye (In humans this crossing is only partial). He cut optic nerves on the right and caused them to re-grow on the left side making the left side of the brain connected to the left eye and the right side connected to the right eye. After the operation, amphibians would move to the right to avoid a large object, whereas before the operation they would move to the left in the same situation (Sperry 1945). This change proved that the nerve's location in the brain made a difference in how the animal saw the world.

The Great Neuroscience Debate: Localization, Holism, and Their Combinations

Human beings always find different ways to interpret discoveries. As more was learned about the nervous system, people began to form differing opinions about whether functions could be located in a specific area of the brain, or whether a unified system was involved. Descartes had made a strong argument that the brain acted like a machine with many parts, but that the rational faculties had to be unitary and global, and could not reside in the brain with its dual hemisphere structures. Not until Franz Joseph Gall's popularizing of phrenology did people begin to entertain the possibility that mental capacities could be localized in specific areas of the brain. To this day, the debate between globalization and localization of structure and function continues to have advocates on both sides. Most recent evidence seems to point to both being involved in a complementarity way, as this book will explain.

Champion for Localization of Structures to Functions: Franz Joseph Gall (1758–1828)

Gall was a controversial physiologist who created a system of matching brain structures to functions, known as phrenology (Gall and Hufeland 2010). The assumption of phrenology was that specific mental functions were located in certain brain areas. Personality, emotions, and cognitions were all thought to emerge from a localized brain area. Gall believed the shape of the head reflected the shape of the brain. Therefore, the mind could best be understood through study of brain size, shape, and weight, rather than philosophy or religion. He based the system on what he considered carefully performed experiments that showed a direct relationship between mental "faculties" and brain "organs." There were 27 brain organs, each associated with a mental faculty. These organs corresponded to bumps on the skull, and so a large bump in a certain area meant that a person had a particular personality quality. The popular culture embraced phrenology. Thousands of people consulted their local phrenologist to learn about their personality strengths and how to develop them (Fig. 1.12).

The academic and scientific community would call into question the research on phrenology, and the theory would prove to be incorrect. We now know, for example, that a larger brain is not necessarily a smarter one. After all, dolphin brains are much larger than humans, and yet, they lack the large cerebral cortex that gives people unique abilities such as language. Despite the rejection of phrenology per se, the approach did inspire further research into the possibility of brain to mind localization.

Fig. 1.12 Phrenology map

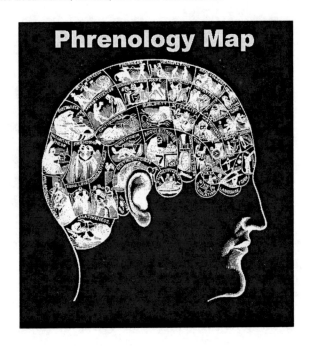

Equipotentiality and Holism of Jean Pierre Flourens (1795–1867) Challenges Phrenology

One of the people responsible for disproving phrenology was Jean Pierre Flourens, an eminent, highly respected French physiologist and university professor. He held many high positions and was credited with doing pioneering work in the use of chloroform for anesthesia (Fig. 1.13).

Flourens refined a surgical method of ablation, which removes small slices of the brain in animals. He worked mainly with birds and rabbits. In his experiments, he carefully observed what functions were lost following recovery, thereby mapping brain structures and functions. His purpose was to test Gall's claims that certain brain areas were involved in specific functions. One of his early ablations was of the cerebellum, which Gall claimed to be the "organ of amativeness." Flourens found that when the cerebellum on both sides of the brain was severed, the animal was unable to coordinate its movements and lost the ability to balance. Ablations of the brainstem, particularly the medulla oblongata, caused the animal to die. When he cut the cortex, the animal lacked motivation, perception, and mobility. He concluded that functions correlate with large brain areas, but he could not find local distinctions. The cortex regulates higher order functioning, the cerebellum is involved in movement control, and the brainstem corresponds to life-giving functions like breathing and circulation. He also found that the cortex was equipotential because intact areas close to an ablation would take over the

Fig. 1.13 Flourens

ablated area's functions. He became a spokesperson for a global, interactive, and plastic correspondence between brain areas and functions that offered strong evidence against Gall's localization theory.

A More Sophisticated Conception of Localization

Paul Broca (1825–1880) is credited with discovering a localized area for language in the left hemisphere. But if it had not for certain events, we might never have thought of the language area as *Broca's area*. This story shows how the play of human personalities often motivates the world's greatest discoveries! (Fig. 1.14).

The ideas from phrenology of localization continued to be championed by physicians such as Jean Baptiste Bouillaud (1796–1881) who had patients with specific aphasias from particular injuries to the brain. Bouillaud's son-in law, Ernest Aubertin (1825–1895) took up his father-in-law's cause for the localization perspective with the Society of Anthropology. Broca, its founder, was a young surgeon, interested in physical anthropology. He sought to compare the head size of different cultures and races. Aubertin claimed to have a patient who had lost his ability to speak and yet understood what was being said. The patient was likely to die soon. He offered this challenge to the Society: If an autopsy did not show damage to a specific area in the frontal lobe, he would renounce his view. Broca agreed to see the patient and perform the autopsy. The man could only utter the word "Tan" which became the name by which he is remembered. The only other word he uttered was the curse, "Sacre nom de Dieu." And yet, he knew that he had been hospitalized for 21 years, and could indicate it by holding up the fingers of his hand. He seemed to have a sense of humor and showed through his movements that he had retained his ability to think and understand. Broca performed the autopsy and indeed, did discover a damaged oval-shaped section in the left frontal lobe. A few months later, Broca had a second patient with the same symptoms whose autopsy revealed damage in the identical area. This area became known as

Fig. 1.14 Broca and Wernike's areas

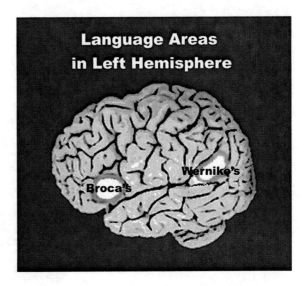

Broca's area and the inability to speak that is associated with this damage was named "aphasia," after Plato's term for being at a loss for words (Fancher 1979).

Another pivotal bit of evidence for localization of function came from Germany around the same time. Two young physicians, Gustav Fritsch (1838–1927) and Edvard Hitzig (1838–1907) found that stimulating a certain strip of area in the back of the frontal lobe of a dog produced movements. Different areas of the strip produced movement in different parts of the body. Their research showed that the cortex could be activated by electricity and that there was a topographical representation of the body for movement in the brain. This famous study provided new evidence for localization of functions in the cortex.

Surgical methods became more refined and other discoveries were made. In Scotland, David Ferrier (1843–1928) found that stimulating an area in the occipital lobes of monkeys made their eyes roll back and forth as if they were seeing things. And when he cut this area, the monkeys became blind. He also found that stimulating the motor strip in the cortex did not produce sensations, but that a strip further back in the brain, now known as the sensory strip in the parietal lobe, did.

By 1874, a new conception of the cortex delineated centers devoted to movement, sensation, seeing, and hearing. The theory was that we take in information and store it in what were called *association areas* located close by to the particular center. Carl Wernicke (1848–1905) developed this theory with aphasias. He pointed out that Broca's area was located at the seat of the motor strip, calling it a motor aphasia because the individual lost the ability to speak, a motor activity. He discovered another kind of language disturbance, where individual retained the ability to say words, but lost the capacity to comprehend meaning. These individuals had damage to a different area, known now as *Wernicke's area*, located in the temporal lobe, just below the sensory strip. These patients hear what is said,

Fig. 1.15 Lashley

but cannot respond meaningfully. Wernicke distinguished Broca's motor aphasia from his newly discovered sensory aphasia.

But localization would not prove to be that simple. Both Broca's and Wernicke's areas were on the left side of the brain. However, babies born with damage to these areas in the left side of the brain developed normal speaking and comprehending abilities. The brain seems capable of plasticity, utilizing other areas to do the job. In addition, these children showed less spatial awareness, possibly because the rewiring left less cortex area available for typical right-hemisphere skills (Fig. 1.15).

The pendulum swung back toward holism following the experiments of the behaviorist Karl Spencer Lashley (1890–1959), one of the early neuropsychologists, and his student, John Watson. If the association areas theory was correct, memory should be stored specifically in a certain area. Lashley did a large research project to see what would happen if he ablated a specific association area. According to the theory, ablating a specific association area not correlated with movement should not disturb the animal's motor memory of the maze. But instead, he found that ablating many areas interfered with the animal's performance. He claimed that these results proved that memory is not localized, but tends to be more global. Today we know that memory is not just one thing. There is short-term and long-term memory, and different areas are involved. The later chapters will describe these discoveries.

Fig. 1.16 Huglings Jackson

The Integrative Approaches of Hughlings Jackson and Alexander Luria

Hughlings Jackson (1835–1911) was a British neurologist who worked with epilepsy. He proposed that people should think of higher level mental functions not as just one process but rather as a combination of simpler skills integrating together. So, speech involves hearing, discriminating sounds, motor movement of the mouth and tongue, along with complex higher cognitive skills. When someone has brain damage causing speech deficits, it can be due to damage to one or several of these simpler skills. Thus, behavior should be understood as an interaction of many areas of the brain, an equipotential and holistic view. But at the same time, each area has its specific functions as well, a localization of functions. Jackson's theory allows for a more complete understanding of how the mind and brain work both locally and holistically, a position we developed further in *The Dao of Neuroscience* (Simpkins and Simpkins 2010) (Figs. 1.16 and 17).

Alexander Luria (1902–1977) was a Russian neuropsychologist who had a profound influence on how we view the mind and brain. He held degrees in education, medicine, and psychology, giving him a broad background. He was also strongly influenced by Lev Vygotsky (1896–1934), the founder of an integrative cultural–historical psychology. He studied the full range of human capacities, from the psychological deficits that develop in undereducated minority groups and brain damaged individuals with aphasias, to an extraordinary individual with unlimited memory ability. He developed an integrative theory of the brain, compatible with Jackson's view. The central nervous system consists of three functional units: the brainstem for regulating arousal and muscle tone, the posterior (back) of the cortex for processing sensory information, and the frontal areas of the cortex for executive functions. Whatever we do engages all three of these units together as a unity, and yet each unit has its own unique functions. So, in Luria's theory, as in

Fig. 1.17 Luria

Jackson's, we see an integration of localization with holism. Luria offered a pluripotentiality model, since any one area of the brain could be involved in many or few behaviors. He observed that higher level brain skills could often compensate for lower skills lost from injury. For example, a young child had surgery to completely remove the left hemisphere. You would expect that this child would have tremendous deficits in language and movement. However, at age seven, he spoke fluently and walked well. The brain had rewired, using the right hemisphere for language instead of the left (Zillmer, Spiers, and Culbertson 2008). By engaging alternative functional systems, lost skills can be reclaimed. This is a primary principle, which we develop in this book, to guide in enlisting alternative functions to regain lost abilities or foster new ones.

Modern Models: The Neuron Doctrine Versus the Reticular Theory

As we entered the twentieth century, researchers began putting together all that had been learned. In the early years of the new century, people believed that the elements of the nervous system took two forms: nerve fibers and nerve cells. The nerve fibers conducted the impulses and were the transmitters of automatic reflexes as well as sensations and perception. This view would prove to be incorrect with the advent of a new imaging technique created by Camillo Golgi (1843–1926) in 1873. Golgi's method used silver to stain cells (see Chap. 3). For the first time, researchers could literally *see* a fundamental neuronal unit. Now, great minds could begin to put together a more accurate model of how the elements of the brain

Fig. 1.18 Golgi

and nervous system working together. Interestingly, two great men would use this new staining method, but what they saw led them to opposite conclusions! (Fig. 1.18).

The Neuron Doctrine was a theory proposed and popularized by Santiago Ramon y Cajal (1852–1934), a Spanish pathologist and histologist, who many consider the father of modern neuroscience. Cajal's doctrine describes the structure and function of the whole nervous system. At around the same time, Golgi, the inventor of the silver staining method, proposed a very different approach to characterizing the whole nervous system. He called it the Reticular Theory (Fig. 1.19).

Cajal's Neuron Doctrine was based on the concept that neurons are distinct, discrete, metabolic units with cell bodies, axons, and dendrites. Although they interact with each other, neurons are separate. These anatomical units follow a Law of Dynamic Polarizing that governs how they communicate with each other, by sending electrical signals. Thus, each neuron is an information-processing unit that takes in signals, processes them, and sends another signal out. This theory emphasized the idea that the nervous system is made up of separate units interacting together through signals.

What Golgi saw when he used his staining method was a highly interconnected network, or as he called it, a diffuse nervous network. So, he developed a competing model of the nervous system. Although Golgi and Cajal shared a Nobel Prize in 1906, the two scientists became bitter rivals.

Golgi's fundamental disagreement was not so much over the structure of the neuron and the ways that neurons interact, but rather that the Neuron Doctrine was reductionist and atomistic. Golgi and others who held the Reticular Theory felt that the Neuron Doctrine overlooked the holistic sense of brain function. They believed that the entire nervous system including the brain, nerves, and spinal chord should be seen as a continuous reticulum with functions that result from a collective

Fig. 1.19 Cajal

action of the whole system, including the neurons. However, the validity of Golgi's theory was denied, and the Neuron Doctrine predominated. The Neuron Doctrine has continued to develop and evolve to this day (Cimino 1999). Readers will see many of the assumptions of this doctrine at work in the other chapters describing brain structure, function, and change.

Cajal predicted that neuroplasticity would be impossible in the adult brain. But a growing body of evidence shows the brain able to rewire in many ways. These findings have led to criticisms of the Neuron Doctrine for being too limited. As a result, holistic perspectives are being resurrected today, since they can be helpful in explaining the new discoveries in neuroplasticity. Broader models are under discussion (Guillery 2007), and new theories are being researched such as distributed cognition (Hutchins 1995). We encourage you to consider a kind of plurality of views, with the multipotentiality of brain tissue. Theories that allow for localization and holism working together can enhance the theoretical picture of the Neuron Doctrine, thereby adding to the understanding of our well-coordinated working nervous system.

The Discovery of Neurotransmitters

Most clinicians today know that neurotransmitters play an important role in many psychological problems. And although drug therapies may not have solved the problem of mental illness, they have offered many people some relief, especially when combined with psychotherapy. Neurotransmitters are one of the keys to how neurons send messages that tell another neuron to fire or not.

The Neuron Doctrine had explained that neurons were separate units. But if they were not touching, how could one neuron tell another neuron to fire or not? The space between them, known as a synapse, would seem to need some kind of vehicle, like a boat, to transport the electrical signal from one neuron to another (Fig. 1.20).

Fig. 1.20 Leowi

Neurotransmitters are chemicals that carry the message across the synapse, from one neuron to another. But this idea was not immediately understood. At first, people thought that the signals at the synapses were only electrical. Otto Loewi (1873–1961) earned a Nobel Prize in 1936 for a famous experiment that gave scientific evidence for the idea that communication at the synapse was not just electrical but also involved a chemical reaction. The story of how it happened is another fascinating example of unconscious intelligence.

Loewi had a strong belief that chemical reactions were involved in synaptic communications, but he could not think of any way to prove it. On Easter weekend in 1921 he dreamt of an experiment, awoke very excited, quickly scribbled down the idea, and went back to sleep. But when he woke up the next morning and tried to read his notes, he was unable to decipher his illegible handwriting. He spent what he called the worst day of his life trying to remember the experiment. He went to sleep that night discouraged and frustrated. But much to his surprise, he had the dream again! Thankful for the second chance, he woke up and immediately went to his laboratory in the middle of the night. The experiment he performed that night would forever change the world of neuroscience.

He placed two frog hearts in separate saline solutions. The first frog heart was dissected with the vagus nerve intact. The vagus nerve controls the heart rate. The second heart was dissected without its vagus nerve. Loewi electrically stimulated the vagus nerve of the first heart and the heartbeat decreased. Then he took some of the liquid from the solution surrounding the first heart and placed it into the solution surrounding the second heart. The heartbeat rate of the second heart decreased similarly to the first. These results indicated that a chemical released by the vagus nerve from the first heart had caused the change in the second heart even though it had no vagus nerve. Chemical messengers were signaling how the heart should beat. This research gave sound evidence for the idea that transmission between neurons involved chemical reactions, not just electrical impulses. The unknown chemical was isolated and found to be acetylcholine, the first neurotransmitter to be discovered (Kandel et al. 2000).

Conclusion: We Enter the Modern Age of Neuroscience

Now the stage was set for a more comprehensive understanding. Neuroscience would develop as a field of combined experts in fields of psychology, neurology, medicine, and even philosophy. The 1990s were declared the decade of the brain, and interest grew exponentially. The rest of this book will unfold the modern story. We invite you to engage in the ideas and make them your own. As you think of your clients not just in terms of their thoughts, feelings, and behaviors, but also as having a functioning brain in a body, you will be able to incorporate the great neuroscience discoveries from throughout history to add new and valuable dimensions into your work.

Chapter 2
Unlocking the Key to Neuroscience Terminology

Embarking on the study of a new subject always involves a period of gaining a set of tools. Although some of the tools can seem complex before you know them, they become helpful once you gain familiarity with what they are and how and when to use them. What might be slowing you down in your learning is that the words used to help navigate through the brain may seem more confusing than what they are describing. Thus, the first step in navigating the brain's systems is to gather the tools you will need to understand the nomenclature that is commonly used to describe the brain. This chapter helps you build the skills you need.

Think of the terminology used to describe the brain like a map. There are many different kinds of maps, such as road maps, elevation maps, some in 2D and others rendered in 3D. Each map uses a labeling system with a logical rationale at its source. The brain pictures that are commonly shown in neuroanatomy books draw on a number of different mapping systems. What is daunting for beginners is that these rationales are rarely explained. Instead, the viewer is expected to be familiar with the system. Another source of confusion is that different mapping systems are used to refer to the same area. Once you become familiar with the typical maps you will realize how each one contributes to giving a more complete picture of the brain and nervous systems.

This chapter explains the logic behind some of the most commonly used terminology and clarifies the different labeling systems that are commonly used. You may be surprised to discover that with the many maps to guide you, the path lights up to reveal fascinating sights along the way.

Viewing the Different Brain Maps

The brain is depicted in pictures, diagrams, and in various types of brain scans such as fMRI and PET. The body exists in 3D, but these technologies show the brain in 2D pictures. Thus, any number of different perspectives might be used.

C. A. Simpkins and A. M. Simpkins, *Neuroscience for Clinicians,*
DOI: 10.1007/978-1-4614-4842-6_2,
© Springer Science+Business Media New York 2013

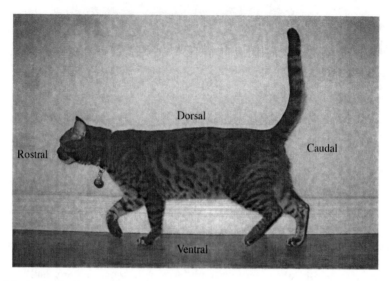

Fig. 2.1 Anatomical Directions

Exterior Perspectives

To view any 3D object from the most revealing perspectives, one would typically begin by inspecting it from the outside, looking at the front, back, top, and bottom. The brain is depicted exactly in this way but with different labels. The four directions are commonly described as *rostral, caudal, dorsal,* and *ventral.* This labeling system is simple to picture with an animal, such as here (Fig. 2.1).

One axis, the rostral-caudal axis, stretches across from the head to the tail. The head or front of the cat is rostral and the tail or far end is caudal. The other axis, called the dorsal-ventral axis, runs perpendicular (or orthogonal) to the rostral-caudal axis. Here, the top of the cat's back is dorsal and its feet are ventral.

When translated to the human brain, this terminology has to alter a little. This is because the rostral-caudal axis bends 90° around during embryogenesis so that the spinal cord is under the brain, not horizontal to it as in many four-legged animals. Thus, when compared to a cat where the axis is straight across, the labels do not always match up in quite the same way (Fig. 2.2).

Here is how the human brain is labeled using this terminology. The rostral-caudal axis places the rostral in the front of the brain (toward the forehead) while caudal is at the back, close to the neck. The dorsal-ventral axis follows the vertical direction of the brainstem and spinal cord. The dorsal side is toward the top of the head and the ventral is lower.

A second system of labeling is also commonly used and in many ways is more familiar and thus easier to apply: Superior (the top), inferior (the bottom), anterior (the front side), and posterior (the back side). Medial and lateral gives one more type of description, where medial is toward the center and lateral is out to the

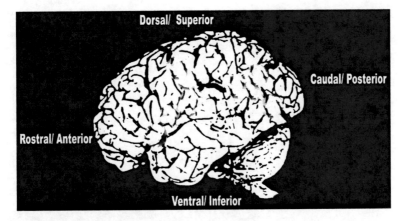

Fig. 2.2 Brain Directions

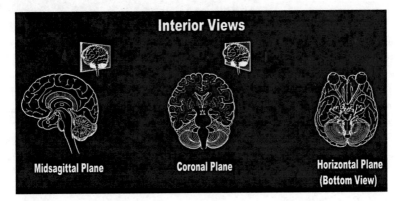

Fig. 2.3 Interior views of the brain

sides. Thus, the medial–lateral axis, looks horizontally, from the inner midline (medial) to the outer edges (lateral). If two areas are on the same side they are ipsilateral, and if on opposite sides they are contralateral.

Interior Perspectives

All of the brain systems cannot be accessed simply from the outside: We need to see what is underneath and inside as well. Thus, another typical way to view the brain is by slicing through, like slicing open an apple to see the core. Conventions for viewing the inside of the brain and the body are fairly universal and consistent with three slices (Fig. 2.3).

A vertical slice from front to back is a sagittal plane (Latin for *arrow*) view. When the sagittal slice goes through the center in this direction it is referred to as the midsagittal plane. A vertical slice going from side to side across is a coronal

Fig. 2.4 Broadmann areas

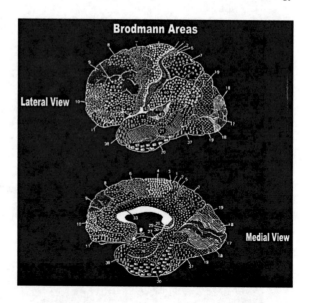

(Latin for *crown*) or transverse plane. Finally, a horizontal section parallel to the floor is the third perspective. This same set of conventions is used for viewing the interior of the human body as well. When labeling brain scans of sagittal and coronal views, you will often see a label of dorsal at the top and ventral at the bottom of the picture, whereas the horizontal plane will place a label of rostral on the side that shows the front (forehead side) and caudal on the side that pictures the back of the brain. These combinations help to clarify what view is being shown.

Brodmann Areas

An early system for labeling the cortex, the outer layer of the brain, was created by Korbinian Brodmann (1868–1918). He divided the surface of the cortex into 52 different sections based on the organization of cells, or cytoarchitecture. He published a map in 1909 (Brodmann 1909/1994). He thought the anatomical arrangement was similar in all animals, but this idea turned out to be incorrect since animals have much smaller cortical structures than humans. Scientists have worked on refining Brodmann's map for a century. The areas are referred to as Brodmann's areas along with the number of the area. For example, Broca's speech and language area is located in Brodmann areas 44 and 45; the primary somatosensory cortex is referred to as areas 1, 2, and 3; the primary motor cortex is area 4; and the primary visual cortex is area 17. Brodmann's areas are no longer considered precise or complex enough to account for the many connections between hemispheres and within hemispheres (Zillmer et al. 2008). But the system is still used frequently when referring to an area of the cortex, and so it is helpful to have an understanding of this mapping system (Fig. 2.4).

Terminology for Parts of the Brain

Different areas of the brain are given names, somewhat like other structures in the body such as the lungs or stomach. Sometimes a structure has a straightforward name that reflects the shape of the structure, such as the olfactory bulb, which is an organ with an elongated, rounded shape, or the amygdala, (Latin for *almond*), which has a curved shape much like an almond.

These structures can be located in the brain and have a distinct form similar to a small organ. But other structures are less physically differentiated. Instead, cells located in particular areas perform unified functions. These groups of neurons that are clustered together are given names to distinguish them. They are sometimes referred to in a way that makes them sound like they function separately and yet, they might be better understood as a system. For example, the basal ganglia are often referred to as if they are one thing, and yet they are a group of structures that function closely together, involved in movement. The limbic system is another group of structures that act together as part of our experience of emotion.

Several different types of terminology are used to refer to such structures. One commonly used word is *nuclei (*plural form), or *nucleus,* (singular). A *nucleus* refers to a group of nerve cells located in the brain that has a specialized function such as, for example, the caudate nucleus, one of the structures in the basal ganglia. The word *ganglia*, (plural of ganglion), is another term often used to denote a densely packed group of neurons connecting the spinal cord and peripheries to areas in the brain. Ganglia are located on either side of the spinal cord and are involved in the stress response and the fight- or- flight stress response. There is an exception, where the word *ganglia* is used to distinguish an area inside the brain, the basal ganglia, the group of nuclei involved with controlling movement. Some believe this is a misnomer and should be called the basal nuclei because it is located within the brain itself. But since it is linked to movement, which involves the spinal cord and PNS, the name makes sense. When areas of the basal ganglia do not do their work, certain subcortical movement disorders result, such as Huntington's disease and Parkinson's disease. ADHD is also linked to the basal ganglia.

Gyri and Sulci

The cortex has many folds with high and low areas. These highs and lows are also given labels to distinguish them: *Gyrus* (plural *gyri*), ridge, and *sulcus* (plural *sulci*), valleys are two other terms that are used to refer to the bulges and creases. Sometimes a particular high or low area is given a name, such as the cingulate gyrus, which is a strip of cortex in each hemisphere involved in emotion and directing attention, or the central sulcus, which divides two of the lobes. Various gyri and sulci are given names and referred to as such (Fig. 2.5).

Lobes is another term used to describe the cortex in different sections, as will be described fully in Part III. The four lobes are frontal, parietal, temporal, and occipital.

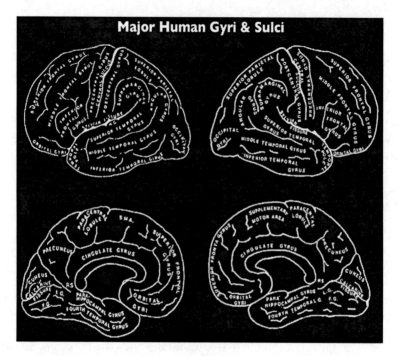

Fig. 2.5 Gyri and sulci

There are two sets of lobes, in the right hemisphere and in the left hemisphere. Although the functions are similar, there are distinct differences between the right and left sides.

These lobe divisions are helpful for getting a general sense of the location for certain functions. For example, the control of vision is located toward the back of the brain in the occipital lobe. The sensory areas are found in the parietal lobe. The frontal lobe performs higher level processing whereas the temporal lobe is involved in auditory functions. Lobes are easily seen without a microscope, as distinct sulci ridges. *Lobule* is another term used to refer to parts of a lobe. Lobules are also clear divisions, but they are only visible if viewed under a microscope. See Chap. 9 for a picture of the four lobes.

White Matter and Gray Matter

Another way to distinguish parts of the brain is by the grayscale shades of the tissues. When we look at brain tissue fixed in formaldehyde under a microscope, some parts appear white while other parts look gray. The white areas have been labeled *white matter* and the gray areas, *gray matter*. Cell bodies and dendrites, the branches from the cell body that bring input signals to the neuron, appear gray under the microscope, and therefore, the cell bodies and dendrites are the parts of the neurons that make up the gray matter. Myelinated axons are the white matter. The myelin is a covering on many axons, the long fibers that carry output signals from one neuron to another. The myelin functions like insulation on a wire. When it is fixed in formaldehyde and looked at under a microscope, it appears whitish in color, so these deeper areas of the cortex are known as the white matter. The brain is made up of white matter and gray matter. Gray matter is found along the outer surface of the cortex and cerebellum, following the hills and valleys in the tissue. It is also found deep inside the brain and around the spinal cord. The white matter is located deep in the brain and cerebellum and in the superficial parts of the spinal cord.

Conclusion

Familiarize yourself with the terminology, and you will have an easier time finding your way through the literature about the brain and how it relates to cognitions, emotions, and behavior. The time spent on these preliminaries opens a rich new world of potential for your therapeutic work.

Chapter 3
What is the Mind and Brain?

As a clinician, you work with minds. Your training and years of practice have attuned you to the subtle nuances of people's thoughts, inner sensations, and feelings. You work with their view of their behavior and what they think about those who are significant to them. In all of this work, you have probably taken for granted that your client has a mind and so do you.

But neuroscience questions this seemingly obvious assumption. Even though common sense tells us that we have a mind, what is it? Science can establish that we have a body and a brain, but how can it prove with certainty that there is a mind? And if you think about it for a moment, if we ask you to show us your arm, you can do it, and if we ask you to show us your brain, you can point to your head. But if we say, well, now show me your mind, you probably would have trouble doing so.

In fact, people have been trying to find the mind for millennia. In a famous story about the transmission of Zen from its great founder Bodhidharma to his first disciple, Hui-k–o, more than 1500 years ago, this same question was asked. Hui-k–o felt troubled, much like our clients, and asked his teacher, "Please make my mind be at peace."

Bodhidharma replied, "Show me your mind." Hui-k–o could not find his mind to show to Bodhidharma. At that moment, Hui-k–o found enlightenment.

The questions about mind and brain are enormously important for us as therapists. The answers influence how we work with clients. We are always searching for techniques that will be most effective in changing the client's mind. As more is learned about the relationship between mind and brain, we gain new methods to incorporate into our practice. Neuroscientists, who are learning more about the brain, ask important questions about the relationship between mind and brain and we invite you to consider the issues now.

This chapter explores the relationship between the mind and brain. The relationship points to the very nature of consciousness and our sense of self. At the end of the chapter, you will find ways to integrate all of this into your practice.

C. A. Simpkins and A. M. Simpkins, *Neuroscience for Clinicians*,
DOI: 10.1007/978-1-4614-4842-6_3,
© Springer Science+Business Media New York 2013

Theory of Mind

René Descartes (1596–1650) began the discussion by separating mind and body. When he said, I think therefore I am, (Descartes 1984), this set the mind apart. Neuroscience findings about the brain have convinced many philosophers that the material or physical brain forms the foundation for the mind. Yet there still remains this sense we have of a mind. So, what is it?

One of the most prominent pragmatic explanations is known as Theory of Mind (ToM). ToM is based on the idea that we attribute mental states, including beliefs, desires, and intentions, to ourselves and to other people. ToM assumes that other people have a ToM like our own. Because we believe that everyone has mental states like ourselves, we attribute reasons for people's behavior. This meta-perspective explains how individuals reason about other minds. The basis for this reasoning involves higher level conceptualizing found in cortical brain areas, a top-down process. We generate representations in our own brain. These mental representations arise from our experience of self and others and from participation in a culture. Our mental representations of the world may or may not actually correspond to the world. In fact, this is called a ToM because the representations are not always observable. Jerry Fodor (b. 1936), one of the foremost proponents for functionalism covered below, is a spokesperson for mental representations. He believes that representations are expressed as a language of thought (LOT). Since our language ability can be located in certain brain areas, we have evidence for existence of LOT and ToM. Psychotherapists often work at the level of a person's ToM. As clinicians certainly know, sometimes a ToM leads to better under-standing of others, but often when people have disturbance, they have adopted a distorted ToM.

But where does the ToM come from? Traditional views hold that we form attitudes and beliefs about the world, an assumptive system, which influences how we perceive. Cognitive therapy addresses this system, helping to make it more realistic and rational. A recent discovery in neuroscience offers a second possi-bility. Mirror neurons have been found in a number of different areas of the brain. They are a special kind of neuron that fires when we do something. But these neurons also fire when we observe someone else doing that thing. Thus, we may understand others because our brain simulates the action. The action of mirror neurons suggests that higher level processing such as ToM could develop later. This simulation theory offers a new dimension to the debates about how we understand others (see Chap. 18 for more on mirror neurons).

Cortical Midline Structures and the Sense of Self

Neuroscientists have found an area located in the front of the brain known as cortical midline structures that become active when we are processing information that relates to our experience of having and being a separate "self" (Bermpohl 2004).

Fig. 3.1 Cortical midline structures

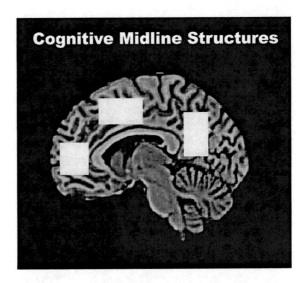

Neuroscientists call this kind of processing, self-referencing, central to generating our sense of self. A number of subprocesses might be involved, including how we represent these stimuli and then monitor them as they are being experienced. We also evaluate them—do these apply to me or to someone else? And we integrate self-referential stimuli together. All of these processes could somehow be correlated with how we come to have a distinct sense of self, and to feel quite capable of distinguishing our own self from others (Fig. 3.1).

Consciousness and the Brain

Given that we have a sense of ourselves, then we must have a way to know that fact. Dictionaries tend to define consciousness as being aware of our own existence, sensations, thoughts, and surroundings. Any person can tell you that they are conscious. And yet, as we think more deeply about it, consciousness eludes us. Meditation methods teach us how to observe the stream of consciousness. Through the process, we can get in touch with it, but the experience is not as rational and logical as one might expect. A single definition can be elusive, but neuroscientists are finding new evidence for providing a clearer understanding of it. Antonio Damasio, a prominent researcher in the neuroscience of emotions provides this helpful definition: "Consciousness is a state of mind in which there is knowledge of one's own existence and of the existence of surroundings" (Damasio 2010, p. 157). These conscious states of mind have content, similar to how the great phenomenologist, Edmund Husserl proposed that all consciousness is intentional toward some object (Husserl 1900, 1970). Many brain processes contribute to these conscious mental states, including areas that regulate alertness (the brainstem), body sensations (insula and parietal lobe of the cortex), emotions (the limbic system), and cognition (the frontal areas of the cortex). With so many brain areas involved, clearly consciousness permeates our existence in profoundly pervasive ways. Neuroscience proposes that consciousness is somehow generated by, through, or in interaction with these brain processes. But how?

Physicalism

Clearly, what happens in the mind has some relation to what is occurring in the brain. But what is the relationship? One prominent approach to the question of this relationship between mind and brain is materialism, also known as physicalism. This philosophical view takes the position that the physical reality of the world is the only true or actual reality. All that we can know of is substance, and that substance is physical or material. The rest is speculation.

In some form, this thesis began with Thales (624–546 B.C.) of Greece, the traditional founder of Western philosophy. Thales believed that the universe is made of a primal substance, water, giving us a compass to navigate the seas of the world. By understanding the material nature of something, we can know all about its properties and behaviors. These are merely manifestations of the material substance as it is.

Many people, especially in the West, believe that at the heart of the universe we will find some kind of ultimate substance. The world is made of matter. Objects and events are just patterns of matter. Matter is what is fundamental. Similar to the pixels on a screen that from a distance, form patterns that we see, the mind forms patterns from the matter the world is made of. To Eastern philosophers, matter is not apart from the mind's patterning: it is inseparable. Mind forms patterns *with* the matter the world is made of.

In the early application of materialism to the brain, the primary substance of mind was the brain's neurons. Without a brain, there would be no human mind. Today, the different positions among the materialists vary on the relationship between the mind and the brain (Place 1999), but at its core, there is always the sense that the brain is fundamental in one way or another.

The Mind is the Brain

Identity theory holds that the mind literally is the brain. The brain is not parallel to or separate from the mind. Dualists may conceptualize the brain and mind as separate and unrelated, or else separate but linked or related, but to the identity theorist, these positions cannot be proved.

Mental processes and brain processes are the same, and therefore are identical. The identity itself is their nature. Although many variations can be argued, with subtle differences, the variations share in the identity. For example, an old ship's hatch cover is made into a coffee table on a base. But it is really the same thing. Whether used as an old hatch cover, or used as a coffee table, it is still the same object, just used for different purposes. Similarly, the mind is the brain, and the brain is the mind. They both are the same thing. The mind is the brain, when thinking.

Our concepts and names for things sometimes obscure what our common sense tells us it really is. If an object has all the attributes that another object has, it is the same object. We can know what an object or substance is by the properties it shares with other objects in the same category. But categories can be deceptive, paradoxically. For example, a certain star shines in the morning and in the evening. So, people can point to the star in the morning and exclaim, "There is the Morning Star!" And in the evening, if they look up in the sky to the place where that certain star was, they could point to what they see and say, "There is the Evening Star!" In a sense, they might seem to be seeing something different. There is a Morning Star, and there is an Evening Star. But these two apparently separate stars are actually the same. They are both one star, a planet we know as Venus; which is actually only one star, seen at different times over the 24-hour cycle. So, things do not necessarily fit with common sense.

Beyond Identity: Eliminative Materialism

A materialist position that takes issue with the identity theory is the eliminative materialist, (Churchland 1988; 1995). The eliminativist believes that the theories of mind and mental representations are simply folk psychology, a taken for granted set of assumptions that are passed down from generation to generation, learned at grandma's knee. The concepts and terminology we use to describe the mind are like phrenology and other old-fashioned beliefs, superstitions, and folk tales. Our common-sense psychological framework is a false and radically misleading conception of the causes of human behavior and the nature of cognitive activity (Churchland 1988, p. 43).

There is no mind outside of the sensory experience activating brain processes. Such mistaken concepts and language are deceptive. The only true mental reality is the neuronal connections of the brain. Therefore, it is unlikely that we can find a one-to-one correspondence or identity between the material structures and functions of the brain with the concepts, thoughts, and beliefs of folk psychology. Since only the physical or material world is real, there is no purpose in pointing out and distinguishing properties using false terms; only false conclusions will be reached. Instead, we should eliminate these older frameworks and embrace a purely material explanation beginning at the chemical/neuronal level. This philosophical position is optimistic about connectionism as an explanatory theory, dismissing other approaches as based on illusory premises.

Functionalism

Granted that the neuronal connections may be important in helping to shape our minds, we also know that thoughts, feelings, and behaviors can have an equally potent influence on the neuronal connections, by either strengthening or weakening the connections. Most clinicians have seen how mental events can influence functioning. One of the theories that may feel most familiar to the therapeutic community is functionalism, which fits seamlessly with the ToM. Fundamentally, functionalism holds that the identity of a thought, wish, feeling, or other mental state depends on its function within the whole system of cognition of the individual (Fodor 1983). Material, such as the brain, is important to the system of the organism only in terms of its function or role in the system as a whole. Thus, a particular thought can be realized within a state of mind or consciousness, from a functional relationship among the parts of the system. But thought is not just happening because of any one part in itself: Parts work in conjunction with each other.

Similar states can be realized from different brains or even different species of brains, as long as they function in the same way. Thus, the same thought can be realized in multiple ways, in various brains, so long as they are organized in the same functional way. The identity of the material that contains, creates, or conveys the thought is not the central determiner; the function is.

So, for example, if a DVD or a VCR tape of your favorite movie is played in a machine that is capable of playing both types of media, the same visual presentation of the movie can be given, even though from different sources. The functional relationship gives information from the parts of the movie, even though they are presented in different media. Similarly, two different individuals can realize the same mental state in their separate brains in response, since they function in the same way. Two people deeply in love can serve as an example of how different individuals with their own unique brain organization, can share in the same feeling. In fact, different species of animals can realize the same responsive state if their organism functions in the same way. The function is the important thing, not just the components. This has broad implications for the philosophy of mind.

Eastern wisdom helps to clarify the idea. The usefulness of a cup is in its emptiness (*Tao te Ching*, Book 11, Wilhelm 1990). So, emptiness is the source, the potential for the functionality of a cup. We do not get our sense of what the cup truly is from its substance alone. The form follows its function, to offer an empty space for use.

Imagine a group of people standing on a boardwalk overlooking the Pacific Ocean at sunset, watching the sun setting. One turns to the others and exclaims, "That was a beautiful sunset!" And they all nod, agreeing. Something was shared among them. They all know what the beautiful sunset was, though they have different backgrounds, and separate and different brains, in different bodies. The functionalist would hold that sharing was possible regardless of these individual factors because the experience of nature's beauty was functionally identical.

Mind Only Theory

We tend to take for granted that the world we experience is real and lasting. We also assume that the same world will be here tomorrow. But therapists know that the mind can have a powerful influence on one's reality sense. People can be very certain that things are negative, as they clearly perceive them to be. Paradoxically, without intending to, they often bring that very reality about. So, the angry woman finds annoying people, objects, and events all around her, a world filled with hostility. The fearful man sees threatening people, objects, or events at every turn, and justifiably finds potential for threat at any moment. The world that we experience is the world we live in. And in turn, the world we live in becomes the world we experience.

The mind-only view drawn from Buddhism (Simpkins 1997) is that everything we perceive is in our minds and is only mind. Our thought makes the world seem real. Meditation, not reason, is the best method to see through the illusion of existence that mind creates. No external reality, no substance actually exists. The basis is non-substance, emptiness. Nothing exists outside of Mind.

In Daoism, the fundamental heart of the universe is uncreated, changeless, unformed, and manifested in opposites of Yin and Yang when things take form. There are many other variations of Eastern views, of course. But the commonalities are that whether the inner heart of things is Mind or Dao, something is experienced, and it has mental qualities.

Before quantum theory, science naively assumed that the world we perceive could be observed objectively and measured accurately, and so we could be certain that the world was there as our measurements indicated. Scientists thought they could build understanding by improving the methods of measurement and developing better technologies. But quantum physics raised paradoxical questions that showed the limitations of accurately observing and measuring. We cannot be so certain that our measurements truly reflect the reality we are trying to measure. We make an error if we naively assume that the world as we perceive it really exists (Albert 1992).

If an object as perceived is really there, we should be able to test its physical attributes. For example, we could weigh it, lift it up, and put it down to feel its weight, compare it with another object of similar weight, or contrast it with an object of a different weight. We could submerge it in water and measure how much water is displaced. But all these tests do not necessarily prove the constant reality of the object. They are inferences about the object. As one of the West's great philosophers, Martin Heidegger (1869–1976) said, we can't find the rockness in a rock by smashing it into little pieces (Heidegger 1962). The true essence or nature of a rock eludes analysis.

Furthermore, our tests simply give us information about the temporary reality of the object as it appears at that moment. There is no direct proof, without any doubt, that the object we measure today is identical to the object we measure tomorrow.

Time, by a succession of instances, gives us the convincing illusion of continuous existence, and as a result, we create the perceived world with our consciousness.

The objects of our mind appear fixed, permanent, and lasting. The illusion of a constant object is partly created by our consciousness filling in the gaps. When we listen to a series of varying tones over time, we hear music flowing. Our mind flows the notes together as a reality, remembers the melody, and even gives it a name and, such as the well-known tune "Row, Row, Row Your Boat." Watching a series of pictures varying slightly or even greatly, if presented in sequence, flickering at a certain range or speed, we see moving scenes. The continuous flow of appearances to which we respond seems constant. The mind joined the separate pictures together to give us a unity through time, experienced by us as a movie, with narrative meanings. This is our consciousness, constructing existence.

Concepts also contribute to the illusion of reality. We label the brain and its parts. We recognize the limbic system and correlate it with emotional reactions. But this label is really an arbitrary convention, sustained by the mind. The limbic system is an abstraction from the real, ever-changing neuronal connections that continue to transform from day to day, and moment to moment. And even the neurons themselves are always in flux. For example, we come to experience the limbic areas of the brain as the system that we label "the limbic system" and think of it as solid and lasting. But there is no actual fixed limbic system outside of the momentary interrelationship of various parts of the brain functioning emotionally. The limbic system is empty of any permanent substance or reality as an organ. But at each moment, the limbic system is there, as the brain functions. And the mind is there, in the brain functioning. They are a unity: empty of independent lasting reality, yet existing dependently as a moment-to-moment reality, in relationship to each other.

Balancing Mind and Brain in Treatment

Karen was a young woman in her twenties who had survived a terrible car accident. Her mother had been driving while she was sleeping in the front passenger seat. Karen suffered back pain and complete amnesia for the event, but otherwise had recovered physically. Her mother did not fare so well. She was critically injured, and years later, was still languishing in a coma on life support. Karen came to therapy to help her reclaim her memory of the accident. She believed that if only she could remember, things would be better.

She had forgotten the course of events just before, both during and following the accident. She was haunted by frightening images and scary nightmares. She felt intense guilt, continually replaying in her mind, if only she had been awake, perhaps she could have done something to prevent the accident. She could not stop thinking about her poor mother, lying pale and unmoving in the hospital bed. She had difficulty making it through the unchallenging job she had taken rather than pursuing her higher education.

Karen had suffered a trauma to her brain. But in addition, over the years since the accident, her mind was filling in the blank spots with disturbing thoughts and feelings. As therapy progressed, it became clear that training her memory was not the place to start. Instead, she needed to explore how mental processes were filling in memory gaps.

She worked in therapy to accept that she might never recover her memory of the exact order of events, that there might always be a gap in her memory. But, she could take control of her thinking. In facing the unknown squarely, she could accept the real tragedy of the situation. She recognized that it was not her fault; and that even though she had recovered and her mother had not, that did not mean she was personally guilty. As she began to take more responsibility, she reasserted control over her life. When the doctors advised the family that her mother was in an irreversible coma, she was faced with the decision whether the family should assent to removing her mother from life support. She considered the financial and personal responsibility, and how her mother might have felt. Then she accepted the loss of her mother, and felt ready to let her go. We heard later that Karen had returned to college and was pursuing a meaningful career.

The mind and brain are in an intimate interplay. Here we saw one of the brain's natural mechanisms, to fill in gaps. In the beginning of therapy, filling in gaps worked against the client. At the termination of therapy, this natural mechanism helped her heal. With warm support, Karen was able to make a better adjustment for a brighter future.

Conclusion: An Integration of Mind and Brain

Clearly, mind and brain are intimately linked. Whether you feel drawn to a Western materialist perspective; whether you accept identity theory, functionalism, eliminative materialism, or an Eastern perspective such as Buddhist, Daoist, Zen, or Mind-only, you will learn how you can work with the mind to affect the brain or the brain to affect the mind of your clients. The relationship goes both ways, and later chapters will show how to utilize these influences therapeutically.

We take a middle way that includes the different perspectives together, like the waves and particles that particle physicists conceive as matter. As you read this book, you will see illustrations of how brain and mind are involved in an interdependent relationship that is explained by one or more of these models. Without our brain, we could not be thinking, feeling, doing beings, and yet our brain cannot be understood fully without the feelings, thoughts, and behaviors that arise from it. Mind and brain are embodied, part of a larger world, continually interacting with others. Your understanding becomes more complete by recognizing the interaction between mind and brain, and thereby broadening, your options for clinical techniques to foster healthy mind–brain change in your clients.

Part II
Neuroscience Methods and Technologies

In recent years, advances in methods and technology have boosted progress in our understanding of the brain and mind. One of the early avenues that proved to be fruitful for understanding the brain was to study people with brain damage, particularly when the exact place of injury is known. A single individual, carefully studied and tested, can reveal a great deal. The method is to compare deficits in functioning from a known point of damage with how a normal person functions. Researchers can reason backwards to what specific sections of the brain are likely to be involved. Chapter 4 describes some of what has been learned from pivotal cases such as Phineas Gage, H. M., and others.

Key technologies and are described in Chap. 5. New equipment has allowed neuroscientists to light up the brain and see inside the black box. For many decades, the brain was been studied in real time through EEG. But the readings have been general and difficult to decipher. Now with more advanced statistical methods, researchers are better able to filter out the noise to hear the real signal. Imaging equipment such as fMRI makes it possible to take an indirect picture of the inside of the brain. The images reveal activations and changes that could never be seen before.

With the new information gathered from imaging, researchers have been able to create mathematical models to help interpret the data more accurately. Chapter 6 gives you some of the frequently applied mathematical models that help characterize the complexities that are being described. When you consider the billions of neurons, each with their multiple connections to other neurons, and many working simultaneously, it makes sense to turn to mathematics, which conceptualizes many interactions together. This chapter will reveal how these mathematical models work. Then, even readers who might not consider themselves to be "a math person," will come to appreciate why mathematics has been so helpful to the field of neuroscience and to you as a practitioner for gaining clearer pictures of the mind-brain interactions.

Combining all these new approaches, neuroscience draws open the curtain that hides the brain, revealing a vast field of potential just waiting to be seen and explored.

Chapter 4
Learning from Brain Damaged Individuals

One of the early ways for learning about the brain was to study people who have had brain damage. Researchers took note of what the patient could not do. Then by studying the absence of function, what is not, they could infer by comparison what the specific brain area in a healthy brain is probably doing.

The best clinical evidence for how the brain influences function has been found when researchers could study patients who had suffered a known type of injury or gone through experiences from a deliberate lesion to a specific part of the brain. Damage can affect functioning in various ways, either specifically or globally. The effects are carefully correlated with the damaged area, teaching us about what areas of the brain influence behavior and experience. Thus, from what is not, we learn about what is.

The Case of Phineas Gage

One of the earliest documented cases where much was learned from studying the loss of function took place as a direct result of a known brain damage in the famous case of Phineas Gage. The dramatic transformation that Gage underwent from his injury became a pivotal case in the quest for understanding how vitally important the brain is to personality.

Gage was a 25-year-old construction supervisor for a railroad company with the job of overseeing the laying of railroad tracks. He was well-liked, competent, and well-adjusted. The year was 1848 and Gage was hard at work, blasting through some rocks. He carefully set an explosion of powder, but this time, it backfired. The blast sent an iron bar flying into his left cheek, through the front of his brain and out through the top of his skull.

Gage was immediately rushed to the hospital. But much to the surprise of everyone, including his doctor, he seemed relatively unaffected and was quickly released.

C. A. Simpkins and A. M. Simpkins, *Neuroscience for Clinicians*,
DOI: 10.1007/978-1-4614-4842-6_4,
© Springer Science+Business Media New York 2013

At first, he appeared to recover completely. But over time, changes emerged. Gage was transformed from the responsible, hard-working person his friends, family and co-workers knew, into a foul-mouthed, irresponsible, stubborn individual who could not hold a job, stick to a plan, or maintain a relationship. His personality changed radically for the worse. He lived for only 13 more years, dying in 1861 after a series of seizures (Damasio 1994).

The bar had penetrated through Gage's prefrontal cortex. The prefrontal cortex is located in the frontal lobe, and is involved in high-level cognition, goal-directed behavior, planning, and sequencing. Damage to this area has broad effects on personality, as this famous case illustrates. The tragic story of Phineas Gage is an extreme illustration of frontal lobe damage. But the kinds of problems he developed are typical of people who suffer frontal lobe damage. Although the IQ does not alter much, these patients suffer from certain symptoms such as shallow emotional affect, reduced responsiveness to pain, and an inability to carry out plans. As one patient said, he could see all sides of a decision, but had no way of deciding which option would be best to actually implement (Damasio 1994).

H. M. and New Understanding of Memory

Henry Gustav Molalson (1926–2008) is better known by his initials, H. M. He was a man who underwent elective surgery to deal with his debilitating epilepsy. His surgeon, William Beecher Scoville, had found the epilepsy to be located in his left and right medial temporal lobes and decided that the best way to prevent the epilepsy was to remove two-thirds of his hippocampus, his parahippocampal gyrus, and his amygdala. Even though a small portion of his hippocampus was not taken out, it became completely non-functional because the endorhinal cortex was also destroyed. The endorhinal cortex is located in the temporal lobe and gives most of the sensory input to the hippocampus. Part of his anterior lateral temporal cortex was also taken out.

Little did anyone know in 1953 the serious effects this operation would have on H. M.'s memory. The surgeon immediately recognized the deficits following the surgery and began testing H. M. He shared his findings about the resulting memory losses with the scientific community. H. M. willingly participated in research for the rest of his life, thereby helping to further scientific understanding about the brain. (Scoville and Milner 1957). From the many years of study with H. M., we have gained strong evidence for the idea that some functions of memory are localized, performed with specific areas of the brain.

Following surgery, H. M. could not form any new lasting memories. In other words, he was unable to commit new events to long-term memory, suffering from what is known as anterograde amnesia. He also was not able to recall events in the period right before the surgery, retrograde amnesia.

At that point in time, people believed that memory was one kind of process. What H. M. taught the world was that the brain has different memory systems.

H. M. was told to trace a star shape. His view of his hand was blocked, so that he could only see his hand reflected in a mirror. This experiment is known as the mirror tracing task. Every time the task was presented to H. M., he believed that he had never seen it before, and so instructions were given as if it were the first time. Normal subjects perform badly on this task at first, but improve with practice. H. M. performed similar to normal subjects, learning the skill. What researchers concluded from this groundbreaking study was that H. M. was able to form memories involved in learning a new skill, how to do something, even though he had no memory that he had learned anything. This skill-learning memory is now known as procedural memory, which differs from memorizing facts (semantic memory) or remembering what you did yesterday (episodic memory). These findings from H. M. along with other work in this area have led researchers to clarify distinctions of many different types of memory such as short-term and long-term memory, declarative and procedural memory, episodic and semantic memory. Recent studies have made further refinements within these categories. For example, a study that compared H.M. to patients with more severe medial and lateral temporal lobe lesions found that H.M.'s ability to remember how to apply grammar and syntax remained in tact, whereas the more severely damaged patients lost these abilities as well (Schmolck et al. 2002). Different types of brain damage impair different functions, adding more evidence for the search for distinctive brain-to-function connections.

As a result of knowing exactly what was removed, coupled with H.M.'s willingness to give himself over to extensive research for more than half a century, the world's knowledge about such crucial functions as memory, emotion, and attention have been greatly fostered. Upon his death, he has donated his brain to science . Researchers at the University of California, San Diego, are slicing the brain into sections and digitizing the images, to make them available to other researchers and students worldwide. Truly, H. M. made a great and lasting contribution!

Split-Brain Patients

Another rich source for learning has come from the study of split-brain patients. The brain is divided into two hemispheres with many structures that seem to be repeated on both sides. Neuroscientists have studied patients who have had damage to one side or the other, as well as those who have had the connections between the two hemispheres, the corpus callosum, severed to help contain epilepsy. Some of the functions have been found to be localized to one side or the other, but not both.

Hemisphere specialization has been observed since the time of the ancient Greeks. Hippocrates (460–377 B.C.) noticed that injuries to one side of the head often resulted in impaired function to the opposite side of the body. He concluded that the human brain is double (Crivellato and Ribatti 2007).

But later when the brain was looked at under a microscope, the two halves appeared to be structured the same. So, it is not surprising that throughout most of medical history, scientific researchers believed the two halves of the brain functioned similarly (Kimura 1996). Over the past 100 years, the details of the different functions of the two hemispheres have been revealed.

Early scientific research performed in Karl Lashley's (1890–1958) laboratory demonstrated hemisphere and region-specific visual learning in pigeons (Levine 1945a, b, 1952). This idea that learning could be restricted to one hemisphere without active or immediate transfer to the opposite hemisphere was studied further in higher animals (Sperry 1961). Meyers and Sperry ran experiments on a cat at the University of Chicago in the early 1950s that revealed differences in the two hemispheres.

Human subjects offered valuable data for understanding hemisphere specialization following the work of Bogen and Vogel (1962), who were the first to perform successful neurosurgical commissurotomy of the hemispheres by sectioning the corpus callosum and anterior commissures to help with intractable epilepsy. These surgeries also permitted observation of how subjects would react to having the two hemispheres separated. In the early years, neurologists evaluated that the left side, which seemed to control language and complex cognitive abilities, was the most important side.

> The left hemisphere became the one to have, if you have only one. Indeed, neurologists were fond of citing case reports of individuals who were born without a right hemisphere, or who had lost their entire right hemisphere in an accident or through surgery, who nonetheless coped successfully with the business of living. (Gardner 1976, p. 353)

Evidence gradually emerged to indicate that the right hemisphere might be important for intelligent functioning, as well. Research in England during World War II revealed that right hemisphere damaged patients were deficient in their ability to organize spatially. They had trouble finding their way back to their rooms on the wards and even had problems putting on their clothes. The right hemisphere played a part in visual skills, depth perception, gestalt formation, as well our sense of touch and proprioception, the internal body sense. Although the left hemisphere was superior in verbal processing of information, the right hemisphere was superior in managing visual–spatial tasks (Gazzaniga 1973).

Using behavioral tests, Sperry (1974) demonstrated that each hemisphere had its own way of conceptualizing and responding to stimulation. Each left and right hemisphere has its own private chain of memories and learning experiences that are inaccessible to recall by the other hemisphere. In many respects, each disconnected hemisphere appears to have a separate mind of its own (Sperry in Springer and Deutsch 1981, p. 52).

Another group of experiments was performed at Dartmouth on a different set of patients who had undergone a similar surgery (Wilson et al. 1977). Gazzaniga did extensive clinical histories of these patients as well as careful behavioral experiments, which have contributed significantly to our understanding of hemisphere functioning (Gazzaniga 2000). All of these studies form a basis for new theories describing how each hemisphere has unique cognitive properties.

Functions of Brain Hemisphere Interpreters

Today, differences between the two sides of the brain are well documented. What has emerged is a multifaceted tapestry of mental processes. Each of the cerebral hemispheres has its own strengths: the left being specialized for language, speech, and problem solving, and the right being specialized for visual–spatial and attentional monitoring. In a similar respect, each has its own limitations: the left in perceptual functions and the right in cognitive functions (Gazzaniga 2000). Currently, it is a virtual dictum that the left and right hemispheres can conceptualize and respond differentially and specifically to sensory information (Novelly 1992, p. 1128).

Gazzaniga has proposed that although we have a highly evolved laterally specialized system, we get our sense of a unified consciousness from the left-hemisphere interpreter. To Gazzaniga, the left-hemisphere interpreter is "the glue that keeps our story unified and creates our sense of being a coherent, rational agent" (Gazzaniga 2000, 1320).

Another researcher, Corballis (2003) suggests that the right hemisphere has its own interpreter as well. His research shows that some aspects of visual processing may occur at a higher level. In this sense, the right hemisphere can be considered more visually intelligent than the left, and therefore Corballis postulated a "right-hemisphere interpreter" dedicated to constructing a representation of the visual world. From these varying accounts, it is clear that human intelligence may have more than one source in the brain.

Hemisphere Preference

Bogen and Bogen (1983) hypothesized that individuals have a tendency to use one hemisphere and its mode of thought more than the other. Some differences among the population as to how they react may be related to hemisphere preference. Hemisphere tests (Harris 1947, 1974) show that individuals tend to rely on one hemisphere more than another, making some resources more readily available than others. Hemisphere preference has been used as a way to characterize how people tend to function. Looked at as a group, studies of both right and left hemisphere functions seem to indicate that each has its strengths and weaknesses, so reliance on one over the other may lead to certain psychological tendencies. Therefore, clients might be helped best when therapeutic techniques are utilized to take these hemisphere asymmetries into account.

When One Function is Lost

Damage from brain injury is often not as dramatic as the case of Phineas Gage, H. M., and split-brain patients. Often, when a patient sustains damage, the damage does not lead to general blunting or global personality change. Instead, these patients suffer a specific loss. This has given researchers confidence in specifying correlations between a certain part of the brain and a particular function.

Temporal lobe damage on both sides can produce an inability to recognize faces, known as prosopagnosia, or face blindness. The word comes from the Greek prosopon (face) and agnosia (lack of knowledge). The term was first introduced in 1947 to describe the case of a 24-year-old woman who had suffered a gunshot wound to her head. She lost her ability to recognize faces of family and friends. She was even unable to recognize her own face when she looked in a mirror. These patients have no problem recognizing colors or reading, nor do they have any problem with their eyes. But they cannot recognize their girlfriend or boyfriend, son or daughter, husband or wife. Even when they look at themselves in the mirror they will say that this person does not look like me. What eludes them is the recognition of who it is in the mirror. These patients can identify the voice, gait, or mannerisms. This strange and very specific deficit has led to pinpointing the medial inferior part of the temporal lobe as being responsible for face recognition.

However, specialization is never a completely simple matter when the brain is concerned. A recent study found that there is more than one form of face blindness. These researchers found distinctions in the type of face recognition problem as ranging between associative types of problems and perceptual types of difficulties. For example, people vary in how well they can pair different pictures, or being able to perceive similar traits in two identical pictures of faces or matching pictures of faces to names. These authors proposed a detailed descriptive categorization that could be helpful for clinicians: apperceptive, discriminative, and identifying forms of Prosopagnosia (Garcia and Cacho 2006).

Another syndrome that sometimes results from head injuries and comas is Capgras Syndrome. Patients seem to recover completely. They appear to be lucid and clear thinking. They can read, write, and talk. They seem normal in every respect except that they hold a delusional belief that a close acquaintance, such as their mother or father, sister, brother, husband, or wife, is an identical-looking imposter. When asked why this person would want to pretend, the patient often fabricates a reason, such as that my mother went to China and instructed a woman to watch over me. In the past, such disorders were given a Freudian explanation, but more recent neuroscience findings explain it as a problem in brain communication. The message goes in through the visual areas, to the temporal lobe, to the inferior temporal cortex. Normally it goes to the amygdala, a part of the limbic system involved in the salience of emotional significance of what we see. This message is processed by the amygdala and limbic system, for an appropriate emotional response. But for these patients, the message never reaches the amygdala, and thus the experience of emotional salience never occurs. These reactions can be measured by a galvanic skin response change in the sympathetic nervous system. When we see someone who we care about, the sympathetic nervous system is activated. But in these patients, no such reaction occurs. This is a striking example of how a lesion between the amygdala and cortex interferes with the normal autonomic response.

What Neuroscience Tells Us About the Conscious and Unconscious Mind

Therapists are accustomed to working with different levels and types of awareness. Freudian and Ericksonian theories proposed separate types of consciousness: conscious and unconscious. Both are operating simultaneously and affecting how people think, feel, and behave. Neurological syndromes have given strong evidence for the fact that such subtle differences in the quality of awareness are occurring in many of the activities we perform. Some things we do in life have a deliberate, aware, declarative, or what some might call a conscious form. But there is also a spontaneous, unaware, non-declarative, or what some might call an unconscious form. These distinctions may seem subtle, but when you think about them in specific cases, the difference becomes obvious. Neuroscientists may conceptualize unconscious states as non-conscious or perhaps as unaware states of functioning.

Walking, running, smiling, or yawning are a few examples of activities that can be performed deliberately or spontaneously. Consider the simple act of smiling. We can deliberately smile when the photographer says "smile", but more often, we find ourselves spontaneously breaking into a smile when something strikes us as funny. These two types of smiling are controlled by different parts of the brain: The cortex is involved in the control of deliberate smiling whereas the basal ganglion is necessary for a spontaneous smile. But how did researchers discover these distinctions?

Clinical examples have helped to reveal the answers. A patient suffered a stroke that paralyzed the left side of the body because his right hemisphere was damaged. This patient could smile on command, but only on the right side. This patient suffered from apraxia, a neurological disorder characterized by a loss of the ability to carry out learned or purposeful deliberate movements while retaining the ability to perform these same movements as part of a spontaneous response. When the patient's wife walked into the room, he would smile fully, with both sides of his mouth. This was because his basal ganglia functioned normally. The basal ganglia, which controls spontaneous movements including the smile reflex, had not been damaged by the stroke. But the cortex, which is involved in deliberate action, had been damaged (Ramachandran 1997).

This case might not provide enough evidence to establish the role of the basal ganglia in smiling. But combined with another patient with a different point of damage, the correlations become clearer. This patient did not complain and seemed fine in every respect. When asked to smile, she could produce a normal smile on both sides of her mouth. However, when she smiled spontaneously, she only smiled with half of her mouth. This patient had suffered damage to one side of her basal ganglia, but her cortex was normal on both sides (Ramachandran 1997).

Evidence is built up by investigating other types of problems. A similar example is seen with the difference between marching and walking. When we march to a command of the signal of a whistle, the movement resembles walking, but is quite a different activity for the brain. The command to march goes to the cortex as a higher level, learned process. We are not evolved for marching, so we make mistakes. However, when simply walking normally, the basal ganglia is involved. Damage to the basal ganglia leads to problems in the normal walking gait, such as in Parkinson's disease.

These examples show a difference between deliberate, higher level cortical activity and spontaneous processing deeper down in the brain structures, such as the basal ganglia involved with automatic movement.

Skillful application of these natural capacities to do something deliberately or spontaneously can be utilized by therapists to help clients with difficulties. For example, hypnosis can elicit spontaneous movement. People in hypnotic trance often experience movements and experiences as happening spontaneously, outside of their deliberate control. Trance phenomena such as hand levitation, where the hand raises seemingly by itself, are typically produced during hypnosis. When we look at the brain activations of a person who is experiencing hypnosis, we can see that the cortex is deactivated whereas the lower areas involved in movement are stimulated (Rainville et al. 2002). Therapists can use this hypnotic effect to help patients bypass limits in one area to activate potentials in other areas.

Conclusion

Taken together, all of these cases and others like it help neuroscience to put the pieces of the puzzle together, to learn from what is and what is not, for a deeper understanding of the whole brain.

Chapter 5
Neuroimaging Technologies

A boon to understanding the brain has been the advances in imaging technologies. Most people have probably seen the pictures of the brain and may have thought that they were seeing something like a photograph. Unfortunately, technology has not gotten to the point where we can literally take a picture of the brain. What we have instead are indirect measures which, when combined with advanced statistical measures, allow us to make estimates that seem to be approximating the brain's structure and function. This chapter will offer some fascinating details about the development of imaging technologies and what the main technologies used today actually do.

The First Look: Staining Techniques

Early staining techniques created by Camillo Golgi revolutionized the ability to see neurons and detailed cell processes. Golgi's technique could be used when looking at neurons under a microscope. Known as the black reaction, this method of staining cells allows the dendrites and axons of neurons to show up under the microscope. Staining continues to be a useful method for studying the cellular foundations of the brain (Fig. 5.1).

Another technique with a long history is EEG, which measures general activations in the brain is the method of EEG. New data processing techniques have made brain wave data even more useful. Since the 1970s new visualization technologies have been developed as well. One of the first methods was CT, computerized transaxial tomography. Visualization technologies evolved further in the 1980s with MRI, magnetic resonance imaging. Each of these technologies gives a piece in the puzzle. Although the pictures are not like a camera photograph, they reveal a great deal when coupled with computational modeling methods.

C. A. Simpkins and A. M. Simpkins, *Neuroscience for Clinicians*,
DOI: 10.1007/978-1-4614-4842-6_5,
© Springer Science+Business Media New York 2013

Fig. 5.1 Staining

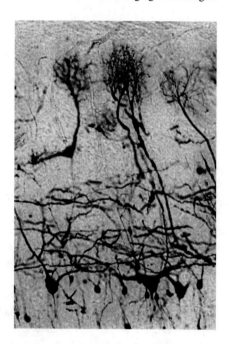

All of these imaging methods, shine light on the inner workings of the brain, both structure and function.

From Early X-Rays to CAT and CT Technologies

The discovery of X-Ray technology by Wilhelm Conrad Rontgen (1845–1923) was a great contribution that won him the Nobel Prize in 1901. This early brain imagining method evolved into the modern CT or CAT scan. A die is injected into the blood that will increase the contrast of the image. The individual is placed into a machine that scans the head. X-rays are passed through and recorded by detectors on the opposite side. The CT scanner rotates slowly around 180° taking measurements at a number of angles. Finally, a computer constructs the image of the brain (Fig. 5.2).

CT provides a static perspective of brain structures, but neurons have an active metabolism with a flow of blood known as cerebral blood flow, CBF. By measuring the CBF, neuroscientists can study the dynamic qualities of brain activity. A method was developed where a radioactive isotope, xenon 1.33 (^{133}Xe) is ingested. It emits a low gamma radiation that remains in the bloodstream. The blood carries the isotope to the brain where its rate of clearance from different regions is quantified as an rCBF. A special technique for imaging rCBF is known as single-photon emission computed tomography (SPECT). SPECT can make an image in 3-D of a radioactively labeled contrast agent. SPECT is an inexpensive alternative to the

Fig. 5.2 CT scan of a tumor

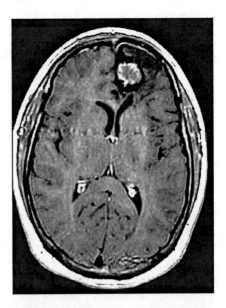

Fig. 5.3 Pet scan

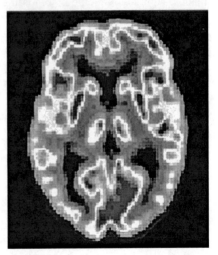

more recent PET. However, a disadvantage is that the isotope stays in the brain for up to 2 days, and thus only a one-time exposure is possible using this method.

PET

Positron emission tomography (PET) is a newer visualization method. Similar to SPECT, PET uses a radioactively labeled tracer substance that is administered intravenously. These radioactive isotopes have a short half-life in the brain and can be directed to a specific target. Each radionuclide is sensitive to particular physiological parameters such as blood flow, glucose, oxygen metabolism, or receptor density of neurotransmitters at certain regions of interest (ROIs) (Fig. 5.3).

Glucose mixes with blood in the brain. The more active an area of the brain is, the more glucose it receives. Therefore, glucose can be a good indicator of the dynamic qualities of brain activation. Researchers have found that measuring glucose metabolism is a better measure of neuron activity than CBF. A more recently used label, O^{15} may be an even better indicator of mental activity (Zillmer et al. 2008).

After the injection of the isotope, a laser shines near infra red light through the scalp. The scattering and absorption depends on the activity in the brain: the more active areas pick up more of the radioactive tracer. As the tracer decays, it emits a positron that moves a short distance until it collides with an electron. The collision causes the emitting of two photons that travel in opposite directions. Detectors all around the scalp register the energy from the traveling electrons. When multiple detectors pick up energy at the same time, the computer assigns it to the same origin and calculates the position of activity as a neural hotspot. PET scans involve millions of calculations from the detectors, estimating their origins. A mathematical subtraction method, comparing to a control scan, is used to isolate blood flow patterns as they relate to mental tasks.

PETs have been used to detect levels of activity in people who have suffered some kind of insult to the brain from tumors, strokes, or brain injury. But they are also used as a research tool to detect brain activity while performing mental tasks, to compare males and females, and for people with mental disorders.

MRI and fMRI

MRI or magnetic resonance imaging is another type of scanner that works on a different principle than X-Rays or PET. Nothing is injected into the body as in PET and no X-Ray is directed through the head. MRI is based on some principles of physical chemistry: Atoms rotate depending upon the number of electrons they have and their atomic weight. Most people have probably seen a simple demonstration of this principle when a physics teacher placed a magnet near some metal flakes. All the flakes jumped into alignment with the magnetic field. Instead of using metal flakes, the MRI machine applies a magnetic field to hydrogen atoms. Hydrogen atoms have one electron, giving them an odd-numbered atomic weight with a distinctive axis of rotation. The MRI machine applies a powerful magnetic field through a superconducting electromagnet. The power of this electromagnet is 60,000 times greater than the magnetic filed of the earth's magnetic field (Kandel et al. 2000). This magnetic field pulls in one direction (Figs. 5.4 and 5.5).

Water molecules H_2O are made up of hydrogen and oxygen atoms. Virtually every area of our body and brain contains water, and therefore the MRI can measure tissues throughout the brain. The machine actually measures the protons

Fig. 5.4 MRI of love

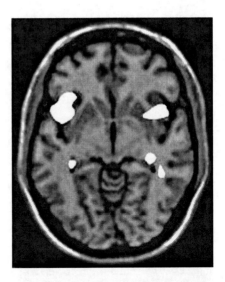

Fig. 5.5 MRI of fear

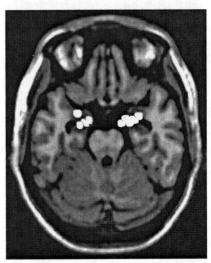

and spin of the nuclei of hydrogen atoms. Under normal circumstances with no magnetic field, the hydrogen protons spin randomly around the axis, with a net sum of zero among all the protons. When a strong magnetic field is applied, most of the spins of the protons pull up into line in the vertical direction.

The magnetization vector cannot be measured directly because, like trying to measure the weight of an object by just looking at it, we need to do something to the object to measure its weight. We can get a measurement of the magnetization vectors by flipping them into a transverse (horizontal) plane using a strong radio-frequency (RF) signal. The RF is applied to the protons at right angle to the vertical direction, and only the aligned hydrogen protons will flip into a horizontal plane.

Fig. 5.6 fMRI composite of
listening to music

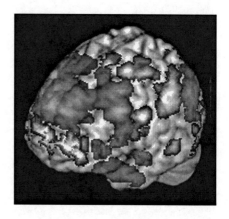

As soon as the RF is turned off, the affected protons spin back to their original
vertical orientation that was controlled by the magnetic field. This "spin-back"
induces a current in a receiver coil that is amplified and measured. Computer
visualization shows the density of hydrogen that has produced the spin-back in
areas all around the brain. What results is a picture of the brain or any other part of
the body that is being measured.

Two main types of measurements can be made from the data. First, we can
determine the density of the tissue based on the variation in the number of protons
returning the RF signal spin-back in different parts of the tissue being measured. A
second type of measure is known as T1 and T2, the magnetic relaxation times. T1
is the time interval needed for the protons to become magnetized by the electromagnet in the vertical direction. T2 is the time interval needed to return to the first
magnetized position after being oscillated by the RF. Protons with shorter T1
values, solids, appear white on the MRI scan and send a more intense signal. The
T1 values that take longer, fluids, appear dark. Images that show T2 values provide
different intensities. They give better contrast for lesions and tumors that show up
brightly. T2 also indicates more oxygenated regions.

Recently, 3-D images are being created based on MRIs using mathematical
algorithms that align slices of MRI data into its three-dimensional shape. MRI has
been used clinically to help with diagnosis and is being used more often as the
technology becomes more available.

fMRI

New advances have found that frequencies emitted by different molecules such as
glucose metabolism and changes in blood oxygenation yield different images.
fMRI uses an idea similar to MRI, but looks at blood oxygenation level-dependent,
BOLD, signals. When the brain is engaged in processing, more oxygen flows to the
area involved. The oxygenated areas show up in T2 contrasts. Then, statistical

methods are applied. The brain is divided up into a grid where each slot gets a unique signature. Fourier transforms are used to decompose the responses into magnitudes and phases over a range of frequencies. Then the responses are color-coded and a 3-D image with activations can be generated (Fig. 5.6).

When compared to MRI by superimposing the images, we can get a mapping of structure and function together for fMRI. These data are helpful, because it can give continuous results of brain activity over a time interval. Subjects can perform experiments while being measured, which provides brain images of cognitive and emotional processing in action. The fMRI does an accurate job of pinpointing where the signal is coming from at a system level across the entire brain at once, and it does so at a high spatial resolution. But fMRI is slow, and so exact real-time correspondence of activity to measurement is approximate. fMRI studies can be combined with EEG, which gives immediate, real-time results though they have lower spatial resolution.

EEG

Electroencephalograms (EEG), is one of the oldest methods for measuring brain activity. It records the electrical signal generated by neurons when they fire. Hans Berger (1873–1941) was an Austrian psychiatrist who is credited with being the first to record EEG from people. He based his device on the EEG technology that was used with animals. Berger is also credited with being the person to discover alpha brain waves. He was an assistant to Swiss psychiatrist and neurologist Otto Binswanger (1852–1929), who was the uncle of Ludwig Binswanger (1881–1966), the existential psychiatrist. Berger did research on lateralization of brain function in a lab with Vogt (1870–1959) and Brodmann (1868–1918).

EEG has been a useful research tool because of its ability to give precise real-time data. Although the output cannot be pinpointed to specific areas, the timing is accurate. Thus, combined with imaging technology, EEG has added a great deal to our understanding of the brain in action.

When undergoing an EEG measurement, a net of approximately 256 electrodes is placed on the scalp, like wearing a cap. The electrodes record small voltage differences from the scalp. Using a reference electrode on the earlobe or tip of the nose as a baseline, a voltage difference is measured. The voltage differences are a summation and averaging of post-synaptic signals. Waves are plotted from the frequency and look as follows (Fig. 5.7):

Beta waves are recorded when subjects are awake and excited. These waves are seen when people are involved in analytic problem solving, decision making, or processing of information. There are several different levels and types of beta waves. SMR (sensory motor response) waves (12–15 Hz) come from relaxed focused attention. Beta 1 (15–20 Hz) occurs with heightened focus. Beta 2 (20–38 Hz) is seen with anxiety and/or heightened alertness. There are also some variations in the patterns of beta activity. For example, the amplitude of beta is low

Fig. 5.7 EEG

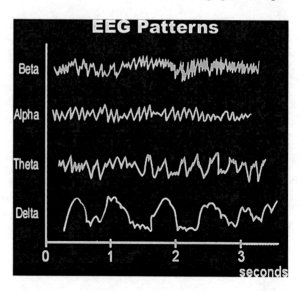

with multiple and varying frequency when people are active, busy, or even anxious. Rhythmic beta with a dominant set of frequencies may occur from drug effects and with certain pathologies. Gamma waves (26–70 Hz) are at the high end of beta. They are fast patterns of brain waves involved in perception and consciousness. Sometimes they are considered part of the class of beta waves. Gamma is seen during REM sleep. Gamma waves have also been found in the dual effect of meditation, where subjects are both deeply relaxed and highly alert at the same time.

Alpha waves (8–12 Hz) occur when awake while also feeling relaxed and calm. The alpha state may be a bridge between the conscious and unconscious. People who are in alpha can act more efficiently and calmly if called upon to do something.

Theta waves (4–8 Hz) are found in light sleep and during the transition between wakefulness and sleep. These waves are often recorded during creativity, intuition, daydreaming, meditation, and when involved in inner focus of attention. They tend to reflect activity from the limbic system.

Delta waves (0–4 Hz) are the slowest waves and occur during deep sleep. They are also found during experiences of empathy and often reflect unconscious activity.

ERPs

Researchers have developed another tool from EEG, known as evoked potential, EP or event-related potential, ERP. When people are presented with a stimulus, the brain has a specific electrical response. ERPs can measure complex psychological

Fig. 5.8 P300 comparison

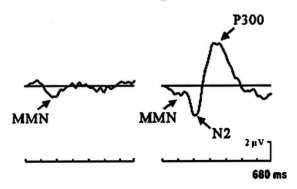

No Response and
P300 Response

processes occurring in the brain as a reaction to a stimulus. But this response tends to be hidden by the generalized EEG activity. Computer technology removes all the averaged EEG responses, leaving a brief, 500-millisecond ERP trace. ERPs are useful for doing experiments with visual, auditory, and sensory stimuli. The specific regions on the skull where these pathways are processed is measured.

A typical ERP experiment involves what is called an "oddball" task. Subjects are presented a number of similar stimuli with an occasionally different stimulus mixed in. They are supposed to press a button whenever they detect something different. An ERP has positive and negative components. When the stimulus is detected (the oddball), there is a brief latency period with no response, known as the N 2. This is thought to represent the moment the person recognizes that this is the oddball stimulus. Then comes a spike that indicates the response. Young adults will show the spike at around 300 ms, known as the P300. When a stimulus is not attended to, there is little change in the ERP signal (Fig. 5.8).

The response time becomes longer as people age. Characteristic ERPs have been found for brain responses to different stimuli. For example, P200 is associated with visual stimuli and P600 is related to hearing or reading grammatical errors in language.

Other Visualization Methods

MEG picks up magnetic fields at the scalp using superconducting wires that are cooled with liquid helium. The measurements can detect currents in the neurons. Like EEG, MEG gives a high temporal resolution for accurate readings of timed responses to stimuli. Although the spatial resolution is better than EEG, it is not as accurate as fMRI. MEG is a large, expensive device, and so it is not usually available.

Transcranial magnetic stimulation (TMS) offers another dimension to measurement. It can temporarily inactivate or activate a section of the brain using rapidly changing magnetic fields. The pulses cause a group of neurons in the cortex to depolarize and discharge an action potential. These activations cause a temporary alteration in brain activity that shows up on a PET or fMRI. Subjects sometimes feel a corresponding change especially when the pulses are in the motor area. In this way, researchers can study what happens when they alter a specific brain region.

Repetitive TMS, called rTMS has longer lasting effects and has been used experimentally to increase or decrease cortical activity. Although the exact mechanism is not clearly understood, researchers believe it may alter LTP, long-term potentiation, a process involved in learning at the synaptic level, and LTD, long-term depression, also at the synapse (Fitzgerald et. al 2006). rTMS is being used as an experimental treatment for major depression (Gershon et al. 2003) and for Parkinson's Disease (Strafella et al. 2006).

Conclusion

Neuroimaging technologies give us information about the structures and functions of the brain. We can now visualize where the brain has been damaged to better understand how the anatomy of the brain correlates with deficits in thoughts, feelings, and movements. More sophisticated experiments combined with fMRI and EEG are helping to know what the brain is doing as people perform cognitive, emotional, and behavioral actions. And technologies such as TMS allow us to alter the brain temporarily and observe what changes in behaviors. As you come to understand what these technologies reveal, you will gain useful information for devising techniques that can enhance brain structures and functions.

Chapter 6
Modeling the Brain

Human beings have always speculated about the nature of things. These speculations have taken form as models. Though we cannot directly know the brain, models can allow us to approximate it. Therefore, neuroscience develops models of the brain.

General Approaches to Modeling

People have been building models to explain reality since the beginnings of civilization. Commonly, a symbol system is used to represent what is experienced. Methods such as rituals and ceremonies were developed in ancient cultures as representative models to connect the inner person to the greater universe. Mandalas are an example of a kind of ancient symbol system that characterizes interrelationships of the individual with the greater universe through pictures and shapes.

Another role that symbolic systems have often played through history is as a way to help control and predict the world. Symbols are seen as signs of how the universe is ordered. People can learn to read these signs and make predictions. Then, when incorporating these predictions as guides to action, certain actions will be better for bringing about a desired result in the real world. Both these roles of symbols are being incorporated into neuroscience models: to represent the brain and to understand, predict, and direct it for healthy living.

Top-Down and Bottom-Up Models

There are two general categories of approach to model making in neuroscience. One is to start at the foundation and build up (bottom-up) and the other is to begin by formulating what we are looking at and then analyze it down to its smallest

C. A. Simpkins and A. M. Simpkins, *Neuroscience for Clinicians*,
DOI: 10.1007/978-1-4614-4842-6_6,
© Springer Science+Business Media New York 2013

parts (top-down). Proponents of top-down theories of the brain hold that first we need to conceptualize what the mind/brain does, and then, later we will be able to find out how it might implement it. These theories conceptualize that the process starts with the representations and higher-order processing. Thus, when we try to catch a ball, a top-down theorist might say that first we have the recognition that we want to get over to that ball to catch it. Then we try to catch it.

Bottom-up conceptions begin with the basic neuronal units and move up to simple connections, and then more complex networks and systems until everything comes together as a unified, emergent activity. So, with our ball example, the sensory input is received through the eyes, with light hitting the retina, traveling through the visual system as the information is processed. Then it must be somehow sent over to the motor system, where it transforms into motor movement. We run with arms extended and hands open to catch the ball.

Neuroscientists have often been reluctant to engage in top-down theorizing, preferring to gather as much data as possible first, working bottom-up, so that the empirical facts will lead to supportable conclusions (the traditional inductive method of science). But whenever a scientist decides to do one experiment rather than another, to choose one hypothesis over a second hypothesis, he or she is operating with bias due to an implicit theory, one that guides the choosing of this hypothesis or experiment over the other (Popper 1965).

Many neuroscientists have come to recognize that an explicit guiding model is helpful for directing research that will yield meaningful results (Churchland 1986). Ultimately, neuroscience is hoping to find a unified theory of the brain that will help to guide research and integrate together the data from the study of brain systems, from the most high-order reasoning down to basic drives for food, sex, and sleep.

Today's computational theories tend to combine these two approaches in various ways. Recognizing the brain's complexities, these newer models are sophisticated. They look for the integration of both directions working together seamlessly and harmoniously. But they do so in different ways. Top-down, bottom-up models are still used at times to help clarify a trend, but more often, theorists are thinking in terms of fields, matrices, and complex interconnections.

Mathematical Modeling Defined

Mathematics has been a helpful tool for working with the complex, multi-factored dynamic data the brain presents. Typically, in the Western social sciences, and especially in psychotherapy, verbal models have been developed to explain human individual, interpersonal, and group behavior. More recently, science has developed sophisticated ways to model using mathematics as an alternative to verbal concepts.

Mathematical models are computational simulations, usually involving computers, which can be used for the study of physiological phenomena and functions or dysfunctions of the brain, organs, or behavior. These models can be formulated to describe different levels, ranging from microscopic models of neurons, for example, to macroscopic models of addictive behavior, as an example of a macro-context. A mathematical model yields equations with solutions that approximate the actual behavior of the brain, organ, or behavior under certain conditions.

When the word *mathematical* is combined with the word *model* one sees a broad spectrum of reactions. For some, the combination seems to endow the model with an aura of truth and exactness. Whereas, many in the helping professions, who are more accustomed to the use of concepts as the building blocks of theories, may feel vague uneasiness with the unfamiliar lexicon. To those unfamiliar with mathematical modeling, we encourage an open mind. Mathematical representations can shed light on dark territory, when used well. And to those who tend to turn to math first we caution that mathematical models, as with all types of scientific models, should always be considered as approximations of reality, never the reality itself.

One limitation of scientific research in general is in the need to assign dependent and independent variables. Typically, in designing research projects, all the dependent variables are kept invariant in order to light up the variable of interest. Mathematical models often have to keep some of the variables invariant, or even disregard them altogether as if they are not even there. Such assumptions and simplifications of certain parameters often become necessary in order to make the equations feasible to calculate. So, in this way, mathematical models, much like traditional studies, are always somewhat of an approximation of reality, not the reality itself. However, an advantage of mathematical modeling is the level of complexity that can be addressed. With the advent of differential equations and certain newer mathematical systems, such as linear algebra, many more factors can be varied at once, to allow for more complex and realistic experiments that come closer in resembling our subtly multi-textured realities.

Just as the map is not the territory it describes, mathematical models are also just representations, not reality itself. However, with the intelligent application of scientific theory, technology, and mathematics, combined with the insight and intelligence of the human mind, we can make better and more useful approximations.

Bayesian Inference: Estimating the Likelihood of Our Models

One problem for science continues to be the confirmation of evidence. When we make an observation that seems to confirm a theory, how do we know that this observation is reliable and valid? The classic example of this dilemma is the raven. Suppose we have a theory that all ravens are black. We have gone out into the wild and observed many ravens. Every raven we have seen is black. Although we might make a theory that states, "All ravens are black," how certain can we be that being black is a true characteristic of being a raven? There is always the danger in the inductive method of observations leading to theories that we just have not encountered the evidence that would disprove the theory, such as seeing a white raven. But how certain are we that a white raven is not living just around the next corner? And as a second concern, when do we know that we have enough evidence to support our hypothesis?

Bayesian Inference offers a new way to answer these questions about confirmation and evidence. The basic idea is that science is an iterative process where we are engaged in gathering new information about an area of interest, continually updating our knowledge about it. Old data is integrated with the new data. Bayesian inference offers a formula for quantifying a probability that guides us in either raising or lowering the likelihood of the hypothesis. Thus, we have a way to accommodate for new evidence to update our theories.

Thomas Bayes (1702–1761) was a Presbyterian minister. But he also was a mathematician who worked on probabilities. During that period of history, people were conceptualizing probabilities as forward looking, that a probability would tell you how likely it would be to choose a white ball from two containers holding a specific number of black and white balls. Bayes turned the problem around and offered an inverse probability solution. What Bayes solved was: Given that you chose a white ball, what can you say about the white and black balls in the containers?

Bayes' idea was that probability is similar to a partial belief rather than simply a determiner of frequency. One of his basic principles was to start from an initial probability P_i of any hypothesis, h. We will call that $P_{initial}$ of h (the initial probability of the hypothesis) or more simply, $P_i(h)$. Then, as you acquire new evidence, e, you will update your initial probability for a final probability based on e. This idea can be expressed as the following equation: $P_i(h) \rightarrow P_f(h) = P_i(h|e)$.

The right side of the equation, $P_i(h|e)$, can be stated in words as the initial probability of the hypothesis h, given the evidence, e.

Bayes offered a way to calculate the probability of the initial hypothesis given the evidence as a function of the probability of the initial hypothesis:

$$P(h|e) = P(e|h)P(h)/P(e),$$

where P is the probability, h is the hypothesis, and e is a piece of evidence. The formula reads thus: the probability of the hypothesis given (or in the light of) the evidence is equal to the evidence given the hypothesis times the probability of the hypothesis divided by the probability given the evidence. The idea is that as evidence comes in, we tend to update our beliefs. Thus, as we gain more evidence about our hypothesis, we will update our belief about the probability that the hypothesis is true.

Typically, neuroscientists take a scientific realist perspective, believing that we can uncover causal mechanisms about how the brain works, and that the aim of neuroscience is to uncover these hidden structures. Bayesian inference and other mathematical modeling methods allow us to estimate how likely it is, given the evidence we gain from EEG and imaging methods such as PET and fMRI, that our hypotheses about these structures is correct. Further inferences are made when scientists hypothesize about the connection between these brain structures and our cognitive, emotional, and behavioral processes. One way to deal with these sorts of issues more realistically is by considering them as likelihoods that will be continually updated as new evidence emerges. Bayesian inference is one way to do that.

Our Stochastic Being: Estimating What Comes Next

Human beings behave stochastically. What does this mean? The word *stochastic* comes from the Greek, to aim or guess. A stochastic system is non-deterministic. For human behavior, this implies what most will agree is true about people: We cannot always predict with absolute certainty how someone will behave next. Only some of our actions are predictable. This is because there is always a random element at play. So, in this sense, human behavior can be seen as stochastic. Psychological theory predicts many of the behaviors that people are likely to exhibit under certain circumstances. But there are always those random, unpredictable elements that make the exact prediction impossible.

Many of the functions of the brain are being modeled as stochastic systems. For example, when working with a stochastic system, we try to discern the deterministic components of the system while eliminating the non-systematic, random aspects. One way that the random quality of stochastic processes can be described is by using probability distributions. In this way, even though the system's future may be indeterminate, we can start from a known initial condition. Then, even though there are many possible ways behavior might go, we find that some directions are more probable than others, given the starting point.

Andrey Markov (1856–1922) was a Russian mathematician who devised an explanation for how stochastic processes behave. This behavior is known as the Markov property. When a system has a Markov property, it means that given the present state, future states will be independent of the entire history. Future states come about from a probabilistic process, not a deterministic one. To follow the process as it travels along its way is called a Markov chain. To understand a Markov chain, imagine the classic example of a random walk. Perhaps you have been walking through the grass in an open field. Each next step can go in any direction. Previous steps do not influence the next step you take. There is an equal probability that you will step in any possible direction.

This process is expressed mathematically. The equation for a Markov chain with a sequence of random variables, x_1, x_2, x_3, which have the Markov property that the future and past states are independent of each other, is stated as this mathematical formula:

$$P\left(X_{n+1} = x \mid X_n = x_n, \ldots, X_1 = x_1\right) = P\left(X_{n+1} = x \mid X_n = x_n\right).$$

So, when using a Markov chain to predict what will happen when you have a sequence of independent and identically distributed possibilities, the probability for the next state, x at a time, n + 1, will depend only on the current state. In order to predict what the system will do next, the idea is that you do not need to know the entire history; you just need to know where you are.

Modern brief therapy has turned toward a more stochastic, Markov-like approach by starting with where the client is here and now. Earlier psychological models saw behavior as determined. Therapy then proceeds by helping people to examine and alter the causes–effect relationship that took them on a causal path to illness. The stochastic view does not assume determinism as simply cause and effect. The past, even if it has been filled with obstacles, is not the determiner of where the client will go next, and so emphasis is not past-gazing but rather, forward-looking. We have all known people with disturbing histories who go on to transcend their circumstances to accomplish great things. History, from this view, is not a determiner of where you are going, only of where you are now. Where you go from here, your future, is an open field of choice.

Self-Regulation: Finding the Optimal Path

If our future is not determined, how do we navigate through life so seamlessly? How can one organ, the brain, help direct our actions without being predetermined as to where to go next? We can walk with ease and make instantaneous adjustments to climb up steps or avoid a stone in our path. We can accurately reach for a piece of popcorn while watching a movie in a dark movie theater. We can pick our loved one out of a crowd and recognize the sound of a friend's voice. These seemingly simple tasks present daunting challenges filled with uncertainties and multiple possibilities. And yet, our brains know how to do them in an optimal fashion without training or anyone there to tell us how. Just the simple act of walking, which may seem trivial, is an extremely difficult task for a robot to do. In addition, though walking over one type of smooth constant terrain is a very difficult problem for a robot, it is even more difficult for a robot to dynamically adapt to changing terrains, movement patterns, and disturbances (which humans do with ease). Recognizing an object or inferring what one should do next in a new situation has required some of our brightest minds to implement in a machine and at a much more rudimentary level than a young child can do.

Neuroscience is tackling these problems, trying to understand how the brain might be achieving these skills with such ease. One of the recent theories with some potential answers is optimal control theory. Neuroscientists are developing models of the brain that are based on the idea that the brain works optimally (Simpkins, de Callafon and Todorov 2008; Simpkins and Todorov 2009).

Optimality is based upon a simple and yet profoundly useful concept: "Determine the best strategy for controlling a dynamic system" (Stengel 1994, p. vii). Many neuroscience models are based on the idea that the brain is operating according to this concept of optimality. Affect regulation methods in therapy draw on this core idea: To guide clients toward developing the best strategy for controlling their own emotions and sense of well-being.

The Bellman Equation and Dynamic Programming

Richard Bellman (1920–1984) was an applied mathematician who is considered a founding father of optimality and modern control theory. He invented a dynamic programming equation in 1953, known as the Bellman equation. This equation offers a method to find a measure, called the value function, which gives the optimal value of a problem as a function of the state variable of that system, x. The function V(x) (value function) describes the cumulative or instantaneous optimized value of a particular decision. The state-dependent set of optimal choices is called the policy function. He went on to formulate a principle of optimality: If the policy function is optimal, then whatever the initial state and decisions are, the remaining decisions will be an optimal policy for that state.

Here is the equation Bellman created to describe an optimal policy for a function, V(x):

$$V(x) = \max_{u \in C(x)} [F(x, u) + \beta V(u)], \quad \forall x \in X.$$

Here β, $0 < \beta \leq 1$. represents the discount factor, x is the state vector, u represents the optimal policy function, $C(x)$ are the constraints for the policy (such as energy or strength), and $F(x, u)$ is the instantaneous reward function.

You may wonder how this equation relates to psychotherapy. Bellman's equation has been applied in reinforcement learning where it can be used to predict what the optimal action should be. Reward becomes the action taken that is based on the highest expected return. We see these different factors at play with our clients, who are operating in a certain state, with various constraints in terms of their environment or personality, and the rewards they seek in life. By thinking in terms of these variables, you can unlock keys to foster better choices for your clients.

Optimality is based on making the decision, which either minimizes cost or maximizes reward. Cost and reward are related in that they are complimentary for the defined problem. For example, think of the best route between two points, such as your house and your work. One might immediately think that the best route is the shortest one. But you might make an optimal choice for yourself that is a slightly longer path because you know there will be less traffic. So, instead of minimizing the distance you decide to minimize time and maximize ease. Thus, the costs are a longer route, but the benefits are quicker time and easier driving. Weighing out the costs and benefits helps to find the optimal solution for your "state" which in this case is to want a pleasant drive home. The brain makes optimality calculations all the time (Todorov 2006). It quickly calculates the adjustments needed for the most optimal solution to the situation.

There are two main types of optimality, finite horizon and infinite horizon. In finite horizon, you build an optimal trajectory over time whereas in infinite horizon, the present state maps the optimal decision. If we characterized therapy in terms of these two types of optimality theory, brief therapy could be characterized as a finite horizon approach that achieves the most optimal state possible in a finite amount of time. By contrast, an infinite time-horizon therapy might be seen as having no expected completion date but rather as striving to make continuous adjustments, which are expected to bring about the best improvement possible in the patient over the long-run.

Stochastic Optimal Control

Bellman's formulation was deterministic and continues to have many applications, especially when the system is definable, such as in machines. But the human being and the brain present us with undefined elements. Life is filled with random disturbances making it difficult to optimize all the factors with certainty. A new form of optimal control has been developed that is known as stochastic optimal control to deal with uncertain systems. We can apply this to human beings as well. As was discussed earlier, a stochastic system is non-deterministic. Thus, a stochastic controller that seeks optimality estimates the system's dynamic state using probabilities and statistics.

The method involves refining your policy (method) so that for each step or iteration, the policy will tell you how to translate into more optimal actions. Thus, feedback is essential. Eventually, every action you take, given an initial condition, will take you on an optimal trajectory to the goal. The idea is to create a method that maps a particular set of states, whatever the current state is, into actions. This map can then guide behavior into the appropriate action to take for the expected reward. This program is done in an iterative fashion, step-by-step, beginning from a blank starting point and continually optimizing by lessening errors and increasing successes over all the possible decisions until the optimal map is found. Therapy works in an iterative way, with clients coming in each week, taking feedback as they update their adjustment until they find a happy, healthy path for their life.

Computational Programs for Modeling the Brain

The brain is an amazingly complex system with its many billions of neurons and supporting cells, interfacing in complex systems of interaction. Bayesian inference, stochastic systems, Markov processes, and optimal control theory have been used to model the brain. We have included a few simplified examples of computational models that will have adaptations to understanding of brain systems involved in psychological problems: the connectionist theory and the dynamical system theory.

Connectionist Model

Connectionist theories hold that neurons make connections when they receive an input and send it to an output. The connections between inputs and outputs are varied and unique. And these connections are interconnected in various ways. The many interconnections are occurring in parallel, at the same time.

Connectionist theories study the connectivity relations between these complex collections of inputs and outputs.

Tensor Theory

One connectionist theory known as the tensor theory found that these connections are much like a mathematical tensor (Pellionisz and Llinas 1982). A tensor is a way of describing something that can be expressed as a multidimensional array, or a space, relative to a choice of a basis for that particular space on which it is defined. These spaces are made up of scalars (single points in space), vectors (lines that move in a certain direction), and matrices (groups of vectors describing 3-D space).

The tensor theory was first formulated based on the cerebellum. Much is known about the structure of the cerebellum, with its limited types of neurons and simple input and output processing. We also know about the functions of the cerebellum. What the model tries to explain is how the component parts of the cerebellum work to bring about the functions it performs. Modeling how one brain system, visualized as a 3-D space is transformed into another 3-D space, accomplishes this goal of linking structure to function. The relationship between these two spaces can be performed by mathematical operations found in linear algebra. Matrix multiplication and transpose are examples of operations that show a direct relationship between mappings from one space into a completely different space. The representations of separate brain system are positions in these spaces and the brain computations are the coordinate transformations from one space to another. Generally stated, the tensor network hypothesis states that a neural network implements whatever is its function as a connectivity matrix to transform input vectors into output vectors. There are two aspects to this theory: that even though each neural-network system has its own basis in space, the different neural networks are in direct relationship to each other. And the second idea is that these systems work in parallel.

In the diagram below of a transpose, a particular type of transformation, with numbers taken from linear algebra, the rows become columns and the columns become rows. You can see that the configuration changes, and yet there is the relationship between the numbers that is preserved. Transposes can prove useful when the new positioning is easier to work with or understand.

```
0 1 2 3          0 4
4 5 6 7          1 5
                 2 6
                 3 7
```

These computational theories of the brain have evolved into a parallel distributed processing or PDP which is the style of computing that transforms one pattern into another by passing it through a large configuration of synaptic connections (Churchland 1988).

Dynamical Systems Theory

The dynamical systems theory (van Gelder 1995) offers another computational model. It is based on the idea that cognitive systems are a realization of the physical systems, somewhat like the connectionist view. But it adds several other components: time and place. Cognitive systems should be understood as acting in the real world in real time.

This situation comes from evolution. Even though the computational potentials of the brain are vast, they are finite. Over the eons of time, some of the tasks were off-loaded, or shared by the body and the environment to make the best use of the brain's resources. This theory is echoed in the writings of the great philosopher, Martin Heidegger (1889–1976). Heidegger believed we never understand ourselves outside of how we are situated in the world. We are always beings-in-the-world (Heidegger 1962). Dynamical Systems Theory holds that any accurate picture of the computational neural system that truly explains brain processing must include the brain's interactions with the world and our body as it occurs through time. According to this theory, even though it is helpful to look *inside* the brain to the neurons, neuroscience should not ignore the constraints and resources, which come from the body and environment on the *outside* (Clark 1997). Experience-based neural plasticity can be explained by such a theory. Nor should we ignore the moment-by-moment changes that inevitably occur as we travel through time as well moving in space.

Dynamical systems theory uses difference equations to model human behavior in discrete time and differential equations to model it in continuous time. As our brief introduction to calculus (see appendix) explains how equations that contain derivatives, known as differential equations, reveal change over time.

We can think of cognition as a multidimensional space of all possible thoughts, feelings, and behaviors. A particular thought, feeling, or behavior is a possible path (or function) that can be followed under certain environmental and brain possibilities, captured by sets of differential equations. These equations are used to represent the person's cognitive trajectory through the highly dimensional state space. This theory uses such terminology as state spaces, points, cyclic and chaotic attractors, trajectories and deterministic chaos.

Conclusion: Looking to the Future

As our models, technologies, and statistical methods improve, we will forge ever deeper into the recesses of the brain to uncover its secrets and activate the vast potentials that lie hidden, just waiting to be discovered. And as we do so, we find ways to translate models into practical techniques for practice.

Part III
Neurobiology from the Single Neuron to Complex Networks

As the brain is explored with the help of new technologies, modern neuroscientists are discovering many ways that individual units, structures, and systems work in concert as a network to produce functions such as language, memory, attention, emotional attunement, and interpersonal relationships. The organization of the nervous system can be appreciated as an exquisitely orchestrated symphony, where all the elements come together to make a masterpiece. Form and function interweave. The many parts are precisely attuned and interrelated to bring about something that is far more than just their sum. Understanding the brain in terms of the structural and functioning systems will be helpful when working therapeutically to alter entrenched patterns.

The chapters in Part III begin with the building blocks of neurons and follow a path through neurotransmitters in Chapter 7 and to brain structures in Chap. 8. Chapter 9 discusses the pathways combining the many structures and neurotransmitters and Chapter 10 shows how they all come together as neural networks. Cognitive, emotional, and social systems of the brain interact, with interrelated functions. These interactions on multiple levels can be utilized for facilitating therapeutic change, as later chapters describe.

Chapter 7
Neurons and Neurotransmitters

Therapists work with behavior, thoughts, and feelings. In a sense, we see the culmination of many different influences that bring a person to therapy such as problems in the larger environment, difficulties in interpersonal relationships, troubling inner experiences, and a malfunctioning nervous system. We know that our experiencing is linked to the nervous system, and that experiencing the environment influences the nervous system. We also know that the nervous system is regulated by neurotransmitters and neuronal interactions, and so you can learn something more about what your clients are going through when you know what happens at the microscopic level.

This chapter shows how neurons react under normal circumstances and what they do when something goes wrong. We provide an overview of the structure of neurons and how they communicate with other neurons at the synapses by sending signals. Neurotransmitters activate or inhibit these signals. Our therapeutic methods influence this neurotransmitter action, as do psychopharmacological drugs. What goes on at the neuronal level gives you new maps for your therapeutic journey with your client.

The Impulse for Action Begins in the Neurons

The nervous system is made up of neurons, nerve cells that extend throughout the entire body. Neurons are supported by glia, which act as support for the neurons. The neurons interlink, forming a network of electrical activity that correlates as an underlying component for everything that we do, think, and feel.

Neurons are the functional units for communication. There are currently believed to be between 90 and 100 billion neurons, with at least 80 billion involved in cognitive processes. Each neuron interconnects with hundreds of other neurons. If you lined the neurons up side by side, they would span six hundred miles!

C. A. Simpkins and A. M. Simpkins, *Neuroscience for Clinicians*,
DOI: 10.1007/978-1-4614-4842-6_7,
© Springer Science+Business Media New York 2013

Fig. 7.1 The neuron

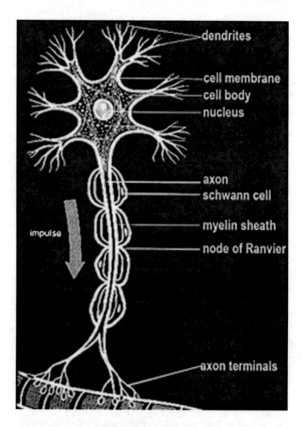

Neurons are made for the specialized processing of information. They have four distinctive parts: dendrites, cell bodies, axons, and synapses. Each part has a specific role to play in the process. It is triggered to play its role and acts as part of a web of interactions in a network. Neural Networks, described in Chap. 10, provide a model for how learning to takes place, tendencies form, and habits become entrenched. At the foundation of these larger neural network systems, the neuron plays its role (Fig. 7.1).

The Role of Each Neuron Part: Dendrite, Cell Body, Axon, and Synapse

The dendrites (from a root word for *branches* in Greek) serve as input zones, taking in information from other neurons. The dendrites come from one side of the cell body. Like the branches of a tree, the further away from the cell, the smaller they get. The ends of the branches can receive either excitatory or inhibitory inputs. Tiny spines, resembling the shape of lollipops, extend from the ends to accept the excitatory inputs. Inhibitory inputs are received at the neck of the spines.

Dendrite spines have become the focus of research because of their ability to alter themselves. These spines on the dendrites are one of the sources of experience dependent neuroplasticity (Grutzendler et al. 2002; Trachtenberg et al. 2002). Therapeutic interventions can foster excitation or inhibition that leads to these kinds of neuroplastic changes in the connections between neurons.

The cell body is the integration zone consisting of a nucleus and cytoplasmic organelles. Here is a brief account of what occurs in this integration: The DNA resides in the nucleus. Attached to the nucleus is the endoplasmic reticulum with ribosomes involved in translating the DNA into proteins. The mitochondria are the energy producers that fuel the process by producing adenosine triphosphate (ATP) that the cell uses for energy. The Golgi apparatus packages neurotransmitters and proteins together and sends them off in vesicles that it creates. These packaged molecules roll down microtubules in the cytoskeleton of the cell to be transported. Thus, through this processing, the input signals from the dendrites are combined, altered, packaged, and then sent out to the axon for communication to other neurons.

An axon extends out from the other end of the cell body. It acts as the conduction zone, like a single wire carrying an electrical signal from one neuron to the next. Many are insulated by a myelin sheath made of glia cells (see section below). Similar to insulated wire, the myelin sheath on the axon keeps energy from escaping and saves energy. This insulation looks like little segments with separations or gaps between the segments. These gaps are important for the regeneration of the signal which speeds transmission of the electrical impulse without having to increase the diameter of the axon. Two types of cells provide this insulation: oligodendrocytes are in the brain (CNS) and Schwann cells are in the peripheral nervous system (PNS).

Finally, we have the output zone at the axon terminals. Here are the synapses where the important communication between cells occurs. The word *synapse* is Greek for clasp together, which is, in a sense, what the synapses do. The terminal end of an axon makes a connection with the spine of a dendrite. The Law of Dynamic Polarization predicts, dendrites generally receive synaptic inputs and axons generally send synaptic outputs. The dynamic polarization process of inputs, integration, and outputs involves a kind of computational procedure that helps to account for how neuroplasticity takes place. Chapter 13 describes the fascinating process of neuroplasticity in detail.

When the synaptic gap between two neurons is close enough, the electrical signal can simply leap across and keep going. But more often, the gap is too large for this to happen. With the larger gaps, the electrical signal is converted into a chemical, known as a neurotransmitter, and the chemicals swim across the gap where they are then converted back into an electrical impulse.

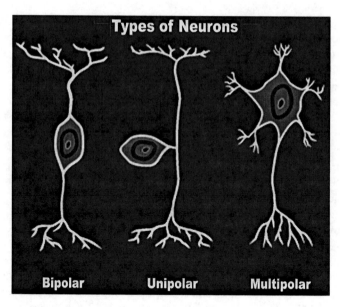

Fig. 7.2 Types of neurons

Different Kinds of Neurons

We see nature's creative diversity in the many sizes, shapes, and structures of neurons. These differences allow neurons to perform varied functions. The larger neurons have elaborate dendrites and axons, or they might cover longer distances, while the smaller ones simply have an axon and a few dendrites (Fig. 7.2).

Muscle neurons that regulate muscle movement are multipolar and are the most common type, with many dendrites and a single axon. They send signals from the central nervous system out to the muscles. Sensory neurons are unipolar, with a single dendrite at one end and a long single axon at the other end, allowing sensory input to be received from the periphery of the body and sent to the nervous system. Their unipolar structure makes it possible for the specificity of sensory information that we find. For example, the sensation of temperature separately from texture information. Interneurons, also known as associative neurons, are bipolar, connecting signals between neurons. They send signals between motor neurons and sensory neurons (Fig. 7.3).

Glia Cells

Neurons are not the only type of cells found in the brain. Glia (singular: glial) are the second type of brain cells. Glia, originally named *neuroglia* cells, are ten times more numerous than neurons. Rudolph Virchow (1821–1902) gave the name

Fig. 7.3 A glial cell

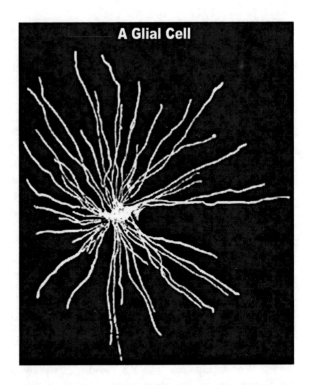

Neuroglia, or "nerve glue," to these specific neurons in 1859. He was a German doctor who has been credited with many discoveries about pathology. He is also known for his cell theory (Hoe et al. 2003, p. 61). Glia cells perform support nursing functions for neurons such as picking up waste, bringing food, providing nerve growth, and as one of our professors called it, "cuddling," (Pineda 2007) which holds neurons in place. They are also the cells that make up the myelinated insulation for the axons, helping signals to move more quickly and efficiently.

Glia cells may be far more important than one might expect. In a study comparing samplings from Einstein's brain with 11 normal men, Einstein's brain was found to have more glia cells throughout, and significantly more in the lateral parietal lobe (Diamond et al. 1985). This suggests that Einstein's brain functioned more efficiently than others. Increasingly, researchers are beginning to recognize that glias are also involved in many important neurological functions (Swaminathan 2011). They are essential for the development, repair, and plasticity of the myelin that covers the axons. They play a key role in keeping the environment of the neurons well-regulated, and so are being researched as key in disease prevention (Butt and Bay 2012).

How Neurons Communicate

Part of what makes the brain unique is the way that neurons communicate with each other. Every neuron has electrical properties that influence how they communicate. In a sense, the electrical signals function like a language, passing information along.

Each cell has a characteristic electrical charge when at rest, called the resting potential. For neurons, this resting potential is more negative than positive. What this means is that when inactive, the inside of the neuron has a higher concentration of negatively charged ions, cations, and the outside of the cell has a higher concentration of positively charged anions. Scientists believe this balance is the result of evolution, theorizing that we originally evolved from a watery environment with salt (Na^+Cl^-).

The Sodium–Potassium Pump Helps Move the Ions

Inside the cell, we find large, negatively charged molecules such as proteins that do not readily leave the cell. There are also more potassium K^+ ions inside the cell than outside. Outside the cell we see more positively charged sodium Na^+. Special doors allow specific ions (Na^+, K^+, and Cl^-) to flow in and out. Usually the doors are tightly closed. The doors open only when there is a disturbance. But like a leaky house, Na^+ tends to seep in even if the doors are shut. If left unchecked it would diminish the cell's negative balance, making the cell less likely to generate electrical signals. The neuron has a mechanism that prevents this from happening called the sodium–potassium pump. This pump rapidly pumps out around three sodium ions for every two potassium ions it takes in, keeping the equilibrium negative in the interior of the cell. This pumping action takes energy to maintain a balance between certain physical and chemical forces. Here we see a principle of the Yin-Yang opposites at work, where sending out more positively charged anions and letting in less, maintain a negative charge inside the cell.

Forces that Alter Electrical Balance

Different concentrations of ions produce certain forces. One type of force comes from osmotic pressure. Molecules tend to want to distribute evenly in an area. Everyone has seen this when steam from boiling water rises and diffuses into the room. Similarly, a high density of ions wants to flow into lower density areas. The osmotic pressure for sodium is to move out of the cell, whereas potassium moves into the cell.

A second type of force is electrostatic pressure. The old saying, opposites attract, is true at the molecular level. Particles with opposite charges attract and similarly charged particles repel. The combination of the osmotic pressure and electrostatic pressure creates *an equilibrium potential*. When the two forces are in balance, the cell is at rest. Most of the cells in the cortex have a resting membrane potential (V_m) of -70 mv (millivolts). Other areas of the brain have slightly different resting potential values. When a change occurs, the electrical potential alters and may become more negative or positive.

The Source of Communication: The Action Potential

Part of what distinguishes neurons from other cells is the way they undergo changes in their polarization to become more positive or negative. When the electrical balance changes, it becomes a signal. Herein lies the source of communication between the neurons.

These electrical changes can be slow and gradual (*graded potentials*) or large and sudden (*action potentials*). Graded potentials polarize the neuron in a succession of small steps, driving the cell gradually toward a firing threshold. When depolarizing goes from negative to less negative, it is excitatory. Once excited, the neuron might go from -70 to -60 mv. By contrast, hyperpolarizing drives the cell to become more negative, such as from -70 to -80 mv, and is inhibitory. With both excitatory and inhibitory graded potentials, cell voltage eventually returns to equilibrium unless the cell receives more inputs.

The movement of sodium ions into the cell activates an action potential. When a certain threshold of depolarization is reached, usually around -40 mv, a nerve impulse is triggered in the axon at the place where the axon originates from the cell body, known as the *axon hillock*. This action potential creates a sudden, brief millisecond-long change to a positive charge in the inside of the cell. This spike to positive is called the *action potential*. Action potentials have an all or nothing property: They fire at full amplitude or not at all. Following the spike comes what is called the *after-potential*, where the interior drops to a more negative level than the resting potential. The cell then returns to its normal -70 mv equilibrium.

Many different types of channels let the ions pass in or out. Neurotransmitter molecules trigger *ligand-gated* channels that open or close in response to neurotransmitter chemicals. Others channels are sensitive to the voltage around them, known as *voltage gated*. Cell membranes can also stretch to create *stretch-gated* channels. Some gates are transient, opening only briefly such as Na^+ channels, while others have sustained openings such as with K^+. Some are very sensitive while others are not.

From Electrical to Chemical Communication Between Neurons

When the electrical signal reaches the synapse, it might be able to flow across if the distance between neurons is close and conditions are optimal. But more often, the distance is too large and something else is needed to communicate across the synaptic gap. Nature provides a second way for signals to be transmitted, by turning electrical signals into chemical signals. It does so by the use of neurotransmitter molecules.

The ends of the axons are filled with neurotransmitter molecules, just waiting to be released. Protein molecules hold neurotransmitter molecules together. As mentioned earlier, there are many different types of channels between cells. Here is how one type, the voltage-gated channel, works. Ca^{++} comes down and the voltage-gated channels open up to start a process called electrocytosis, which removes proteins that are anchoring the neurotransmitters. The neurotransmitters then migrate to the end of the terminal and lock onto proteins located at the end. As soon as they lock on, the presynaptic channels open and the high concentration of neurotransmitters spill out into the synapse where there is a lower concentration. They bind to receptors on the dendrite of the neighboring neuron that are specific to the neurotransmitter being sent. Once across the synapse, the signal becomes electrical activity again. Thus, the signal goes from electrical to chemical and back to electrical.

Neurotransmitters Carry the Messages

The neurotransmitters are chemicals that carry the message across the synapse, from one neuron to another. When trying to understand what neurotransmitters are and do, keep in mind that they are messengers. Some transmit the information to activate and others to deactivate. They carry the signal across the synapse to tell the next neuron whether to fire or not. This signal, acting in conjunction with other neurons, carries a meaning beyond a simple unitary electrical signal. Through networks of neurons excited and inhibited in patterned ways, meaningful signals tell the organism to run away or freeze, to feel pleasure, to cry, or to go to sleep or digest its food. Many of the activities of living become engaged one way or another as a result of the signals transmitted across the synapse.

Other Chemical Messengers

Neurotransmitters are not the only chemical messengers in the body. There are four main types of chemical messengers. Besides neurotransmitters, there are also hormones, neurohormones, and paracrine signaling. Hormones are chemicals that are secreted into the blood from endocrine cells. They influence cells at various locations around the body and tend to act quickly by traveling through the blood. Neurohormones are chemicals secreted into the blood by a neuron. These neuro-hormones alter other cells around the body. Paracrine signaling involves the secretion of chemical regulators that have a local influence on the cells close by.

Neurotransmitters act differently from these other chemical messengers by working through synaptic connections between neurons. Several criteria have been established for what makes a chemical classified as a neurotransmitter. It must be synthesized and released from neurons and have biochemical machinery within the presynaptic neuron to do so. The chemical must be released in response to an electrical signal and produce a physiological response in the postsynaptic target. Known antagonists of the transmitter can block the postsynaptic effects. Also, there must be mechanisms to terminate the action of the neurotransmitter such as chemical deactivation, recapture, reuptake, or diffusion.

Neurotransmitters

The imbalance of neurotransmitters in the brain has been linked to a number of mental disorders. Psychopharmacological drugs have proven to be helpful for schizophrenia, mood disorders, and anxiety. Psychotherapy also alters the balance of neurotransmitters, and evidence seems to indicate that the combination of drugs and therapy work best with serious disorders. Part VI deals with psychological problems and describes how specific neurotransmitters are involved. This section provides what you need to know to understand the action of neurotransmitters that are most relevant to psychotherapy. You can keep these actions in mind when devising your treatments, and later chapters provide methods and techniques.

Neurotransmitters are produced by the neurons. Drug therapies do not add anything that is not already potentially available within the brain/body system. Rather, they stimulate or inhibit neurotransmitter activity toward a better balance. Psychotherapy alters the balance of neurotransmitters as well. Some treatments involve deactivating an inhibitor, which brings a net effect of stimulation, or activating an inhibitor, which results in greater calm. (See Chap. 19 for how psychotherapy alters the brain). By readjusting the interactions among these opposite elements and forces, a new pattern is formed, to addresses the problem with a better balance.

Different Types of Neurotransmitters

Research has discovered hundreds of different neurotransmitters. But certain ones are seen throughout the brain while others are local. Neurotransmitters are of particular interest to clinicians, so we focus on them here.

Generally, neurotransmitters are divided into types according to molecular size. The two smallest types are amino acids and biogenic amines. Another group, neuropeptides, tends to be larger molecules. Neuroscientists know more about amino acids and biogenic amines, but recent findings indicate that there are many more neuropeptides.

Each neurotransmitter has a distinctive shape, somewhat like a key, and each receptor, the place on the receiving neuron, has a shape, similar to a lock. This allows a specific neurotransmitter to attach to a specific receptor, fitting together like lock and key. Sometimes only one neurotransmitter can fit to a receptor key. At other times several different neurotransmitter may fit and will compete for the same receptor. Although there is specificity, the process of transmission has room for flexibility as well. Some of the neurotransmitters travel along specific and identifiable pathways, such as the dopamine reward pathway, while others such as glutamate are more widely distributed around the brain.

The amino acids glutamate, excitatory, and GABA, inhibitory, provide excitation and inhibition throughout the brain. Another group, the biogenic amines, includes acetylcholine (ACh), dopamine, norepinephrine and serotonin. They are slow and act as modulators. The neuropeptides include endorphins and opioids. There is one other class of neurotransmitters made up of lipids and gases.

Amino Acids: Glutamate and GABA

Glutamate and gamma aminobutyric acid (GABA), are found in every cell of the body. We find amino acids involved in rapid communication between neurons. There are more than twenty different amino acids, but glutamate and GABA are the most common ones. Glutamate is the primary excitatory neurotransmitter and is widely distributed throughout the brain. As an excitatory neurotransmitter, it tends to cause neurons to be more active. It is synthesized as a by-product of glucose metabolism in the Krebs cycle and removed by reuptake of the neurons and glia.

GABA is the primary inhibitory neurotransmitter in the brain. It is extremely common, with up to one-third of all synapse receptors being receptive to it (Zillmer et al. 2008). One of the most prominent GABA systems is made of Purkinje cells that extend out into the cerebellum. GABA also extends into parts of the basal ganglia, the brain area involved in controlling movement. People with Huntington's disease are thought to suffer a loss of inhibitory GABA in the basal ganglia. They have difficulty controlling their movements.

Acetylcholine

ACh plays a role in stimulating the parasympathetic nervous system. It is widely involved in regulating behavior as well as in modulating physiology. So, sensory processing, learning, memory, mood, attention, sleep, arousal, biorhythms, and aggressive behavior all engage the ACh system. The effects of this neurotransmitter are often referred to as cholinergic. This is simply the adjectival form of the term ACh.

The chemical structure is simple. It has a positively charged nitrogen atom with four attached methyl groups. Because of this configuration of nitrogen and methyl groups, this neurotransmitter is classified as a quaternary amine.

$$CH_3 \qquad\qquad O$$

$$| \qquad\qquad\qquad ||$$

$$H_3C - N^+ -- CH_2 - CH_2 - O - C -CH_3 \qquad \text{Acetylcholine Structure}$$

$$|$$

$$CH_3$$

This neurotransmitter is synthesized from acetyl coenzyme A (acetyl CoA) and choline. Choline is found in foods such as vegetables, egg yolk, kidneys, liver, seeds, and legumes. It is produced in the liver as well. Once choline crosses the blood–brain barrier, it can be used to create ACh. This reaction can be reversed, but it tends to shift toward the formation of ACh.

There are two types of receptor sites sensitive to ACh: They are muscarinic and nicotinic receptors. Muscarinic receptors work slowly and last longer. They are located in the parasympathetic autonomic nervous system, found around the heart and the smooth muscles. Nicotinic receptors work quickly for fast neurotransmitter action to facilitate voluntary muscle control. They are found both at the ends of motor neurons activating skeletal muscles and at the peripheral somatic nerves. Nicotinic receptors are also found all around the cortex. Deficiencies in nicotine receptors have been implicated in Alzheimer's disease.

ACh is in three main brain systems known as cholinergic brain regions. The septal projections come from the limbic region involved with hippocampal rhythms and memory. The pons projections are involved in REM sleep. The basal forebrain projections mediate cortical activation and learning. There are other areas that release ACh as well through the interneurons.

ACh is central in many functions from body to brain to mind. Many years of research has shown that ACh is involved in sleep, waking, and associated with arousal and alertness in the reticular activation system (Vanderwolf 1992). Cholinergic neurons play an important part in modulating cortical excitability and sensory processing (Feldman et al. 1997). It also has effects on cognitive functioning in learning, memory, and attention. For example, blocking muscarinic receptors interferes with acquiring and maintaining many types of learning (Spencer and Lal 1983). Cholinergic systems may also play a role in the decline of memory in old age as well as in Alzheimer's Disease.

Monamines: Dopamine, DA, Norepinephrine, NE, and Epinephrine, EPI

Another group of neurotransmitters are called monoamines because they possess a single amine group. Dopamine, DA, norepinephrine, NE, and epinephrine, EPI, are the most common ones.

Dopamine

Dopamine plays an important part in many different systems of behavior and emotions. The activity of DA is directly related to motivation and learning. Dopamine is also involved in motor control. Although the two pathways are separate, DA may help to integrate sensory, motivational, and motor functions together (Feldman et al. 1997).

DA's synthesis, and that of the other neurotransmitters in this group, occurs in several steps. The precursor is L-Tyrosine, an amino acid that comes from dietary protein and is synthesized in the liver. Next comes the formation of dopamine from L-DOPA. Thus, dopamine not only plays a vital role as a separate neurotransmitter, it is also a precursor to NE and EPI. So the order is: L-Tyrosine to L-DOPA, to Dopamine and finally to Norepinephrine and Epinephrine.

There are five receptor types, designated as D1, D2, D3, D4, and D5. They are all part of the G-protein-coupled receptor family.

Dopamine is involved in three main pathways that can alter psychological functioning. The mesolimbic or reward pathway projects from the brainstem to the basal ganglia, limbic system, and prefrontal cortex. We are wired to enjoy life-sustaining activities such as sex, sleep, and food. The release of dopamine through this reward pathway is one of the mechanisms that help to reinforce these vital activities by giving us feelings of happiness, pleasure, and satisfaction when we engage in them. See Chap. 9 for details.

The second pathway, the neocortical dopamine pathway begins in the brainstem and goes to the frontal cortex and limbic regions. This pathway is associated with cognitive functions including learning, regulation of attention, and social behavior (Stahl 2000).

The third dopamine pathway goes from the substantia nigra of the brainstem to the basal ganglia and is involved in the regulation of voluntary motor movement. Decrease of dopamine leads to loss of movement ability whereas too much dopamine brings about disruptions in smooth motor movements. Parkinson's Disease is the result of not enough dopamine, and Huntington's Disease comes from having too much dopamine in this pathway. Treatments for these disorders have involved regulation of dopamine, with moderate success, especially in the early stages (Breedlove et al. 2007).

Norepinephrine

These neurotransmitters were first discovered by Walter Cannon, an important psychologist who devised an early theory of emotion known as the Cannon-Bard theory of emotion (Simpkins and Simpkins 2009). He discovered a substance he called "sympathin" because it was released when they stimulated sympathetic nerves (Cannon and Uridil 1921). Later this substance was identified as norepinephrine (NE).

Epinephrine is often called adrenaline and norepinephrine is frequently referred to as noradrenaline. Both NE and EPI are also part of the fight- or- flight hormones released by the adrenal glands in response to stress as part of the sympathetic nervous system. But they are also found in the brain as well, with important neurotransmitter functions.

Norepinephrine helps to regulate mood, overall arousal, attention, vigilance, and sexual behavior. This neurotransmitter is also involved in the regulation of hunger, satiety, and body weight. Low levels of norepinephrine may be involved in symptoms of depression and anxiety. Norepinephrine is released from neurons in three distinct, relatively small clusters in the brainstem. These areas are the locus coeruleus (LC) in the pons, the lateral tegmental system of the midbrain, and the dorsal part of the medulla. The LC is the most prominent noradrenic nucleus with the largest projections. Projections extend to the cortex, limbic system including the amygdala and hippocampus, and to the thalamus and hypothalamus. Thus, change in just a few neurons in one of these areas can range out to have a widespread influence over large areas of the brain.

Serotonin

Serotonin has a chemical name, 5-hydroxytryptamine(5-HT) and is found in large areas around the brain. It is produced in an area concentrated along the midline called the raphe nuclei in the brainstem and midbrain. The pathway branches out, extending to the cerebellum, to the spinal chord, and out to the forebrain and limbic system.

The effects of serotonin released into a synapse are terminated through a reuptake process, where the serotonin is carried back into the releasing neuron to be reprocessed for later use. Inhibiting the reuptake of serotonin was found to help the symptoms of depression and so serotonin is believed to play a role in clinical depression.

Serotonin also plays a definite role in eating and body weight regulation. Low levels of serotonin have been linked to aggression, particularly impulsive aggression (Brown 1994). A number of studies in the 1990s found a relationship between serotonin dysregulation and aggressive behavior and hostility in adolescent boys with conduct disorder (Unis et al. 1997). Another study found a similar correlation between hostility and aggression in healthy males (Cleare and Bond 1997). Suicidal behavior, which could be thought of as self-directed aggression, is also associated with lower serotonin levels (New et al. 1997). Another disorder that has been linked to serotonin is OCD. We do not have conclusive evidence for abnormality in serotonin for OCD sufferers because other neurotransmitters are involved as well, such as dopamine (Feldman et al. 1997). But evidence for serotonin's involvement comes from the success of medications that increase serotonin.

Neuropeptides

Peptides are a class of compounds that are short chains of amino acids with more than sixty different types. They form one of the many parts that make up proteins. Opioids are neuropeptides that are synthesized from mRNA to slowly release when there is increased demand. They tend to be co-localized with other neurotransmitters such as the monoamines and act as neuromodulators. They are released by repetitive stimulation or in a burst of firing and are broken down by enzymes. There are several different kinds of opioids: B-endorphin produced in the pituitary gland, hypothalamus, and brainstem, enkephalin, produced throughout the brain and spinal chord, and dynorphin, also produced throughout the brain and spinal cord. Opioid receptors are distributed all throughout the brain and spinal cord, especially in the limbic areas.

The opioids can mimic opiate drugs such as morphine, giving a feeling of euphoria and reducing pain. But the neurotransmitter action is not always simple. Neuropeptides often act as co-transmitters that activate pre and postsynaptic responsiveness of other neurotransmitters such as dopamine, norepinephrine, and serotonin.

Conclusion

The brain is capable of reacting and responding all the way down to the chemical and neuronal level. As neurons communicate electrical signals through their sophisticated neurotransmitter systems to relay the messages, we find new ways to understand how the brain learns and is capable of responding to our needs. As a highly reactive and yet at the same time stable system, the neuronal level reveals a combination of complexity and simplicity.

Chapter 8
Brain Structures

Neurons do not act alone, but tend to behave together in structures. We can identify many structures in the brain that tend to perform certain functions. Keep in mind that the structures we describe in this chapter do not work alone. The brain partakes of many paradoxical qualities. In some ways, the brain functions as a unity while also comprising many separate parts. To truly understand the brain and how it affects psychological functions of human cognition, emotion, and behavior, you will need to think of both the parts and the whole. Then you gain a better understanding of mental disorders with helpful ways to heal them.

This chapter describes the key brain structures of the brain, with their functions. You will become familiar with the parts that are activated for thinking, feeling, and behaving, giving you a resource for therapeutic interventions.

Holism and Localization

You can think of the brain from two general points of view, either as wholes and systems or parts and localization. (Ramachandran 1997). Golgi's theory of the reticular activating system, mentioned earlier, exemplifies the holistic view. Theories from the holistic position postulate that subgroups of parts always work together as a unity. The parts view, which Ramon y Cajal exemplifies, considers the brain as made up of many localized parts, which are involved in separate functions. Certain parts are more important for one function, and not another. Through history, both the holism and localization perspectives have added to our understanding of the brain, offering useful paradigms. Each has importance for understanding the brain. Over the years, the dominance of either localization or globalization has prevailed, for a time. But a more mature perspective includes both localization and generalization together, as both are involved. We can learn from both models.

C. A. Simpkins and A. M. Simpkins, *Neuroscience for Clinicians*,
DOI: 10.1007/978-1-4614-4842-6_8,
© Springer Science+Business Media New York 2013

In this chapter the brain is divided into two hemispheres, with most of the organs being repeated on both sides. It can be best understood as three general areas: the lower brain, the internal areas including the basal ganglia and limbic system, and the cortex, the outermost areas.

Lower Brain System

The lower parts of the brain regulate our basic functions such as breathing, heart rate, and other automatic functions. We find a great deal of interaction among the different parts of the lower brain with higher brain areas (Fig. 8.1).

At the base of the brain is the brainstem. It comprises the hindbrain and midbrain. The hindbrain contains the pons, the medulla, and the reticular formation. This area is the transition between the spinal cord and the brain. All axons that pass between the brain and spinal cord go through the brainstem. This area also contains the nuclei of the cranial nerves XI and XII and is important in regulating vital body functions such as breathing and heart rate.

The medulla is located at the bottom of the brainstem and at the top end of the spinal cord. It regulates breathing, blood pressure, heart rate, and wakefulness. Moving up from the medulla is the pons or bridge. The pons connects the two sides of the cerebellum and then connects these sides to the opposite side of the cerebral hemispheres. It also connects the cerebellum to the medulla. The final area in this lower section is the reticular formation. This area quickly filters incoming information before sending it to the thalamus. It is involved in circulation, sleep, respiration, digestion, attention, and arousal.

The Cerebellum

The cerebellum (Latin for *little brain)* contains more neurons than all the other parts of the brain and yet only occupies 10 % of the brain's weight. The cerebellum is highly convoluted. In fact, if unfolded, the cerebellum would be larger than the cortex. It is connected to the brainstem, the hypothalamus, and thalamus.

Like the cerebral cortex, the cerebellum has two hemispheres and lobes with connections between the lobes. Also similar to the cortex, the cerebellum is layered, although it only has three as opposed to the cerebral cortex's six layers. The layers consist of a granule level of tiny cells, a second layer of purkinje cells, and a molecular layer. The purkinje cells are exquisite two-dimensional cells, which are by nature expressing her artistic design with an elaborate tree-like structure, that allows each cell to receive input from 200,000 spines!

The cerebellum regulates neuronal signals to other parts of the brain through loops of interaction. It is involved in motor learning and cognition. Thus, it serves a variety of functions in being a part of the regulation of higher cerebral processes

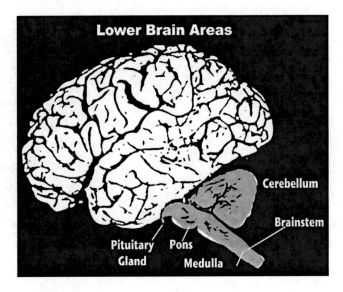

Fig. 8.1 Lower brain areas

in motor planning, cognition, autonomic (involuntary) function, and problem solving. Some of the motor skills it regulates are posture and the command of movement, as well as being important for motor learning. It is also important in cognition for learning, processing words, anticipatory planning, and making time-based judgments.

In regulating motor learning and coordination, the cerebellum helps to combine rapid, skilled movements together such as learning to play the piano or ride a bicycle. It controls and corrects these compound movements through feedback and timing such as trial and error correction or supervised learning networks. During the learning period, the cerebellum is highly active. Eventually, after time and practice, the movement goes from effort to effortless. At this point, the cerebellum becomes less active.

Bassal Ganglia System

Moving up through the brain in the interior, we find the basal ganglia. All of the motor areas from cortex send information to the basal ganglia. The basal ganglia are a C-shaped set of structures, made up of four interconnected nuclei that span from the brainstem to the cortex. The substantia nigra is a brainstem structure in the midbrain. The caudate nucleus, the putamen, and the globus pallidus are located under the cerebral cortex (Fig. 8.2).

The functions of the basal ganglia involve voluntary motor movement and movement coordination. The basal ganglia do not generate motions directly but

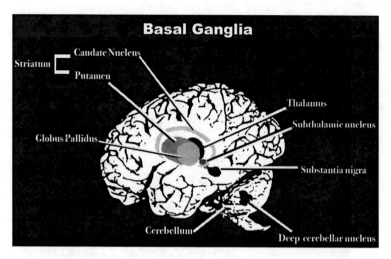

Fig. 8.2 Basal ganglia

rather takes input from the cortex, modifies it, and then passes it back to the cortex via the thalamus. It is also involved with higher order cognition that is involved in moving such as motor planning, sequencing, and maintaining learning. This area is also part of predictive control, attention, and working memory.

Movement that comes from the basal ganglia is automatic, such as when walking or smiling. One patient had damage to the left side of his motor cortex. When he was being tested for loss of function, he was asked to smile. He was able to deliberately make a half-smile on command, with the left side of his mouth moving into a smiling position, but not on the right. However, when his wife walked into the room, he broke into a broad smile on both sides of his mouth. Because his basal ganglia was not damaged, he could smile naturally and effortlessly. But he had lost the ability to make deliberate movements. An opposite kind of damage occurs with Parkinson patients whose basal ganglia are affected by the lower level of dopamine to activate movement. However, they can sometimes learn to temporarily bypass their inability to walk, for example, by deliberately marching. Walking is an unconscious skill controlled by the basal ganglia while marching is a deliberate conscious movement skill, activated by the motor cortex.

The Limbic System

The limbic system, also located in the interior of the brain, has been given much attention by therapists, since it is intimately involved in emotion, fear conditioning, fight- or -flight response as well as learning and memory. James Papez made an early anatomical model of emotion that was called the Papez Circuit, but it was limited in what it included. Paul MacLean (1952) expanded the model, adding

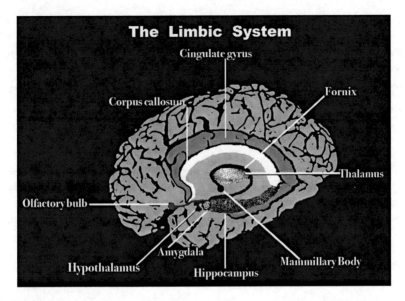

Fig. 8.3 Limbic system

structures and giving it the name *limbic system* that is still used. Today, many structures are considered part of the limbic system, although exactly which structures are included will vary (Fig. 8.3).

Most people agree that certain structures play a central role in the limbic system. These limbic structures typically include the amygdala, hippocampus, cingulate gyrus, fornix, hypothalamus, and thalamus. But these designations as to what structures are included vary. Several other areas are sometimes considered important for emotions, and thus are often included as part of the limbic system. These other structures include the olfactory cortex involved in the sense of smell, the pituitary gland regulating hormones, the mammillary body, the orbitofrontal cortex (part of the pre-frontal cortex), and the nucleus accumbens, important for reward, laughter, pleasure, addiction, and the placebo effect. Keep in mind that all of these structures interconnect and interact together, although some contribute more to one function than to another. With so many varied structures all closely interacting with each other and with the higher cortical functions, it makes sense as to why our emotional life plays such an important role in every aspect of living.

Amygdala

The amygdala regulates emotion. There are two amygdala structures, one in each hemisphere, located deep inside the temporal lobe and connecting to many other structures including the hippocampus and the prefrontal area and the thalamus. These many connections help the amygdala to play its vital part in the mediation and control of a broad range of emotions from love, happiness, and affection to fear, anxiety, rage, and aggression. It plays a primary role in forming and storing memories associated with emotional events and experiences. When damaged, people lose their feelings about other people and situations, such as knowing who a person is, but not knowing whether they like or dislike that person.

Hippocampus

The hippocampus located close to the amygdala in the medial temporal lobe, is important in learning and memory. There are two hippocampus structures, one in the right hemisphere and the other in the left, both shaped like a sea horse. The hippocampus has a number of memory functions. It is involved in short-term memory and its consolidation is into long-term memory. It is also part of episodic and declarative memory and the detection of novel stimuli.

The hippocampus has a three-layer circuit. It receives parallel inputs from the different senses where the sensation enters a sensory buffer. Then the signals are encoded and sent onto working memory and short- term memory. When the memory is actively attended to, it becomes consolidated. The consolidated memory then moves onto long-term memory. The long-term memories can later be sent back for retrieval. Damage to the hippocampus has been shown to produce amnesia. Therapy can help people to retrieve painful memories, and then recon-solidate them in a healing way. (See Chap. 17 to learn how).

Another central function involves spatial learning, such as knowing the map of your living room. A large body of animal and human research has shown how critical the hippocampus is for spatial orientation and spatial memory.

One of the most exciting recent discoveries is that the hippocampus undergoes neurogenesis through the entire life span. Research shows that the hippocampus can grow or shrink depending upon experience, with positive or negative effects on learning and memory. People suffering from clinical depression or PTSD have a shrinking in their hippocampus, whereas those involved in spatial navigation or novel experiences undergo growth in the hippocampus.

Thalamus

The thalamus is located at the center of the brain. It is the main gateway to the cortex, with a back and forth system that regulates what is happening outside the body. Thalamic cells are built for relaying information, since 75 % of the thalamic neurons are relay cells. In fact, all sensory input, except for olfaction, pass through the thalamus.

Specific thalamic nuclei exist for each sensory system. Within each thalamic nuclei are neurons specialized for a specific sensory input such as touch. The touch neurons are specialization further, where some relay painful touch and others are for temperature from touch. Each of these areas carries the information rapidly to the appropriate area of the cortex. The separate bits of information from the thalamus then map onto the appropriate cortex area where the information gets organized, synchronized, and unified. It is in these maps on the cortices where neural plasticity can occur (Hendry et al. 2003).

Hypothalamus

The hypothalamus, located right below the thalamus, is vital for regulating the internal states of the body. It is the seat for the biological clock, the circadian rhythms for sleep and waking, hormonal rhythms, and sexual activity. Homeostasis is also maintained by the hypothalamus, such as the body's ability to maintain a steady temperature.

The hypothalamus is central in the regulation of thirst and appetite as well. An early dual-center hypothesis for the control of eating proposed that there are two centers for appetite: one for hunger and the other for satiety. The ventromedial hypothalamus (VMH) was thought to be the satiety center (Hetherington and Ranson 1940). Lesions to the lateral hypothalamus (LH) seemed to trigger rapid weight loss (Anand and Brobeck 1951). But this theory proved to be too simple. Today the theory of appetite control in the hypothalamus integrates hormones and signaling systems together. The arcuate nucleus, located in the hypothalamus is where the appetite controller resides that is governed by the circulation of hormones. Five peptides are secreted into the bloodstream. They are insulin, leptin, ghrelin, obestatin, and PYY_{3-36}. Two types of neurons in the arcuate nucleus are sensitive to these peptides and signal for either a decrease in food intake or an increase. These signals exert effects on second-order neurons in the VMH and LH, which lead to release of other hormones that stimulate or inhibit appetite.

The hypothalamus also monitors signals to the pituitary to direct the release of hormones involved in stress. This is known as the hypothalamus–pituitary–adrenal-axis (HPA), which helps to regulate the stress response. Thus, the hypothalamus plays a role in emotions that sometimes become intertwined with the other regulatory functions, such as in eating disorders, sleep disturbances, and sexual problems.

The Insula

The insula is located deep within the lateral sulcus that separates the temporal and parietal lobes. The insula is sometimes grouped with the limbic structures because of its important functions with emotions. It is involved with many basic emotions such as anger, fear, disgust, happiness, and sadness. The insula has close interaction with the thalamus and the amygdala: it receives information from the thalamus and sends output to the amygdala as well as to the orbitofrontal cortex.

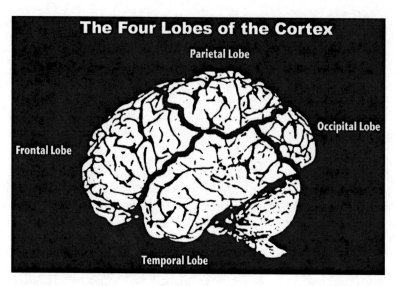

Fig. 8.4 The lobes

The insula has also been shown to play a role in addiction (Naqvi et al. 2007). And mirror neuron researchers have found that the insula is involved in both the recognition and experience of disgust (Wicker et al. 2003). The functional interconnections between the insula and other areas help to account for this.

The Fornix

The fornix that is a C-shaped fiber that begins in the hippocampus on each side of the brain come together as a bundle of fibers in the midline of the brain, arches over the thalamus, and connects to the mammillary bodies of the hypothalamus. The exact function is unclear, although it seems to serve as a connection among these different areas.

The Cerebral Cortex

The cerebral cortex, Latin for bark of a tree, is the outer layer of the hemispheres. The cortex is sometimes referred to as the higher part of the brain. Here we find the source of higher functioning such as cognition, language, speech, memory, and visual processing. The cortex has many convolutions, gyri, and folds, sulci. The folding increases the surface area of the cortex, so that more than two-thirds of the surface is hidden from view.

The cortex contains two primary types of neurons, excitatory and inhibitory. The structure of the cortex tends to be uniform, with six horizontal layers that are

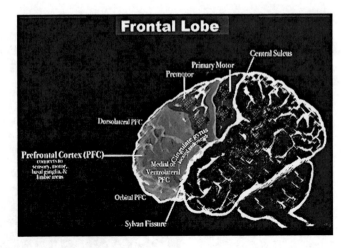

Fig. 8.5 Frontal lobe

recognized by their different cell type and neuronal connections. There are some exceptions to this structure, such as in the motor cortex, which does not have a layer IV.

Each hemisphere is divided into functional sections or lobes. There are four lobes: frontal, parietal, temporal, occipital. The lobes run along the sulci grooves, although some are more evident than others. The central sulcus divides the frontal lobe from the parietal lobe. The sylvan fissure, running horizontally, separates the frontal and parietal from the temporal lobe, which is below. The boundary between the occipital lobe, located toward the back of the cortex, from the temporal and parietal lobes is less visually distinct. Each lobe monitors different functions, although they are all interrelated, interacting together (Fig. 8.4).

The Frontal Lobes

The frontal lobe accounts for nearly one-third of the cerebral cortex. It extends from the central sulcus forward to the anterior, (front) portion of the brain (Fig. 8.5).

The Prefrontal Cortex

At the front of the brain behind the forehead is the prefrontal cortex. Typically, the prefrontal cortex is divided into three subregions: the orbital, also called orbito-frontal, the dorsolateral, and the medial including the cingulate gyrus located deepest within. These regions are now understood as circuits of interaction with other brain areas. The three prefrontal cortex areas follow a general circuitry connecting them to lower regions. Connections move up from subthalamic areas, through the basal ganglia, to the thalamus, and up to different parts of the

Fig. 8.6 Homonculus

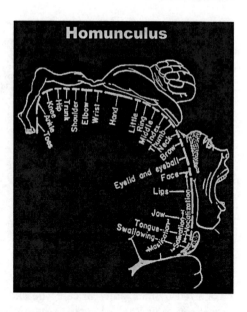

prefrontal cortex. The fact that the prefrontal areas have extensive links throughout the brain shows how interrelated brain functions really are, both specifically and generally, localized and globally.

The dorsolateral area is involved in executive functioning. Executive functions include planning, higher level decision-making, sequencing, and goal directed behavior. Our independent thinking that is not based on the environment is processed in this area. When this area has dysfunction from strokes, Alzheimer's Disease, tumors, or accidents, people are poorly organized, have inadequate memory search strategies, and become dependent on the environment.

The orbitofrontal area is involved in personality characteristics such as empathy, social appropriate behavior, and emotional control. Much was learned about the functions of the orbitofrontal area from cases such as Phineas Gage and other patients with damage to this area (See Chap. 4).

The Motor Cortex

The primary motor area of the frontal cortex is located closest to the parietal lobe, just anterior to (in front of) the central sulcus. It is important for control of movement, known as motor control. This primary motor cortex strip maps to the body, somatotopically, similar to how the sensory strip maps in the parietal cortex. Larger portions of the cortex are devoted to areas we use more, such as the hands and face, whereas smaller areas are given to the feet and back. The Homunculus (Fig. 8.6) pictures the relatively larger areas of the motor strip that are devoted to the hands and face compared to other parts of our body.

There is also a non-primary motor cortex, located in front of the primary motor cortex. The non-primary motor cortex includes a supplementary motor area and a pre-motor area. Damage to the primary motor cortex leads to an inability to perform fine motor movements in the fingers, whereas damage to the non-primary motor cortex brings about difficulty with larger movements such as stance, gait, and initiating voluntary movement sequences.

Mirror neurons, which are involved in understanding and empathizing with the intentions and actions of others, are located in the motor area of the frontal lobe. This system is becoming increasingly more important in social cognition and empathy, as Chap. 18 describes.

The Cingulate Gyrus

The medial or cingulate gyrus, also called cingulate cortex and sometimes called cingulated gyrus or cortex, is located on the inside of the prefrontal cortex and above the corpus callosum, the fibers that connect the two hemispheres. This area is involved in motivated behavior, spontaneity, and creativity. Focus on complex behavior and focus of attention are also processed in the cingulate gyrus. This area is primary for the emotional reaction to pain and the regulation of aggressive behavior. It has also been found to play an important role in maternal attachment as evident in behaviors like nursing and nest building in animals (MacLean 1985).

The Parietal Lobes

The parietal lobe is involved in sensation and perception of touch, pressure, temperature, and pain. It also integrates sensory information from the body to produce a perception or cognition of the sensation. Another function of the parietal lobe is to locate objects in space and map the body in relationship to the world. The anterior end of the parietal lobe is the sensory strip, located next to the motor strip in the frontal lobe. This sensory cortex is mapped somatotopically, similar to how the primary motor cortex is mapped to the body (Fig. 8.7).

Damage to the left parietal lobe interferes with the capacity to understand spoken or written language. When the right parietal lobe is damaged, patients develop visual–spatial deficits, such as navigating even in familiar places.

The Temporal Lobes

The temporal lobe is located in each hemisphere of the brain, near the temples. It is primarily involved with auditory information and is where the primary auditory cortex is located. The left side plays a larger role in understanding spoken language. Some visual processing occurs near the bottom of the temporal lobe, but the primary visual areas are located in the occipital lobe. The aspects of visual

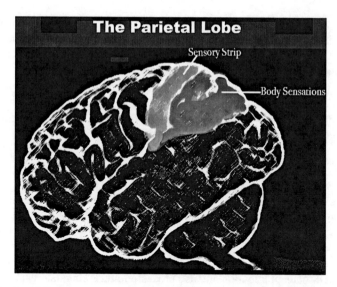

Fig. 8.7 Parietal lobe

Fig. 8.8 Temporal lobe

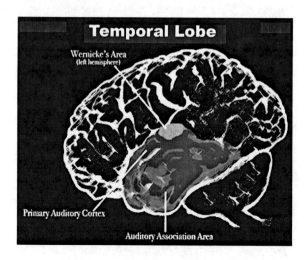

processing that are located in the temporal lobes involve perception of moments and face recognition (Fig. 8.8).

People with anxiety problems have metabolic abnormalities in the temporal lobes and right amygdala, even when they are resting. (Boshuisen et al. 2002). Panic disorders are accompanied by both structural and functional changes in the temporal lobes. The volume of the temporal lobes in patients suffering from panic disorders was found to be smaller than in normal subjects (Vythilingam et al. 2000). Temporal lobe damage on both sides can produce an inability to recognize faces, known as Prosopagnosia or face blindness. (See Chap. 4 for more about temporal lobe damage.)

Fig. 8.9 Occipital lobe

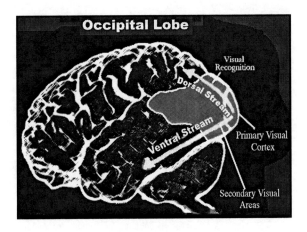

The Occipital Lobes

The occipital lobes are located in the posterior, caudal (back) of the brain. Axons coming from visual input from the eyes pass through the thalamus and are directed to the primary visual cortex located in the occipital lobes. The visual cortex is also sometimes called the striate cortex because of its striped appearance. There are more than 32 zones for visual processing involving such different aspects of seeing such as color, texture, and movement, all located in the occipital lobes. People often think that blindness is caused by harm to the eyes. But blindness, called cortical blindness, can also occur without any damage to the eyes when parts of the occipital lobes are damaged (Fig. 8.9).

Visual information leaves the primary visual cortex through two different streams: the ventral stream, the "what pathway," which travels to the temporal lobe and the dorsal stream, the "where pathway," which goes to the parietal lobe. The ventral stream is involved with recognizing the form and representation of objects whereas the dorsal stream is concerned with motion and where objects are located. There is some controversy about these two pathways, and it probably will prove to be more complicated, but the distinction shows that the brain often separates interacting functions.

Conclusion

From studying the different areas of the brain, one begins to see how the brain is both specialized and generalized. Different parts work together to bring about certain functions. And yet, damage to a specific area can have a very specific effect. The later chapters will continue to develop the interrelationships and functional systems that are involved in areas that therapists work with such as attention, emotion, and memory.

Chapter 9
Brain Pathways

Thoughts, feelings, and behaviors are intimately involved in the flow of pathways in the brain, which are dynamical systems of interaction among structures, energy flow, and chemicals. Through these pathways, the complex collection of neurons described in earlier chapters, becomes expressed. Your growing understanding of the nervous system will give you new insights about psychological processes. When people are not functioning at their best, you can help them shift their nervous system by means of its natural pathways, to re-establish optimal functioning as nature intended.

The Nervous System

The basic components of the nervous system can be seen without magnification. Observe the diagram (Fig. 9.1) and you will notice interconnections between brain and spinal chord spreading throughout the entire body. These interconnections allow signals to flow around in patterned ways. The patterned ways are stable, forming dynamical systems that neuroscience has formulated as pathways, flowing in cycles. When people have problems, they are often experiencing some dysfunction in one or several of the natural pathways. Therapy can help guide the client back to healthy functioning of the relevant pathways.

Central Nervous System and Peripheral Nervous System

The nervous system consists of two main parts: the central nervous system, or CNS, made up of the brain and spinal cord, and the peripheral nervous system, consisting of all the other nervous system parts that lie outside of the brain and the

C. A. Simpkins and A. M. Simpkins, *Neuroscience for Clinicians*,
DOI: 10.1007/978-1-4614-4842-6_9,
© Springer Science+Business Media New York 2013

Fig. 9.1 The nervous system

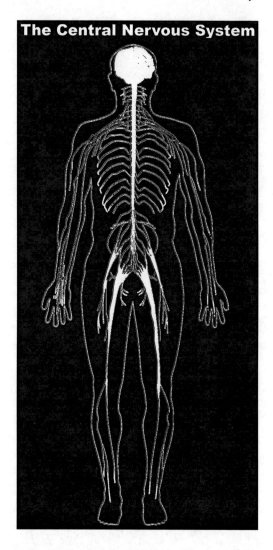

spinal cord. There are three main components to the peripheral nervous system: the cranial nerves, the spinal nerves, and the autonomic nervous system. The components communicate information to and from the CNS.

Cranial Nerves

Twelve pairs of cranial nerves connect directly to the brain. Three of them are sensory pathways, five are motor, and the other four have both sensory and motor functions. They are labeled as their name plus roman numerals, such as olfactory

(I) and optic (II), all the way up to the hypoglossal, (XII), nerves of the tongue. The vagus nerve (X) is used as a measure for parasympathetic nervous system and how well individuals are reacting to stress (Porges 1992). This measure has been used as a helpful adjunct to some of the newer psychodynamic and psychobiological therapies. The Polyvagal Theory helps to explain how the heart, lungs, and nervous system are interconnected and are responsive to emotions, cognition, and our world of interactions with others. The therapeutic applications of this theory are described in Chap. 18.

Spinal Nerves

The spinal nerves, connected along the spinal cord, are arranged at regular intervals, with the name of each corresponding to the part of the spinal cord where they are connected. Thus, the cervical spinal nerve connects to the neck, and the lumbar connects to the lower back, etc. Each spinal nerve has two distinct parts or roots, as they are called, that function differently. The dorsal root projects from the body to the spinal cord and the ventral root projects from the spinal cord out to the muscles.

Support System Through the Ventricles

The brain and spinal cord are protected from rubbing against bones and muscles by a membrane called the *meninges*. More cushioning is provided by the cerebrospinal fluid, CSF. The ventricular system produces the CSF. The word *ventricle* refers to a small cavity or chamber in a body or organ. The Ventricular System consists of four interconnected cavities or chambers found in the brain. These four ventricles contain choroid plexus, a membrane with many small blood vessels in the fluid spaces that secrete the CSF. The cerebrospinal fluid fills and circulates through the ventricular system at a rate of approximately 20 ml/h (Fig. 9.2).

The cerebrospinal fluid has two main functions. It supports and protects the brain and the spinal cord from rapid movements and trauma. It also provides a transport system for hormones and neurotransmitters and a medium of exchange of materials for nutrition and waste between the blood vessels and brain tissue.

There are four ventricles: the lateral ventricles, one and two, are located in each hemisphere. The remainder of the system is located in front of the cerebellum. The third ventricle has a central aqueduct. CBF passes through the fourth ventricle on its way to circulate and surround the outer surface of the brain and spinal cord. Finally, CBF is reabsorbed back into the circulatory system through large veins right below the top of the skull.

Fig. 9.2 The ventricular system

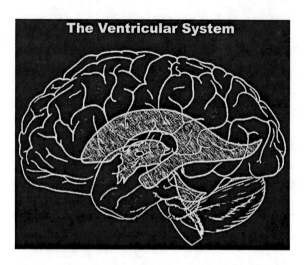

Autonomic Nervous System

Therapy can have a direct effect on the autonomic nervous system. Your interventions can calm it when over-activated and stimulate it when under-activated. Later chapters of the book offer ways to help regulate the autonomic nervous system, so understanding what is happening on the neurobiological level will make it easier for you to choose the appropriate intervention.

The autonomic nervous system is made up of groups of neurons known as autonomic ganglia. The word *autonomic* means independent, reflecting the earlier belief that these nerves were independent of the brain. Now we know that these ganglia are controlled by the CNS and span the central and peripheral nervous systems. However, they are autonomous in the sense that they are not subject to conscious, voluntary control. Treatment modalities such as hypnosis can tap into these systems by activating the involuntary, unconscious.

The neurons of the autonomic nervous system are found in many different locations outside the CNS and around the body. This system is made up of three subsystems: the sympathetic nervous system, the parasympathetic nervous system, explained below, and the enteric nervous system, a local network of sensory and motor neurons that regulate digestion.

Sympathetic and Parasympathetic Nervous Systems. The sympathetic nervous system is found in the spinal cord, specifically in the thoracic and lumbar regions. In a somewhat general way, one can think of the sympathetic nervous system as an activation system. The SNS prepares the body for vigorous action. It consists of two paired chains of ganglia located to the left and right of the spinal cord. It regulates increases in blood pressure, speeding up the heartbeat, and dilating the pupils.

The parasympathetic nervous system is found above and below the sympathetic nervous system. *Para* means "around" from Greek, and this is how the parasympathetic system acts. The neurons are located above and below the neurons of the SNS, but they are not collected there in a chain. Instead, these ganglia are found all through the body, positioned close to an organ, permitting appropriate cycles. For example, during exercise, sympathetic activation constricts blood vessels or inhibits digestion. When the workout is over parasympathetic activation relaxes vessel walls and stimulates digestion.

Different forms of treatment may excite or inhibit one or the other system, partly because the two systems use different neurotransmitters. For the most part, the sympathetic system releases the neurotransmitter norepinephrine, although the sweat glands use the neurotransmitter acetylcholine. The peripheral system releases the neurotransmitter acetylcholine. The two systems work together to regulate, protect, and foster appropriate responses. The systems of activation and deactivation are involved in emotions such as fear and anger, as well as participating in the stress response. Together, these two interacting systems maintain the control that keeps the body in homeostatic balance.

Regulatory Pathways

The brain does not work alone. We know that the central nervous system helps to keep us in balance with the environment. A complex group of mechanisms goes into regulating the organism's ongoing homeostasis both internally and externally with the world. Emotions are part of these regulatory systems that help to keep us in balance.

Sensations of pleasure and pain arise from the patterns of interactions that come together at a particular time. These processes often occur automatically and might seem beyond our control, but there are ways to control these patterns. The larger principle we can discern is that processing of pain and pleasure occurs through distinct pathways. The patterns of signals relate to feelings of pleasure or pain when elicited by stimulation. And since the patterns are part of a dynamical system, appropriate action can be taken. By understanding how these patterns come about and learning how they are regulated, we can begin to take the steps needed for positive change.

The brain has many dynamic regulatory systems. We describe here the systems that have the most profound effect on psychological processes and can be influenced by psychotherapy. One pathway controls positive emotions and drives toward fulfillment, satisfaction, and enjoyment, known as the *reward* pathway. Another dynamic regulatory system, the *fear/stress* pathway, provides the capacity to respond to threat by fight, flight, or freeze if necessary, to help deal with the threat and return to homeostatic balance. We also have regulatory systems controlling appetite and sleep. When any of these systems become out of balance, we see disorders and problems. Thus, all of these systems are important for psychotherapy. Consider these mechanisms when devising treatment methods and you can add another dimension to help strengthen the effects of therapy.

Fig. 9.3 Sensory processing

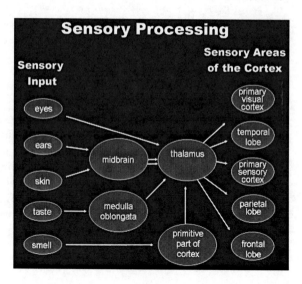

Sensory Processing

The senses provide a window to the world. Our receptor organs detect certain textures, temperatures, sounds tastes, and visual stimuli. All animals have specialized parts of the body that detect stimuli, but there is large range of diversity among the types of detector organs in different animals. Each receptor is adapted to a certain kind of stimuli. Thus, we have different types of receptors for different types of stimuli, each with a range of responsiveness. Depending on the qualities and intensities of these stimuli, our experience is either pleasurable or painful to some degree (Fig. 9.3).

Sensory processing begins with a receptor cell that is specialized to detect a particular energy or chemical. When the receptor cell is stimulated by the right stimulus, it converts the energy into a change in the electrical potential across its membrane. This process is called sensory transduction. The sensory events are represented by the action potentials and coded into a pattern of electrical activity. The intensity and location of the stimulus influence the signal that is sent. The sensory signal travels through a sensory pathway for the particular modality to the highest level of the brain. The signal travels from the extremities, up through the spinal chord, and then through the brainstem to the thalamus where it gets relayed to the cortex. Each sensory system has its own pathway. Within the cortex, we find receptor fields of cortical maps for the particular sensory modality. Attention is what helps us to notice some stimuli and not others. Within the cortical maps we find neuroplasticity.

Fig. 9.4 The pain pathway

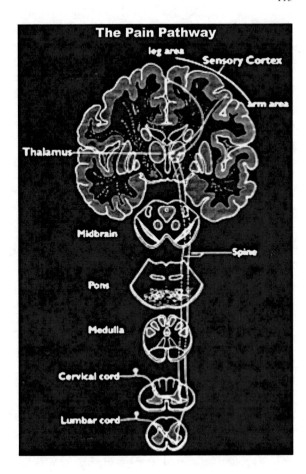

The Pain Pathway

Pain has its own pathway in the central nervous system. Pain signals are carried rapidly along myelinated fibers and slowly along unmyelinated C-fibers. The axons move up the spinal chord and the signal is then sent to the brainstem, through the medulla and pons, which controls pain-related vocalization. Then the pain signals are distributed to the different thalamic areas and up to cortex. The cingulate cortex is activated by pain, especially when people believe the stimulus will be painful (Rainville et al. 1997). The emotional and sensory components of pain, which seem to be associated with different subregions of the cingulate cortex (Vogt 2005) can make a difference in the pain's experienced intensity. Thus, psychological approaches that influence expectancy and the emotional element involved in pain will have an effect on how the brain responds. We can affect our feeling of pain much more than we think using psychological methods which literally encourage a downregulation of the pain response in the brain (Fig. 9.4).

Pain control involves a number of neurotransmitters. The brain contains natural opiate-like substances called opioids. The brain moderates pain similarly to how opiate drugs do, using three types of opioids: endorphins, enkephalins, and dynorphins, as well as certain opioid receptors. Thus, the pain signal that comes through the ascending pain pathways can be controlled and inhibited by the corresponding descending pathway. Interestingly, placebos have been shown to stimulate the same brain regions that opioids do. Both placebos and opiate drugs deactivate the pain regions in the cingulate cortex. In studies where an opioid blocker was given, the placebo did not reduce activity in the cingulate cortex, and thus the placebo was not effective (Wager et al. 2004). But other studies showed that opioid receptors do not account for the entire placebo effect. These studies suggest that the placebo effect probably has additional factors influencing how it works (Grevert et al. 1983).

The Reward Pathway and the Experience of Pleasure

Motivation toward pleasurable experiences is controlled at several points, from the chemical–neural level to higher brain areas such as the limbic system and cortex. The hypothalamus plays an important role in motivation as well (Fig. 9.5).

When we have pleasurable experiences that result from primary life-enhancing actions such as eating and sex, or from fundamental emotions like happiness and love, dopamine is released, and we feel pleasure. This is the reason that this pathway is called the reward pathway, since the dopamine system is critical in behaviors involving pleasure and reward.

The reward pathway can be divided into two different routes: the *nigrostriatal* pathway and the *mesolimbic* pathway. Both are associated with reward and pleasure. Normal feelings of pleasure for behaviors that are necessary for survival such as eating, drinking, and sex travel the *mesolimbic* pathway. When the cortex receives and processes a rewarding sensory stimulus, such as a favorite food, it sends a signal for activation of a part of the midbrain known as the ventral tegmental area, where dopamine is produced. The ventral tegmental area then releases dopamine to the nucleus accumbens, the septum, the amygdala, and the prefrontal cortex. Dopamine transmits the sensation of pleasure. Thus the release of dopamine is associated with feelings of reward and pleasure, whether from sex, food, or any other pleasurable experience, including movement. The nucleus accumbens activates the motor functions while the prefrontal cortex focuses the attention, two processes that are involved in procuring the pleasurable experience. The endocrine and the autonomic nervous systems also get involved. They modulate the reward pathway through the hypothalamus and the pituitary, which tends to self-regulate the individual level of pleasure and vitality each person has. (Squire et al. 2003).

The nigrostriatal pathway is associated with motor movement, releasing dopamine into the basal ganglia for this. Reward is experienced through this pathway. The loss of dopamine in this area has been widely researched (Zillmer et al. 2008) and results in the symptoms of Parkinson's disease, in which motor control is lost.

Fig. 9.5 The reward pathway

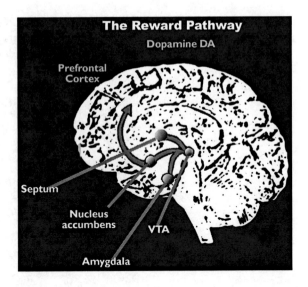

The rewarding effect of drugs is mediated through the reward pathway, affecting dopamine production and release at various stages of the system. This connection between drug effects and dopamine production, release, or reuptake helps to explain why drug addiction is so difficult to overcome, since dopamine is primary in life's enjoyment.

The Fear and Stress (HPA-Axis) Pathway

Our nervous system has a built-in capacity to detect a threat as a brain-body response. Several brain systems involving cognition and emotion are involved. When facing a perceived threat, the amygdala signals the endocrine system through the hypothalamus, pituitary, and adrenal glands, known as the hypothalamus–pituitary–adrenal (HPA) pathway. The hippocampus, where memories are stored, is closely linked to the amygdala, the gateway for processing emotions. Thus, the HPA pathway is not only activated by an immediate threat, but also from remembering a threat from the past (Fig. 9.6).

The fear and stress pathway (HPA axis) regulates how the body responds to threat by modulating the secretion of hormones and neurotransmitters that help to keep the body in balance. There is nothing inherently pathological about the fear pathway. However, even when the threat is gone, people sometimes continue to sustain the reactions as if there were still in danger. Remaining stuck in this pathway without the ability to naturally flow back to a resting state transforms this normal, healthy response into a stress response. Therefore, when the experience of threat is repeated or sustained, the fear pathway produces stress, so the pathway transforms into a stress pathway.

Fig. 9.6 The fear/stress HPA pathway

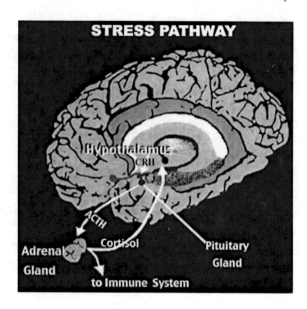

How the Fear/Stress Pathway Reacts

The fear/stress pathway is made up of three key areas: the hypothalamus, pituitary, and adrenal glands, thus the name HPA-Axis. The hypothalamus is a primary center for regulating homeostasis with the environment. The hypothalamus maintains this balance by acting as an integrator. It receives converging inputs from the sensory and autonomic systems that are related to the internal and external environment. Then the hypothalamus responds quickly. Typically, neuronal cells in the brain release their transmitters to other neurons through synaptic connections, which can take some time.

The hypothalamus not only releases neurotransmitters and neuropeptides to certain target areas in the CNS, it also initiates a fast response, sending the chemicals as signals into portal capillaries located in the anterior pituitary gland, which then go directly into the circulatory system. So, as the hypothalamus receives the input, it releases hormones and neurotransmitters that are routed quickly to the circulatory system, giving that immediate alerted reaction to danger that makes it possible to respond quickly.

Several hormones are produced and released to activate the fight- or- flight response to danger and when under a sustained stress. Corticotrophin-releasing hormone (CRH) from the hypothalamus, ACTH, adrenocorticotropic hormone from the pituitary, and cortisol from the adrenal glands signal the heart to race, palms to sweat, and breathing to become shallow. Cortisol helps to release the stored glucose in the body, giving the burst of energy we need to take action.

The normal system has a balancing set of hormones and neurotransmitters to slow down the stress reaction when the danger has passed. Production and release of glucocorticoids is fed back into the brain and pituitary to slow down the synthesis of CRH and ACTH.

Thus, when we are feeling fear, the autonomic nervous system gets involved. The sympathetic nervous system prepares the body to take vigorous action, and parasympathetic nervous system involvement inhibits action. These two systems work together when we feel challenged somehow by situations that we experience as fearful, threatening, or stressful. The heart pumps faster allowing us to run away or fight. At the same time, digestion slows so that we are not distracted by hunger needs.

This circular process of activation and deactivation is usually kept in balance. Our system is well equipped to respond rapidly to a traumatic, frightening, or dangerous event with a large increase in activity along the HPA pathway. And it is equally well equipped to calm down as soon as the threat is passed and return to balance.

Biological Rhythms

Eastern philosophy has long recognized the importance of rhythms in the theory of recurrent cyclical change. Cycles of change are seen everywhere in nature. We live in the midst of seasonal changes and day/night changes. Our attention goes through cycles as well, with periods of alertness followed by periods of rest. We have internal biological rhythms that help us to synchronize and regulate these inner rhythms with the outer environment.

Twenty-four hour circadian rhythms are found in every living thing on the planet, from the single cell organism to human beings. Plants exhibit rhythms when they undergo photosynthesis by day and cell division at night. Even bread mold has daily rhythms with spores being produced every 24 h.

We have many different cycles. Infradian rhythms last longer than 24 h. The female menstruation cycle is one example. We also have shorter cycles, such as ultradian rhythms. At the endocrine-behavioral level, we find 90–120 min-long ultradian cycles embedded within the sleep cycle. Ultradian rhythms also relate to rest–activity cycles during the waking hours. At the cellular level, growth and replication take place in a cyclical manner as well (Rossi 2002).

The circadian rhythm controlling our sleep/wake patterns has far reaching effects on every aspect of life. Human beings undergo a complex circadian cycle, which influences our behavior, sleep/wake patterns, hormone production, body temperature, appetite, and digestion, attention, and alertness.

Brain Areas Involved in the Circadian System

If there is an endogenous circadian clock, where is it located? Curt Richter (1894–1988) is credited with introducing the idea that we have a biological clock located in the brain. Richter's research on animals clarified that the clock is a cyclical mechanism, governing the animal's eating, running, drinking, and sexual behavior (Richter 1927). The exact location of this clock is in a small area of the hypothalamus, the suprachiasmatic nucleus (SCN), located behind the optic chiasm.

The SCN regulates our sleep–wake patterns by sending signals to the pineal gland, which in turn secretes melatonin, beginning two and three hours before bedtime. The release of melatonin makes us feel sleepy. The clock's link to the outside world is through signals of lightness and darkness. The retina has a population of retinal ganglion cells that are different from the rods and cones involved in seeing. These special retinal ganglion cells send signals directly to the SCN (Hannibal et al. 2001). The light of day tells the system to become alert and awake, just as darkness sends the message for sleep.

The SCN's clock mechanisms can be located in the individual neuron of the SCN tissue. Each neuron has a genetic mechanism, oscillating at the molecular level of protein synthesis. Through the process of transcription and translation is a negative feedback/positive feedforward system that turns the clock-genes on or off in each cell. Little is known about exactly how all the cells synchronize together, but patterns of cells firing together in synchrony seem to provide clues. Each cell has a rhythm and when the many cells fire together, the clock-like functions are activated. Neurotransmitters may be involved in communications between cells (Evans 2009).

Are our Circadian Rhythm Internal or External?

Researchers have wondered whether these rhythms are programmed from within our body, endogenous, or driven by the environment, exogenous. Studies lend evidence for the 24h cycle having an internal source. Bread mold taken onto the Space Shuttle Columbia retained its daily rhythms even traveling through outer space (Evans 2005)! But other studies indicate that even though an endogenous clock generates circadian rhythms, these rhythms are also exogenous in the way they synchronize in the environment (Breedlove et al. 2007). For example, when hamsters were given normal 24h day/night lighting, their activity/sleep patterns followed a 24h cycle. But when they were placed in a cage with constant dim light, their activity levels occurred a few minutes later each day. Even though the hamster's endogenous clock seemed to have a slightly longer period than the 24h day/night schedule, they adjusted to synchronize with the shorter external environment cycle (Rusak and Zucker 1979). Furthermore, shifting the light by one hour each day resulted in the hamsters shifting their pattern to match. So, even if the natural body cycle is longer than 24h, it corrects itself to match the 24h environmental lighting. Humans also exhibit this pattern of drifting away from the 24h sleep–wake cycle. A subject, who was isolated from any cues about day and night, ended up getting only 74 full nights of sleep extended over 77 days (Weitzman 1981).

Even though we all synchronize to the 24h light/dark cycle, individuals vary widely. Some people naturally like to stay up late, known as night owls, whereas

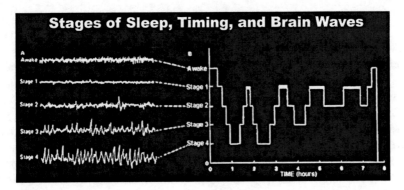

Fig. 9.7 Stages of sleep

those who prefer to get up early are called larks. People usually have a sense of their inner clocks, and can easily tell you if they are a night person or a morning person.

Sleep–Wake Cycle

The sleep–wake cycle is an exquisitely orchestrated set of well-coordinated cyclical patterns that regulate the brain and body. The circadian rhythm, known as C-process synchronizes with a second chemical process referred to as the S-process. The S-process seeks a homeostatic balance. We accrue a kind of sleep debt the longer we stay awake. During the day, adenosine builds up in the attention and arousal centers of the brain. When the level gets high enough, we fall asleep. The levels of adenosine lower again as we sleep, to begin the process anew each day (Fig. 9.7).

Sleep involves brain activity that is different from waking. We have evidence for these differences from EEG that measure brain activity. The first discovery was that there were two types of sleep: Slow-wave sleep (SWS) and rapid eye-movement sleep (REM) (Aserinsky and Kleitman 1953). Slow-wave sleep occurs in four distinct stages, each with distinctive brain wave patterns indicating different brain states and body responses at each stage, followed by a REM cycle. Cycling through the stages takes approximately 90 min, an ultradian cycle, and people commonly go through five or six of these ultradian cycles per night.

In normal waking the brain exhibits a combination of beta and alpha waves, depending on the level of arousal. (See Chap. 3 for a description of EEG and brain waves.) Stage 1 is light sleep, with muscles relaxing and occasionally twitching. This stage is approximately 5 % of our sleep time and shifts us from the shorter beta waves to slower alpha brain waves. In Stage 2, we stop responding to the environment. Breathing and heart rate slows and body temperature drops. This stage comprises nearly half of our sleep cycle. During Stage 2, we see EEG sleep spindles (small clusters) and K-complexes (larger spike clusters). Stages 3 and 4 are the deep sleep stages with slow, rhythmic delta wave activity. The whole brain is highly synchronized, similar to a room of people continually chanting one phrase in unison (Massimini et al. 2005). These stages are the hardest to wake up from and comprise approximately 12–15 % of sleep time. Here we find limited muscle activity.

Once the four stages are complete, something completely different occurs with REM sleep. REM comprises 20–25 % of sleep time. When REM begins, the brain waves suddenly become fast and random, resembling the waking state. But this brain state is different from being awake. The brainstem area is actively inhibiting motor movement, making the body completely relaxed and limp. The eyelids are closed but the eyes are moving rapidly back and forth. Rapid eye movements are accompanied with shallow, quick breathing. Because this period of sleep involves active brainwaves combined with inactive muscles, it is sometimes called paradoxical sleep (Breedlove et al. 2007).

REM is the phase of sleeping with dreaming. There are two types of REM sleep, NREM (non REM) and REM. We dream during both REMs, but NREM dreams contain little content. They tend to be flashes of visual images. Night terrors, which are distinguished from nightmares, occur as a frightening series of images that can occur in the NREM phase. REM dreams are much more elaborate, with emotional content and story plots.

One group of brainwave patterns associated with REM sleep is known as PGO waves. PGO refers to the pons, geniculate (in the LGN-thalamus) and occipital lobe, areas that are involved in REM sleep. PGO waves correspond directly with movements of the eye (Nelson et al. 1983). Each phase of these waves has a unique high amplitude pattern. First, the waves appear in the pons, then move into the lateral geniculate nucleus of the thalamus where they slow down, and finally go to the occipital lobe as a steady pattern (Brooks and Bizzi 1963). PGO waves appear during the earlier stages of sleep when we are deprived of sleep. If sleep deprivation continues, PGO waves find their way into the waking hours to cause hallucinations (Kalat 2007). Thus, dreams are important for keeping the balance between reality and fantasy where it belongs.

REM Sleep and Memory Consolidation

REM sleep has another important function: It contributes to memory consolidation. Researchers have found that memory storage happens through long-term potentiation (LTP) of glutamate neurotransmitter receptors in the hippocampus. This process takes place when theta waves are produced. Recall, theta waves are often recorded when we are involved in deep inner focus of attention, and this is the characterization of REM sleep. Thus, we need sleep to help consolidate memories (Larson and Lynch 1986). This fact helps to explain why students are often disappointed when they stay up all night to study, and then do poorly on the test. Although they have done the work needed to start the memory process, they did not get the sleep time needed to move their learning from short-term memory to long-term consolidation.

Sleep is clearly important for sustaining life and serving many important functions. Psychotherapy can work in conjunction with medical sleep clinics to help people overcome sleep problems by encouraging the natural built-in sleep mechanisms to return to balance.

Conclusion

The nervous system functions in a balance of activations and inhibitions, elicited by different pathways in the brain. We have many pathways serving different functions, to sense what is happening around us, to experience and moderate pain, to react to threat, to experience pleasure, and to be awake and asleep in rhythmical patterns. When people have psychological problems, these systems are thrown off balance. The therapeutic methods you use can shift these pathways back to their natural tendency for balance and smooth regulation.

Chapter 10
Neural Networks: How Neurons Think and Learn

The many pathways of the brain interact with other areas of the brain, functioning together in a network. These networks of interactions account for plasticity, change, and learning. This chapter provides the fascinating story of how the brain can be understood as a neural network.

Psychology has long had theories of learning that have stood the test of research and time. We know that people acquire new knowledge and understanding. But how does learning take place in the brain? What happens at the level of the neuron? The neural network uses a simplified version of a neuron that combines many of them together in specific ways.

As you enter into the description that begins with one simple input, output unit that is correlated with a neuron, you can understand how the brain may be actually engaged in a thinking and learning process. Since the neural network is a good model of the brain, it gives us a glimpse into how structure and function might be coming together to produce our amazing brain–mind capacities.

Network theory has also found its way into psychotherapy. With the advent of family systems therapy in the 1950s, problems and their solutions are often viewed in a larger, more inclusive family network. New possibilities for treatment strategies emerge by focusing on the network of interactions. Therapy can elicit new learning and change in the network, as this chapter and later chapters will show.

Neurons Interact in Networks

According to the neural network model, the brain is a web of interdependent neural connections. Through the interconnections among neurons, signals are sent in patterned ways. Variations in patterns lead to varying phenomena of cognition.

Neural networks simulate how these basic neuronal elements of the brain interact together. They model the link between neuronal activity and mental processing.

C. A. Simpkins and A. M. Simpkins, *Neuroscience for Clinicians*,
DOI: 10.1007/978-1-4614-4842-6_10,
© Springer Science+Business Media New York 2013

Understanding the composition and mechanisms of neural networks will clarify how the brain could be eliciting thinking and learning. This chapter presents the development of neural network theory, some basic principles, and applications to therapy.

What is a Neural Network?

A neural network is constructed to resemble the brain's neurons. It is based on the idea that even though the brain is complex, it is made of simple units, neurons, which individually have simple behaviors. Each neuron behaves similarly to the others. The complicated functions of the brain all arise from these simple units connecting to each other. A neural network tries to makes the component units interact as these simple neuron units might be doing in the brain. Then, if the model is able to produce a cognitive achievement such as recognizing an object or learning something new, the model offers an explanation as to how the brain may be achieving such a process. We can also learn more about how brain structure and function are interrelated.

Sometimes people think that neural networks are like sophisticated computers. But computers differ from neural networks in a key way: Computers need specific and literal steps to follow in order to solve problems. But neural networks, being modeled on the organic brain, can form something new. For this reason, the neural network is a foundational tool for understanding just how human beings learn.

According to the neural network perspective, the brain is a web of neural connections, like Indra's net, described in the introduction where you imagine a net, stretching all around the room with a crystal-clear jewel glittering at every juncture reflects all the other jewels. Similarly, the neurons in the brain interact dependently. Through the interconnections among neurons, signals are sent in patterned ways. Variations in patterns lead to different kinds of cognition. Neural networks simulate how these basic neuronal elements of the brain interact together to bring about cognitive activity. They can learn. And we can learn from them.

The Forming of Neural Nets

The brain learns at the neuronal level. When given an input along with a desired output, the brain learns to produce the desired output. A neural network can simulate this learning process artificially. Then the neural network can do what the brain does when it learns, such as recognize patterns, learn by example, and learn previously unknown things. For example, a face can be represented in many ways, such as pixels on a screen, a drawing of the eyes, ears, nose, and mouth, or a photograph. A neural network can be trained to recognize faces, so that when the network receives a certain pattern of pixels as inputs it can output whether a face is represented or not. With further training, the network can even output the name of the person who matches that face. In a sense, the neural network exhibits mental properties, like a brain.

Early Development

E. L. Thorndike is often credited as being the father of connectionism. He was one of the earliest people to postulate the basic concept of neural network theory. In his 1931 book, *Human Learning*, Thorndike introduced a core connectionist idea that learning is the strength of the connections between neurons.

> The capacity to learn and remember could find its physiological basis in the movement processes of the neurons...By the hypothesis, the neurons move so as to hold some new spatial relation to neighboring neurons...The strength or weakness of the connection is a condition of the synapse...Let us call this undefined condition which parallels the strength of a connection between situation and response the intimacy of the synapse...Thus, certain synaptic intimacies are strengthened and others weakened, the result being the modifiability of the animal as a whole which we call learning...The learning of an animal is an instinct of its neurons. (Thorndike 1931, pp. 57–59)

Thorndike believed that what separates human learning from animal learning is just the numbers of neuronal connections. "A quantitative difference in associative learning is by this theory the producer of the qualitative differences which we call powers of ideation, analysis, abstract and general notions, inference, and reasoning" (Thorndike 1931, p. 168) Human beings have far greater possibilities for neuronal interconnections than animals. Neuronal connections are strengthened through trial and error as the child develops. Thorndike performed careful infant research, convincing him that the variation of neuronal connections between individuals accounts for the differences in intellectual development between people (Thorndike 1931). He cautioned that his ideas were preliminary, but felt confident that his theory would later be proven through experimentation. Thorndike's theory has found empirical verification through the neural network, and has been greatly expanded.

Evolution of Neural Network Theories

The origin of the first neural network is credited to Warren S. McCulloch and Walter Pitts who presented the idea in a famous paper, "A logical calculus of the ideas immanent in nervous activity" (McCulloch and Pitts 1942). They offered the first introduction of the idea of constructing a neural network using mathematical algorithms. An algorithm is a process or set of rules that are used for calculation or problem solving. Their algorithm set out the basic steps for creating a neural network. They assumed that the network would be based on simple neurons called threshold logic units (TLU's). These TLUs were modeled as binary devices with fixed thresholds.

Rosenblatt (1958) furthered the field by designing what he called the *perceptron*. The perceptron was more general than the TLU. It had three parts: A unit of two inputs, an association region, and an output unit. The association region could learn to associate or connect the inputs to a random output unit. But the associational interaction was two-dimensional.

Problems and their Solutions

The enthusiasm was interrupted when Minsky and Papert (1969) wrote a well-argued book entitled *Perceptrons: An introduction to computational geometry* that pointed out some serious limitations. The simple perceptron acting as a neural network worked well with situations of either yes and no or yes or no, but it could not handle what is known in logic as the Exclusive Or (XOR): This or that, but not both. In simpler terms, the perceptron did not have the capacity to distinguish between both and neither.

For example, in a game of soccer, imagine that the ref tosses the ball up in the air. A kicker from each team rushes to kick the ball. If the kicker from one team kicks the ball first, it goes forward. If the other kicker kicks the ball first, it goes in the other direction. But if they both kick it at the same time, it does not go forward for either team. It cannot go in both directions at the same time. Perhaps it slips out sideways. However, both situations are possible outcomes of the ball being kicked. Finally, imagine that the ref does not throw the ball, so neither kicker tries to kick. The simple perceptron could not tell the difference. It could tell when a ball was kicked, but got confused when it tried to determine whether both tried to kick the ball, so it did not move, or neither tried to kick the ball, and again, it did not move, because it was not there. So it thinks the ball was kicked in both cases. There is no difference from the simple perceptron, which has no way to distinguish the two possible states of affairs in this situation. The perceptron gets it wrong.

As the chart shows, if the inputs are both zero, as in the example, the ref did not throw the ball nor did they kick; or zero and one, the ref tossed the ball and one of the kickers kicked, the outputs make sense and the perceptron can tell the difference. But if the two inputs are both 1; he tossed it and they both kicked at once, so it looks as if there is no input, because the perceptron can only recognize one of them. The other should be a 0. The perceptron, due to its logic, cannot output both. It should show that there is input, but it does not. It outputs 0. This perceptron model fails for the XOR case (Fig. 10.1).

The consequence is that the perceptron could not perform certain kinds of pattern recognition, one of the main early applications of neural networks. It was not until the 1980s that cause of the problem was found: the processing structure of a single-layer led to its limits. In order to perform the function of these kinds of pattern recognition, a new processing structure was necessary. Since function

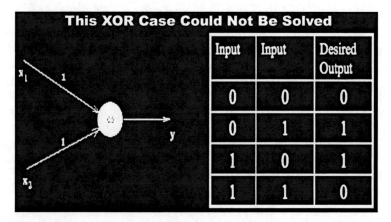

Fig. 10.1 The XOR case could not be solved

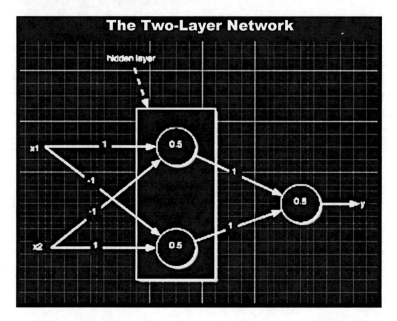

Fig. 10.2 The two-layer network

and structure are related, a change in the structure was necessary to make the function possible, or perhaps, a change in the function of the processing structure. But whether this was possible was not clear.

Adding another dimension: *depth*, gave an improvement to the processing structure. A neural network was constructed with a second layer. The idea of a multi-layered network is that there are other layers, not directly connected to the outside world of inputs or outputs. Instead, the output of one layer provides the input to another (Fig. 10.2).

The network can discriminate by means of the hidden layer. By connecting together multiple TLUs in a two-layer network, the XOR problem can be solved.

In our example, the extra dimension could be for the kickers to pay attention. The kickers have to be paying attention to the ref in order to perform successfully. So, if the ref tosses the ball, and one of the kickers is not paying attention, the other kicker, who is paying attention, kicks the ball and it moves forward. But if the ref tosses the ball, and neither is paying attention, they would not kick, so the ball will not move forward. And if both are paying attention and both kick, the ball will not move forward either, just like the simple example. But there is a noticeable difference. In one case, they do not kick and the ball does not move. In the other case, they kick and the ball does not move. Difference permits discrimination. The second level makes it possible.

This discovery made the perceptron useful. The principle is applicable to therapeutic change, as well. Intelligent change cannot be only two-dimensional. It requires another level of processing, so that distinctions can be made. Therefore, clients who have an XOR-like adjustment cannot tell whether their response is to their actual, outer real world situation, or to their inner conflict, without an added level. The added level can be brought about in many ways. Then the client can distinguish between the two states of affairs: either response to inner conflict instead of responding to outer demands, or response to outer demands while not focusing on inner conflict. The client cannot respond well to both at once. But in order to know the difference, there must be another level of processing, just like the perceptron.

From Biological Neurons to Artificial Neuronal Units

Neurons are the computational building block of the brain's architecture. In reality, each neuron is a complex system of electrical and chemical activity with a wide variety of size, shape, and properties. The single-layer TLU, modeled similarly to the biological neuron, is the building block of the neural network.

The neural network model simplifies the complexities and variations of the brain's neurons by showing what neurons have in common. In general, all neurons take in input through the dendrites, process the inputs in some way, and then send outputs through the axons which then split into many branches. At the end of each branch is a synapse. Signals are transmitted to other neurons at the synaptic connection. The signal either excites or inhibits another neuron that is connected at that synapse. If enough incoming pulses converge on a neuron within a certain amount of time, the neuron fires, transmitting a new impulse into its dendrite and then out through its axon. Learning results from the excitatory and inhibitory changes that take place between neurons at the synapse.

The model of a neuronal unit in a neural network is fairly simple: The unit receives input from its neighbors or external sources and uses this to compute an output signal. This output signal is sent to, or "propagated" to, other units. The second task is to adjust the weights, as will be explained. The diagrams show how the biological neuron is modeled as a TLU.

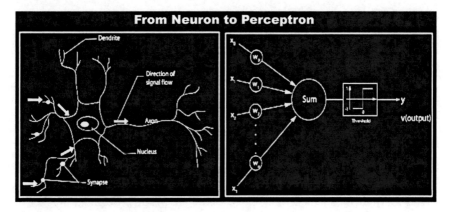

Fig. 10.3 From neuron to perceptron

Connections Between Units

The connections between the neuronal units of the brain can also be modeled in neural networks. The links between neurons are characterized in terms of their synaptic strength known as the bias or weight (W). By connecting the synaptic elements of the neural net in various ways, with differing weights, distinct degrees of intensity of activation become possible. Thus, an input sent to a neuron with a weak connection to the next neuron will have less activation than an input sent to a neuron with a strong connection to the next neuron (Fig 10.3).

The weight of an input is a number that is multiplied by the input to give a weighted input. This number has a specific value that is either positive (+) or negative (−) to result in an input that is either an excitation or an inhibition. These weighted inputs are then added together to help determine whether the neuron will fire or not. In the example of kicking the ball, imagine the kick with the shoes of the kicker. The kick must be firm enough when the kicker kicks the ball to make it move. And if the shoes are soft and light, made of bubble wrap, the ball will not move much when kicked, whereas if the shoe is a good, firm shoe, the ball will move farther. So, the weight affects the consequent result of the connection. In a formal sense, connection requires a rule, known in neural networks as an activation rule.

Activation Rules: The Threshold

An activation rule takes the weighted input data and combines it with the activation that the neuron already has. The simplest activation rule is the threshold. If the amount is larger than a pre-set threshold amount, the neuron will fire. Otherwise, the neuron does not fire. Again, in the example, the kick must be performed with at least a certain amount of forcefulness, the threshold, to affect the ball significantly enough to make it move. The threshold is fundamental to the neural net model, just as the discovery of the concept of threshold was fundamental to the origins of empirical experimentation in Wundt's laboratory in the beginning of modern psychology (Rieber and Robinson 2001).

The computation for the activation can be expressed in this equation:

$$\text{Activation} = \sum_{i=1}^{n} w_i x_i = w^T x$$

The neuronal unit performs a weighted sum of all the inputs (x) multiplied by a guessed value for the weights (w). $\sum$ is the symbol for summation, x is the symbol for the inputs and w_i is a shorthand way of expressing a group of weights without having to list them. So, the term w_i expresses all the weights (weight a, weight b all the way the ith weight) and x_i expresses all the inputs (input a, input b, all the way to the ith input).

If the weighted sum of the x's is larger than the threshold, the neuron outputs a 1 and fires. In other words, the neuron activates. If the sum is less than the threshold, the neuron outputs 0 and does nothing. The equation that shows whether the neuron will fire or not is:

$$y(x) = \begin{cases} 1 & \text{if } \sum_{i=1}^{n} w_i x_i \geq \theta, \\ 0 & \text{otherwise} \end{cases}$$

Other Activation Concepts

Thresholds are not the only way to characterize what will activate the neuron. Sometimes the difference between firing and not firing is not a simple yes or no, but instead forms a more subtle relationship. Such situations are best represented by a mathematical function. An advantage of using a function is that it can use real-valued inputs and generate real-valued outputs.

One way to understand the difference between threshold activation and function activation is to consider the difference between discrete and continuous change. A discrete change is a jump from one value to another one, whereas a function changes continuously over time. For example, the gradual curve of the sigmoid function shows how the change varies with time. The steepness and shape of the curve can be varied as well, helping to visualize how quickly and at what point the changes occur. Psychologists are probably familiar with the linear function, used to model correlations, in which the strongest correlation is a straight line (Fig. 10.4).

There are many ways to model activation functions. The list above shows some of the typically used activation functions, but there are many others. These show how the neuron activates at the threshold.

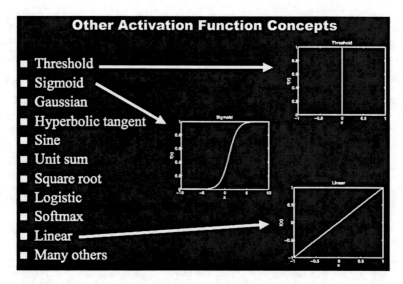

Fig. 10.4 Other activation function concepts

Multi-layered Neural Networks

Each neuron communicating with each other is a slow process, but neurons work together in large networks. Once the neuron has been characterized in this simple manner, the interaction between neurons becomes the focus of attention for building the system. Artificial neural networks (ANNs) utilize this simple conception of the neuron and put the neuronal units together into systems of interconnections. By simplifying the conception of a biological neuron as a simple TLU or perceptron, more complex combinations of interactions can be explored. Multilayered networks are variations and combinations of the single perceptron network. Some typical layered network topologies are shown below (Fig. 10.5)

Feedforward and Feedback Systems

The properties of the connections between units in a network are grouped in two broad categories: feedforward and feedback. In feedforward systems, the signal or data travels in one direction only: from input to output. The processing of the input may extend over multiple layers of units, but there is no feedback. The input is never sent back to a previous layer or neuron, it just moves forward. Feedforward systems are used for pattern recognition and in bottom–up and top–down processing. The simplest example of a feedforward system is the perceptron model mentioned earlier (Fig. 10.6).

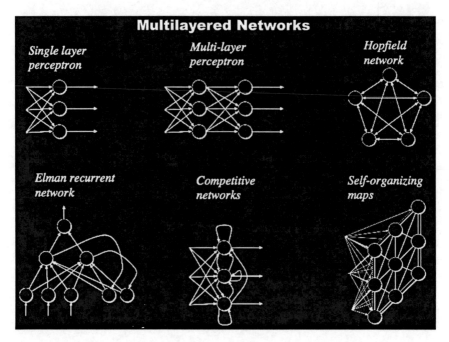

Fig. 10.5 Multilayered networks

Fig. 10.6 A feedforward system

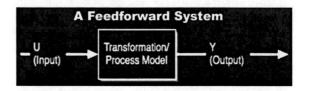

Feedback adds another dimension to networks. Signals can also travel in both directions by adding loops. Then before the information goes to the next neuron, it is fed back through the system, to give the system more information. The feedback influences the system as a controller, so that it can make adjustments. A classic example of a feedback system is the Hopfield network, which is often used to model associative memory. Neurofeedback and biofeedback rely on the feedback model to establish voluntary control of involuntary functions (Fig. 10.7).

Feedback systems are dynamic, continuously changing until an equilibrium point is reached. They will stay at this equilibrium until a new input comes in, requiring a different equilibrium to be found. The reader can see how these networks are sometimes referred to as either interactive, recurrent, or feedback systems.

Both feedforward and feedback have advantages and drawbacks. Feedforward methods offer a simpler computation and easier sensor requirements. However, as might be expected, without receiving any feedback, these models can drift into

Fig. 10.7 A feedback system

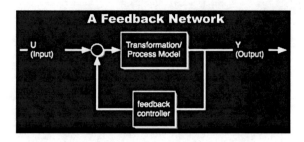

errors with no way to correct themselves. Also, any disturbance to the system tends to throw the system off track. Feedback systems correct errors as they go along. These systems also tend to be robust, which means they are more stable and handle disturbances better.

Models of Learning

Once the neural network architecture is set up, the neuronal system can be trained to learn. Generally, the learning process involves how outputs are processed at the synapse. This is where learning takes place. The strength of the synaptic connection weights is not known. Learning involves computing the values of the weights. The learning takes place from trial and error. The neural network makes an error, adjusts the weights, and then tries again. The process is repeated until the correct value is reached.

Back Propagation

Back propagation is the artificial network's way to learn from experience by correcting the errors. Here is how it works: In the learning situation, an output is produced that needs to be corrected. The learning system gets inputs and sends them forward through the network. The system begins to correct by computing the error between the output and desired output (delta value). This delta value is *back propagated,* or sent back through the layers of the system. Then the weights can be updated in the right direction, so that the errors can be corrected.

Hebbian Learning Rule

In general, learning methods involve adjusting the weights of the connections. Hebb (1949) presented a principle for learning rules that has become a foundation for neural network learning models. His principle is known as the Hebbian learning rule. Simply stated, neurons that fire together, wire together. The result is what Hebb called a Cell Assembly. To apply this principle in a neural network, strengthen the interconnections of units that are active at the same time, and weaken the interconnections of units that are inactive at the same time. Increasing the value of the weights strengthens the interconnections. The amount of increase or decrease is the learning rate.

Types of Learning

There are many learning methods, but they fall into several basic categories: Unsupervised learning, Supervised learning, and Reinforcement learning.

Unsupervised Learning. In unsupervised learning, no feedback is provided to indicate whether the network's associations are correct or not. An output unit is trained to respond to clusters of patterns within the input. The system discovers certain features of the input. Different from supervised learning, there are no pre-given categories for classifying the patterns. The network must discover by itself the similarities in the patterns of the data. Unsupervised learning works with self-organizing networks that possess the ability to infer patterns from the input-only data. Errorless learning, used with brain-injured and schizophrenic patients to teach skills in coping with life situations, is a form of this.

Supervised Learning. Supervised learning, also known as associative learning, sets the weights explicitly using a priori (pre-given) knowledge. A second method of supervised learning teaches the network patterns that allow it to change its weights by a learning rule.

Learning that uses *a priori* knowledge begins with a training set during which known input/output pairs are presented to the network. The network adjusts its weights to produce the desired output. The network compares the actual response with the desired response and gives an error value based on the difference between the two. Following that, the weights are modified to gradually reduce the error.

To summarize, the perceptron becomes "trained" by beginning with a guess for the weights, giving the system an input that will generate an output. The output will be 0 or 1. This output is compared with the desired output using the perceptron learning algorithm equation. If the output is too large, the weights are decreased, and if the output is too small, the weights are increased. Gradually, the system output comes closer and closer to the desired output. In a sense, like people, the system optimizes its learning.

Reinforcement Learning. The idea of reinforcement learning will be quite familiar to most readers. As in supervised learning, training is given. But unlike supervised learning, the teaching does not involve the size of the error. Instead, training gives an indication as to whether the output is right or wrong, or if it is going in the right or wrong direction. Correct outputs are reinforced and incorrect ones are not.

Applications of Neural Networks

Neural networks have found applications in a broad range of fields. There are many different types of neural networks to suit the diverse needs of disciplines. Neural networks help to explore memory, sensory system, and motor functions. And neural networks have been applied to a broad range of problems such as autism, various forms of dementia, chemical dependency, sleep disorders, depression, schizophrenia, and Parkinson's disease, to name a few of the most researched areas.

Neural networks offer a new research paradigm for therapists. They have been used to investigate psychotherapeutic models as well as to learn more about typical disorders. In being able to correlate together multiple and hidden variables, neural networks can also help to better understand the complexities of inner dynamics that have eluded researchers in the past.

Psychotherapy, Brain, Character, and Varied Treatment Modalities

Levine and Aleksandrowicz (2005) developed a neural network theory to capture some of the subtler qualities of psychotherapy and client personalities. They used a neural network to show how psychotherapy moves the client from a less optimal to a more optimal state. Neural networks are an effective means of modeling such movement. The network shifts from one steady state to a more optimal one, which gives the client a corrective emotional experience.

The neural network model combines data from brain regions and neurotransmitters with a character cube built around three dimensions: self-directed, cooperative, and self-transcendence (Cloninger et al. 1993). Each dimension ranges from low to high and includes a broad and descriptive array of character qualities that de-emphasize traditional disease categories and encompass mental health as shown by degrees of happiness and wisdom. Multiple therapeutic methods can also be integrated into the model. Typically, therapists approach clients using one therapeutic method, such as humanism, psychoanalysis, or cognitive behavioral, to give a few examples. Each form of therapy concentrates on certain character dimensions more than others. This neural network model moves from non-optimal states to more optimal ones by appealing to the best combination of dimensions drawn from multiple therapeutic models (Levine and Aleksandrowicz 2005). The ability to integrate the best features of different therapy models will enhance treatment, and neural network research helps to do so.

Neural Networks for Alcoholism

Neural networks have also been applied to specific problems. Ownby (1998) created a neural network model of the diverse factors influencing alcohol dependence. His neural network models allowed him to test his hypotheses about etiology and treatment that are usually unfeasible to test with real subjects. The computational model gives a way to examine relations among neural structures that the various theories consider primary in alcohol dependence to understand the problem more fully.

Ownby created a model of the pathological need to drink by beginning with a neural network that simulated normal desire. Then the network was set up to associate low levels of sensory and motivational cues with a low desire to drink. As the cues become stronger, levels of desire became higher. The network was also set up with corresponding brain reactions from areas typically thought to be involved in alcoholism as well as a network of the neurotransmitter systems and genetic vulnerabilities typically involved. The networks were altered to simulate treatment effects. For example, decreasing limbic system inputs and increasing prefrontal cortex activity simulated the impact of psychosocial therapy, which tends to have a calming effect on the emotions while increasing the use of rational understanding.

A neural network model can simulate symptoms of alcohol dependence. When the network simulated a mild desire to drink, only a strong stimulus increased desire. But in a network with a strong desire, even a weak stimulus elicited a strong desire to drink. The network also simulated the dynamic process involved in alcohol dependence in response to environmental and internal cues. The network correlated with real alcoholic dependence. By simulating the effects of both psychosocial and pharmacological treatments to test their relative effectiveness, the neural network model predicted that the combination treatment would work best (Ownby 1998).

Depression and Learned Helplessness

Neural network modeling methods are particularly suitable for modeling problems such as depression that involve neuro-chemical and neuro-electrical dimensions. They can also be used to answer questions about the interactions between mental processes and neurochemistry. Neural network systems provide a useful way to compare theories that may seem incomparable. For example, Seligman's theory of learned helplessness (Seligman 1975) has been modeled in multiple ways and analyzed along numerous dimensions such as biochemical, motoric, cognitive, behavioral, mastery, control, affective, and many more (Leven 1998).

Memory and Stress

A neural network model was developed to help understand memory under stress. Cernuschi-Frias, Garcia, and Zanutto investigated two types of memories: stressed and unstressed and implemented them on the same neural network. They found that memory retrieval occurred according to a continuous function of how stressed the individual was at the time of retrieval: low stress retrieved unstressed associations and high stress retrieved stressed associations (Cernuschi-Frias et al. 1997).

Conclusion

Neural networks have given us a model for how the brain could be generating cognitive processing. From a simple neuronal unit to such complex processes as being able to learn, make discriminations, and recognize things, neural networks help us to begin to bridge the gap between brain and mind. This model offers great hope for understanding how the brain and mind interact together, to achieve all that the human being can do.

In addition, neural networks are a useful paradigm for therapeutic change. Therapeutic models can be researched by using a neural network. They may prove to be helpful for therapists to do preliminary testing.

Part IV
How the Brain Changes Through Time

Psychotherapists are always trying to help people make changes. How do human beings make a change? How does the environment change through time? What is the nature of change itself? Understanding how the brain undergoes change provides answers to these long-asked questions.

We can think of brain change as occurring over three timescales. Each reveals a different perspective on how the material and functional intertwine as the brain has traveled through time. The most obvious is the evolutionary timescale. Researchers have found that the ecological niche or what the organism does in its environment, along with survival instincts, have influenced how the structure of its brain evolved, as Chap. 11 reveals. A shorter timescale is found in the developmental lifecycle of the individual where environment and genetics intertwine to significantly alter the structure and function of the brain for that individual's entire lifespan. A third timescale has been recognized in recent years: experience-based neuroplasticity. Real-time is immediate, taking place in the present. Recent research points to the discovery that real-time influences can literally sculpt the structure of the brain, even in adults.

Part IV presents the changing brain in these three timescales. First in Chap. 11, the evolutionary theories are described, including some of the most recent discoveries along with ancient theories that continue to be upheld as the new research unfolds. Chapter 12 presents the developmental process of the individual, beginning from inception through the earlier years of brain development including infancy and childhood. Two other phases when the brain undergoes large changes occur in adolescence and again during old age. Chapter 13, on brain plasticity, cites many of the important studies that have led to revising the view of the brain as fixed and unchanging in adulthood. The brain can and does change all through life. We can cultivate change deliberately by the actions we take, thoughts we think, and emotions we feel.

The helping professions are in the business of facilitating change in others. Most therapists and teachers have seen rapid, real-time change happen in people, sometimes drastic, other times subtle. And so, you can embrace these new findings as scientific verification of the potential power for therapy and education done well. In addition, you can look to the specifics for guidance in making our interventions the best ones possible.

Chapter 11
Evolution of the Brain Over Eons of Time

Brief therapy has become more prevalent in our modern, hurried age. And in fact, therapy is getting shorter and shorter with the advent of one-session treatments. Clients want their changes to happen quickly. Given this therapeutic climate, you might wonder how evolutionary change occurring over thousands of years is relevant for brief therapy.

One of our teachers, Milton Erickson liked to say, "Your unconscious is a lot smarter than you are." This same principle can be seen in how the brain evolved. We find in the evolution of the brain many changes that reveal ingenious solutions to new demands. As therapists, we are always helping our clients to discover new ways to adapt to their life situations. Evolution is a showcase of ingenious solutions to impossible situations. Why not learn from nature at its best, having produced the richly varied and beautiful collection of species we have in our world? We invite you to entertain new creative perspectives, as you share in nature's masterpiece of creation through time.

Top-Down Approach to the Brain

We have discussed how neuroscience explores the brain as a bottom-up approach, beginning with the neuron as the fundamental unit and moving up into systems of interactions. But an alternative way is to explore the brain from top-down. Evolutionary neuroscience begins from the evolved functioning brain and deduces down from that to the individual units as they developed through time.

This kind of top-down study helps in understanding the broader context of changes that might not be understood by just looking at our brain as it is today. It also helps for uncovering brain-mind relationships. We learn by contrast as we explore what is different about the human brain from the rest of the animal kingdom. What evolutionary studies reveal is that the kind of brains we have today

C. A. Simpkins and A. M. Simpkins, *Neuroscience for Clinicians*,
DOI: 10.1007/978-1-4614-4842-6_11,
© Springer Science+Business Media New York 2013

has been constructed through an evolutionary process that developed through time. We see that certain structures in our brains are found in all species through time. More complex species evolved as a combination of branching like tree and climbing up a ladder.

Therefore, evolution helps to explain how our brains have come to be the way they are today. Research studying the evolution of different species through the ages uses modern technology to reveal a great deal of the brain's hidden nature, adding depth of perspective. Although we hope our therapeutic effects will occur faster than the eons of time required for evolution, we can learn much from the change that has taken longer!

Three Evolutionary Methods

We have learned about brain evolution through three main types of investigations: Fossils, comparative methods, and brain mechanisms (Fig. 11.1).

Most people will know that much of what we know about evolution comes from studying fossil records. We can observe these ancient skulls, which are preserved as fossils, being made of bone. Scientists have learned about brain size relative to the animal being studied. Brain tissue is usually tightly packed inside the skull, thus, much can be inferred about the proportions of soft tissue even though it is not preserved. For example, early primates had a larger proportion of neocortex than mammals. Skull fossils reveal major fissures patterns that can give clues to the subdivisions of the functional organization of the neocortex in the fossils. Van Essen (2007) theorized that densely interconnected regions tend to resist separation during brain growth, forming gyral bulges, whereas looser, poorly connected regions tend to form sulcus folds.

A second way to explore brain development has been done through comparative methods. It is currently held that complex life on Earth evolved only once. The molecular template that is passed from generation to generation can be modified. Phylogenetic (ancestral) classifications can be made to shows points of divergence from a common ancestor. Similarities in the generations show preservation of the original code, while divergences reflect alterations in the code. These divergences are what have led to the diversity in our world.

The phylogenetic relationships are deduced from comparative evidence. A more recent method for tracing these ancestral relations is known as Cladistics, founded by Willi Hennig (1913–1976). Hennig believed that groups sharing a common ancestor, clades, are likely to share more biological features. Cladistics uses what Hennig called "character analysis," that is observable features of an organism, as a means to reconstruct the ancestral (phylogenetic) relationships. For example, one could investigate a feature of the brain, such as the neo-cortex, and look at species where it is larger or smaller, present or not present. Then, hypotheses can be made about how the groups are related.

Fig. 11.1 A fossilized brain

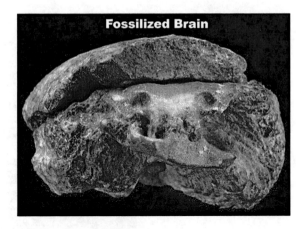

The features that are shared in common come from a common ancestor and are homologous. Homology does not imply an identical structure or function. For example, all mammals have a somatotopic organization in the brain (specific regions of the cortex that are responsible for different areas of the body) but these areas develop to perform different functions. Sometimes, groups that do not share a common ancestor possess similar characteristics. Such characteristics that evolved independently are called analogous. Analogous characteristics come about from similar adaptations and similar functions.

A third source of information for brain evolution is brain mechanisms. For example, in comparing larger with smaller brains, the mechanism of neural connections evolved differently. Larger brains have greater number of neurons located at longer distances from each other. Larger brains had to evolve different mechanisms of neural transmission to maintain communication between neurons, as smaller brains have.

Single Ancestor Assumption

The varieties of species existing today evolved from a single common ancestor, according to the Darwinian view. The evolution of the human brain also is thought to have evolved from a common ancestor. One compelling bit of evidence is found in comparing the brains of mammals. Each of the main structures of the human brain has a counterpart in other mammalian brains. The differences tend to be quantitative, not qualitative (Breedlove et al. 2007). In this sense, we are interconnected and related to all species, at a fundamental, original level.

All species share certain common features. All have a have a central nervous system consisting of a brain and spinal cord, and a peripheral nervous system. The neurons have a ganglion structure and axons with neural conductance.

Fig. 11.2 Ancient vertebrate
fossil

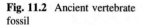

Although the structure of these systems is similar for vertebrates and different for invertebrates, the common roots of all are evident (Fig. 11.2).

Vertebrate nervous systems share even more similarities. They all develop from a hollow dorsal neural tube (See Chap. 12) and they have segmentation with pairs of spinal nerves that extend from each level of the spinal chord. The cerebral hemispheres have hierarchical control over the activity of the spinal chord. The central nervous system is separated from the peripheral nervous system, and functions are localized in the central nervous system.

Intelligent Design Versus Self-Organizing Systems

Power resides where information resides. (Warren McCulloch 1965, p. 229).

The conception of evolution as the motivation for the development of the species and of the brain spawned a debate between intelligent design and the self-organization. These concepts are ancient and timeless. We can learn from both theoretical positions.

Intelligent design, sometimes called the watchmaker argument, has its roots in a belief in God, or a prime mover, as the first cause. A watch that you might come across, with its intricate mechanism, or in effect, anything that is found in nature, must have come to be through some form of intelligence, guiding idea, or purpose that made it. The smallest entity to the most complex system requires a first cause, an intelligent designer who somehow set it into motion.

By contrast, the self-organizing system view of evolutionary theory has a different basis. The idea of self-organizing systems has a long history, dating back to Descartes. For the evolutionary theorist, there is no intelligent designer or first cause. The laws of nature tend to produce organization. Thus, given enough time, space, and matter, organization is inevitably going to come about. And systems tend to be in balance, or to find it.

According to this view, everything we encounter in the world is a self-organized system. These systems began as simple entities, connected to each other according to some kind of an implicit organization, such as the learning rule fundamental to a neural network. The result is an ordered, stable system. And similar to how the unsupervised neural network functions (see Chap. 10), there is no external source required to guide it. It may be simple or complex, but one common feature is that the system is not being guided by an outside source. Properties emerge from the system. In brains, these emergent properties are such things as consciousness—our thoughts and feelings. These emergent, collective properties are always more than the sum of the parts. This idea is applied in many areas such as physics, chemistry, economics, biology, as well as in studies of the brain.

The argument between intelligent design and self-organization is not settled by any means. The self-organized system proponents claim that evolution explains how complex elements could arise at any level of the species. They sight certain tendencies of nature such as natural selection and competition between units as the means by which these systems operate. As Darwin argued:

> Nothing at first can appear more difficult to believe than the more complex organs and instincts should have been perfected not by means superior to, though analogous with, human reason, but by the accumulation of innumerable slight variations, each good for the individual possessors (Darwin 1859/1999, p. 375).

But the intelligent design proponents believe that this does not explain the origin of the single units that make up the systems. How did that first tiny element that self organizes itself into the complex system first come to be? Nor does it explain how these units pick one organization over another. These questions are left open, and the reader will certainly have an opinion.

No matter what view is held regarding the ultimate starting point, the understandings about self-organizing systems can be helpful in learning more about how our brains are organized and function. The achievements of the brain, a relatively small part of our anatomy, are truly a marvel. Investigating this complex development and evolution at many levels may give us a helpful perspective on this magnificence, which we all carry in our skull. The ultimate answers of origins and initiators are left open.

Self-Organizing Systems

Contemporary views of brain evolution are based in the evolutionary principle that the brain is a self-organizing system. Beginning billions of years and continuing through the ages, the original molecular structure changed in response to different factors. It appears that nature did not start over anew with each evolutionary change. Instead, changes occurred as a series of add-ons to the brain systems that already existed. The variations have resulted in the wide range of variations in brains among species. Generally, "the evolution of vertebrate brains can be related to changes in behavior" (Breedlove et al. 2007, p. 165).

The emergent properties that occur from the self-organizing brain system are behaviors, feelings, and thoughts, our very consciousness itself. (See Chap. 3 for the relationship between mind and brain) The basic idea is that the dynamics of the system can tend, by itself, to increase the inherent order of that system.

Self-organizing systems rely on certain basic components. Feedback is primary. When information is positive, it tends to be sent forward through the system, feed-forward. When the information is negative, it goes back into the system to become feedback for making an adjustment.

Self-organizing systems have certain common properties with multiple rules for the system. These systems are dynamic and absent of centralized control (competition). The system also has a global order. Furthermore, the system tends to be insensitive to damage by having redundant parts to help protect it. It also has methods for self-maintenance and repair. The systems have complexity, with multiple parameters and they are built with hierarchies of multiple self-organizing levels that tend to be in a balance between competition and cooperation (Pineda 2007).

How Self-Organizing Systems Evolved Through Time

There are two theories about how this self-organizing brain system evolved: the Ladder Theory and the Tree Theory. Each helps to explain different findings, and so evolution is best understood by combining them.

The Ladder Theory

The older evolutionary theory looked at brain evolution as similar to a ladder, with gradually increasing complexity. This view continues to offer some helpful ways to categorize and compare species. It tends to emphasize quantitative differences. For example, as we move up the ladder by comparing brain weight to body weight, several findings emerge. As brain weight increases, the complexity becomes greater.

Humans are at the top of this ladder, with the most complex brains, although our brains are not the largest ones. Encephalization is a term that defines the amount of brain mass that exceeds an animal's body mass. The idea is that the larger the brain in relation to its body size, the more intelligent, since there will be more processing for association beyond what is needed to regulate the body. Quantifying of encephalization, the encephalization factor, has been used as an index of comparative intelligence. A brain to body mass ratio (encephalization Quotient or EQ) has been used to quantify intelligence.

$$EQ = m(brain)/Em(brain).$$

Dolphins have the highest EQ for cetaceans. Sharks have the highest for a fish and octopuses have the highest for an invertebrate. Humans have the highest EQ for any known animal (Gregory 1988)

Smaller brains have a shorter period of neurogenesis (birth of neurons) than larger brains. For example, a rat brain takes a few days to develop while a human brain needs5–6 months to come to full development. Larger brains have more cells and more variety in cell types than smaller brains. Also, as brain size increases, so does the percent of the brain devoted to the ability to perform complex computations, known as association areas. A total of 80 % of the human brain is devoted to association, while only 16 % of the rat brain performs such tasks.

The Tree Theory

The second theory views evolution like a tree, or even a wide bush, with many branches. The evolutionary tree theory is widely accepted today (Kaas and Preuss 2003). According to this view, there is not a single *telos* or purpose, such as toward greater size or complexity. Although larger brains tend to be more intelligent, the needs of the species vary. Qualitative differences evolved as varied needs arose. Brains structures evolved to assist the organism's adaptation to its needs. For example, an animal living underground might have evolved a smaller brain and a reduced visual system. These adaptations would conserve the animal's metabolism for the most effective and efficient existence in an underground environment. An animal with a large area devoted to vision would be far less successful living underground than the animal with a smaller vision system.

The largest brain is not necessarily the smartest one. A case in point is the dolphin (Fig. 11.3). The dolphin brain is much larger than the human brain, yet dolphin brains do not have as much lamination as human brains. More lamination or layers of varied neurons produces more complex behavior and capacities. A large portion of the dolphin brain is devoted to the regulation of body temperature, a necessary function for their underwater environment, whereas a much smaller area controls cognitive functions, less necessary functions in the life of a dolphin.

Fig. 11.3 Human brain/
dolphin brain

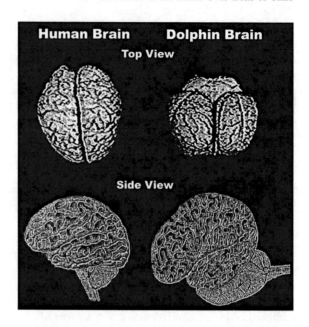

Many of the brain variations found in different species are in response to local
conditions in the environment. For example, humans rely on vision more than their
other senses. Thus, humans need more complex visual circuitry and do, indeed,
have a larger brain area devoted to vision. Rats, which use their sense of smell for
multiple functions, have as large an olfactory area in their brain as their
neo-cortex.

Researchers first thought that the more complex parts of the brain evolved later,
such as the multi-layered visual areas. However, further research has revealed that
highly differentiated laminar organization occurred in some of the first mammals
and perhaps even in more primitive species (Kaas and Preuss 2003, p. 1152).

One way that nature adapted the needs of larger or more complex brains is by
increasing the amount of brain specialization. Larger brains have more areas of
specialization than smaller brains. For example, rat brains have three to five visual
areas whereas humans have 32 visual areas. Less specialization is needed for a tiny
brain to function optimally, since connections are shorter and less complex. Direct
wiring between parts integrates all the activity together. In a larger brain, interactions are more complicated, being further apart and having interconnections that
are more intricate.

Small brains can send a signal from the brainstem to neo-cortex and back very
quickly. Large brains do it just as quickly even though a larger distance must be
traversed. How is this possible? The ingenious solution that nature has offered is
increased myelination of the axons. Myelination is the insulation of axons.
By insulating the axons, signals can travel more efficiently over longer distances
and save energy.

All of these variations help to explain how and why the human brain functions as it does. The ladder and tree theories can be integrated together, to offer a fuller picture of brain evolution by understanding how brain size and brain complexity interact for the best adaptation.

The Crown of Evolution: The Isocortex (Neocortex)

One of the most special achievements of brain evolution is the isocortex also known as neo-cortex or just cortex. The isocortex developed in the higher mammals. *Iso* translates from Latin as "same" and *cortex*, means bark of a tree. The isocortex is special. It distinguishes mammals from the rest of the animal kingdom. The isocortex used to be known as the neocortex, because it was considered the newest evolutionary part of the brain (Fig. 11.4).

One paradigm for evolution of the anatomy of the mammalian brain is the triune brain (MacLean 1967). In the course of evolution, three distinct brains emerged, one after the other, and now exist together in the human brain. First was the reptilian brain that includes the upper brainstem, cerebellum, and thalamus, governing instincts and fundamental vital functions such as heart rate, breathing, body temperature, and balance. Developing out of the brainstem in the first mammals came the limbic brain, with paleomammalian structures, including the hippocampus and the amygdala. The limbic brain was capable of memory and emotional evaluations of experiences, often unconsciously made, that can have a powerful effect on behavior. Finally, the neomammalian structure known as the cortex evolved out of the limbic system. It appeared most prominently in primates and developed fully in the human brain. The cortex has two large cerebral hemispheres, with the capacity for language, abstract thought, and the ability to learn. The triune brain functions interdependently with interconnections among all the parts.

The cerebral cortex has become the most significant structure as it met the adaptation needs more effectively. The cortex expanded in predatory mammals more than herbivorous ones. One evolutionary theory suggests that the origins of human intelligence are intimately linked with the search for more highly nutritious foods, such as meat, and the sharing of meat with fellow group members (Stanford 1999).

The size of the cortex has increased markedly in primates, from the smallest monkeys to human beings. The increased complexity of social interactions has resulted in a large evolutionary advantage. Thus, the theory is that evolution favored growth in the areas of the cortex that facilitates social interactions such as language.

Increased folding of the cortical surface area has allowed for a larger cortical surface that fits inside the skull. It has also led to better organization of the many complex behaviors and cognitive capacities that are regulated by these areas.

Fig. 11.4 Triune brain

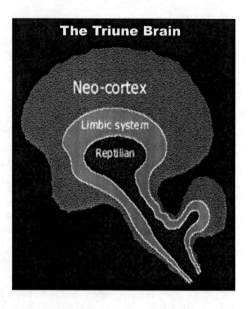

The prefrontal cortex has made the most evolutionary advances in the human brain. This area, dedicated to voluntary motor control in other species, is responsible for our ability to plan and retain information while performing a task. Although there are some who disagree (Sherwood et al. 2005), researchers have found that humans have a larger volume of white matter than other primates. The white matter is composed of myelin-covered axons that communicate with other parts of the brain, providing far greater connectivity to the rest of the brain (Schoenemann et al. 2005).

Conclusion

Evolutionary theory and research has provided significant pieces of the puzzle. Ideas that seemed straightforward, such as that the largest brain is the smartest brain, do not capture the complexity of the brain's functional and material sophistication. As we continue to learn more about our past, we may be able to make a better future. Insects, animals, plants, and humans all share a common nature. All life is valuable and perfect in its own way, and therefore all should be equally respected.

Chapter 12
Brain Development Through the Life Span

The process of human development is one of the most dramatic examples of neurogenesis and neuroplasticity. Consider the amazing fact that we all begin as one single cell, the fertilized egg that grows and matures into the adult brain with estimated 90–100 billion neurons. The early development of the brain provides a template for some key aspects of neuroplasticity and neurogenesis that occur throughout life. So, understanding how the fetus and infant develop will help you better understand brain structures and functions.

Early development unfolds in orderly stages. Some stages are self-regulated and genetically determined, while others are responsive to the surrounding environment. The interplay of what seems to be inborn genetics (nature) with the environmental influences (nurture) shows that with the brain, nature and nurture are ongoing interactive processes from the very beginning. We see this interplay of nature and nurture all through the life cycle. As therapists, we are aware of both influences and work with them sensitively to best help our clients make their changes. Brain development through life reveals dimensions of how the brain undergoes change, giving us potential sources for therapeutic interventions at various stages of the developmental process.

The Stages of Brain Development, from Conception to Infancy

The stages of brain development from conception to early infancy have been divided into a number of phases. Some of these phases are repeated at key times during the life span, such as in childhood, adolescence, and old age. You may find it helpful to visualize this sequence of development as a paradigm for brain development that offers a way to better understand brain structure and function as it transforms through life (Fig. 12.1).

C. A. Simpkins and A. M. Simpkins, *Neuroscience for Clinicians*,
DOI: 10.1007/978-1-4614-4842-6_12,
© Springer Science+Business Media New York 2013

Fig. 12.1 Brain
development from conception
to birth

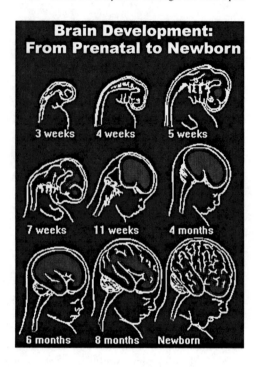

Phase 1: From Conception to Neural Induction

Phases of Early Brain Development		
1. Neural Induction	E18–E24	Genetically
2. Proliferation	E24–E125	determined
3. Migration	E40–E160	
4. Differentiation	E125-postnatal	
5. Synaptogenesis		Environmentally
6. Cell Death/Stabilization		Sensitive and
7. Synaptic Rearrangement		Self organizing

When the ovum is fertilized, it begins as a single cell containing much of the information to guide its development. The cell quickly begins dividing. This process, called *gastrulation*, is the division of non-neuronal cells. The earliest stages are driven largely by genetics. But as the process continues, environment plays a larger and larger role. The growing group of cells forms three layers: endoderm, mesoderm, and ectoderm. The ectoderm layer will eventually develop into the nervous system.

Fig. 12.2 The neural tube

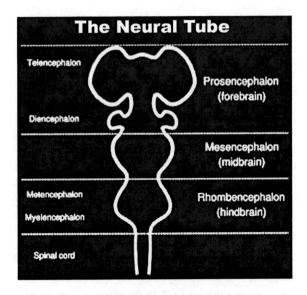

The brain begins to take form on the 18th day. This process, called neural induction, involves three phases. First comes the induction itself. From the interaction of endoderm and mesoderm, a peptide is released that causes the ectoderm to thicken. This ectoderm thickening leads to the development of a neural plate. Second comes neutralization at day 20, when the neural plate becomes a neural groove, leading to the formation of a neural tube. On the 22nd day, the two ends of the tube close off to form the rudimentary beginnings of the brain. Finally, neural patterning begins on the 24th day. This third phase is where we begin to see the brain take shape.

The cells segment as they acquire different identities and dimensions. The neural tube divides into three primary brain vesicles: the forebrain, midbrain, and hindbrain. The primary vesicles divide further to form the secondary brain vesicles. From top to bottom, they are the telencephalon that will become the two hemispheres, diencephalon, mesencephalon, metencephalon, mylencephalon, and spinal cord. All of these vesicles will eventually develop into the adult nervous system structures (Fig. 12.2).

Phase 2: Proliferation

Proliferation is the phase where new cells are produced. Following the closure of the neural tube, the cells divide at a blinding speed of 250,000 cells produced every minute! By the 125th day, around the middle of the pregnancy, the fetus has all its cells.

At first, all cells are stem cells. They have the potential to become any kind of cell. A single layer of cells called the Ventricular Zone forms along the inner surface of the neural tube. Two weeks into the proliferation phase, these cells undergo a change. They lose the ability to become anything, stop dividing, and become neuronal cells.

Phase 3: Migration

The neurons do not stay in the ventricular zone, but begin to move in a process called cell migration, from the bottom level of the neural tube up to the top. Migration is the phase where cells make the voyage to their final destinations. They travel along the surface of a particular type of cell, the radial glial cells, which are the first type of cells to differentiate in the ventricular zone. They provide a primary support system, like scaffolding (or a highway) to help other cells make the journey to where they need to go.

Three main types of scaffolding structures are formed to help other cells migrate. The inside-out pattern is a layered structure with mostly radial migration patterns, essential for the formation of complex circuits that make up the cerebral cortex, hippocampus, colliculus, and cerebellum. A second type of layered structure is formed by a mixture of radial and tangential migration found in the formation of the retina and spinal cord. Non-layered structures make up the brainstem, thalamus, hypothalamus, and other areas in the mesencephalon and diencephalons. Some cells stop partway up the scaffolding, while others climb all the way to the top. It is not known what causes them to stop or to migrate sideways.

The process of migration is complex, involving many types of chemicals called immunoglobulins and chemokines that help guide neuron movement. Sometimes problems occur in migration when there is a deficit of these supportive chemicals. These deficits can lead to such problems as smaller brain size or mental retardation (Berger-Sweeney and Hohmann 1997; Tran and Miller 2003). Pathology can result from imbalances. For example, schizophrenia has been correlated with an over-abundance of chemicals (Crossin and Krushel 2000).

Phase 4: Differentiation

From the 125th day until after the baby is born, cells differentiate into what they will become, such as a hippocampus cell or a cerebellum cell. The environment the cell finds itself in after migrating seems to help it specialize. For example, if you take any cell in the brain stem and put it into, for example, the hippocampus, it will become a hippocampus cell. The tendency to be influenced by our environment takes place at a fundamental level, all the way down to the single neuron!

Once the cell has made a decision, so to speak, it can begin to use or *express* particular genes. This means that the cell transcribes (converts) a particular type of gene to make the exact proteins that it needs. Then the cell will acquire its distinctive features.

Cells differ from each other in their arborization of dendrites. They also differ in connectivity, neurotransmitter molecules, the proteins they express, and their receptor subtypes. A differentiated cell seems to know exactly how to produce these features. Experimenters tested how differentiated cells seemed to know what to do. The decision does not seem to be an absolute one but may depend more on timing. For example, in one experiment, neurons that were immature, when transplanted from one part of the cortex to another, took on the characteristics of the neurons in the new area (McConnell 1992). But neurons transplanted at a later stage kept some of the qualities of neurons in their previous location, only taking on some new characteristics of neurons in the new area (Cohen-Tannoudji, Babinet, and Wasserf 1994). This phenomenon is analogous to how very young immigrant children take on the new language of the country they move to, with little or no foreign accent. But when immigration occurs later in life, the child tends to speak the new language with a foreign accent (Kalat 2007).

Phase 5: Synaptogenesis

The young neurons grow more axons and dendrites, adding synapses in a process known as synaptogenesis. Growth cones are formed at the ends of both axons and dendrites, reaching out with threadlike fibers into the spaces between. These fibers pull the growth cones in a certain direction, bringing the axons and dendrites along in that direction. The axons are guided by chemicals released by the target neuron to draw them either toward or away from the target cell. The axons of a pre-synaptic cell extend downward and flatten outward toward the target cell. Protein scaffolding occurs on the post-synaptic side to help build the connection.

The synaptic connections form rapidly on dendrites and on little bumps on dendrites called spines. These connections are affected by environmental influences after the baby is born. The cell bodies of neurons also grow in size as their dendrites keep developing.

Phase 6: Cell Death and Cell Life Lead to Stabilization

The initial surge in synaptic growth begins to level off and then declines after the first year of life. Eventual cell death is an integral phase of human development, known as apoptosis, from the Greek *apo* (away from) and *ptosis* (act of falling). Every living cell has the potential for death programmed into its genes. Similar to the Eastern conception that life and death both play an important part in the cycle of existence, the process of dying cells has a positive effect, of sculpting and pruning, to help bring about just the right balance.

Apoptosis can occur from an external signal, when the cell binds to a death receptor on another cell. A second way apoptosis happens is triggered internally, when a cell's mitochondria become permeable so that things leak out. Stress can induce such a process. Cells also possess caspase, proteins that can cut things up when they are released. Other proteins usually inhibit caspase. But stress or tumors may activate the caspase, bringing about the cell's death.

Just as there are forces for death, there are also forces toward life. Neurotrophic factors influence cells to grow and develop. One such factor, nerve growth factor (NGF), has been known for 40 years (Levi-Montalcini 1982). NGF counterbalances the dying of cells. There are other neurotrophic factors, a family of cells similar to NGF that each influences various types of cells. And there are some other types of neurotrophic factors that are different from the NGF family, that also help to prevent the death of cells.

Phase 7: Synaptic Rearrangement

Structures are refined as the synapses rearrange themselves through the developmental process. This often takes place following cell death. Thus, we can see how birth and death are interrelated processes that interact toward the development and growth of the brain.

At first, there is a high exuberance, where axons reach out without much accuracy to make contact with all their neighbors. Sorting and adjusting occur as the neurons figure out which target is the right target. This occurs in several ways. Neurons compete for space, growth factors, and targets. Latecomers do not continue to live, because there is no place for them to connect and thrive.

One of the most important influences on whether a synapse is maintained or lost seems to be its neural activity. The Hebbian learning rule offers a possible explanation. After the synapse is established, if one axon is activated and gets a response, the synaptic bond strengthens. But if there is no response, the synaptic bond weakens. Correlated activity leads to dynamic connections. They can change.

Another perspective on how synapses become arranged is by looking at dendrite growth. Dendrite growth during neurogenesis takes place in three phases. The first phase takes place during embryogenesis. Here we find the initial formation of neural networks guided by the neurotropic messenger molecules. Following birth, during the critical period, synaptic connections are refined by sensory and motor experience. During the rest of the life span, neurogenesis occurs from activity-dependent memory and learning (Matus 2000)

The Interplay of Genetics and Experience in Development

Genes are central to the developmental process of the brain. Every neuronal structure and behavior can be influenced by changes in the corresponding gene. These influences from the genes take place within the cell itself, and so they are considered intrinsic factors. But external or extrinsic factors can also influence the developing cell and these extrinsic factors affect the genes as well. Extrinsic factors are such things as nutrients or chemicals that might enhance or interfere with development, even as early as in the womb (Breedlove, Rosenzweig, and Watson 2007). Cell-to-cell interaction at the synapse is another extrinsic way that genetic information can be altered, as Chap. 7 reveals in more detail.

Everyone is born with a genetic constitution also called the *genome* or *genotype*. This genotype is the individual's genetic makeup. Genotype is distinguished from phenotype. The phenotype results from an interplay of the genotype and the environment. Thus, the phenotype contains all the physical characteristics of the individual under certain environmental conditions. Although the genotype is determined at the moment of fertilization, the phenotype changes throughout life. The continual interaction between the genotype and experiences bring about these changes (Rossi 2002). Twin studies have shown that two individuals with identical genotypes develop very different phenotypes, believed to be because of their different external influences. One of the clearest indications that development from the same genes does not lead to the same individual is the fact that identical twins always have different fingerprints (Breedlove, Rosenzweig, and Watson 2007). Also, animal studies with clones who share identical neurons, such as a study with identical cloned pigs, showed variation in behavior and temperament, just like in normal siblings (Archer et al. 2003). Studies of genetically identical mice raised by different mothers, showed significant differences in behavior, particularly in their ability to learn and respond to stress (Francis et al. 2003). All of these studies can serve as a beacon of hope for the potential power of psychotherapy when it fosters new experiences for clients.

Critical and Sensitive Periods

The synaptic patterns can change throughout life, but there are certain times in early development when the nervous system is even more capable of changing in response to experience. Critical periods are time-limited windows of opportunity. At these critical points in the developmental cycle, the brain can be responsive to certain stimuli that are necessary for development of a skill to take place. At these times, neuronal pathways await specific information from experience in order to proceed with normal development. A timely experience is the crucial element for triggering genetics. If the brain does not receive the appropriate stimulus, certain skills become more difficult or even impossible to acquire later. If the needed information is not received, brain areas associated with these skills may be used to perform different functions instead. Sensitive periods last longer than critical periods. During these times, the nervous system is more receptive to certain environmental stimuli for learning.

Critical periods differ from adult learning in several ways. The changes are only possible during a short period, whereas adult plasticity can take place at any point (See Chap. 13). Unlike learning, during the critical period, neurons are able to choose a more permanent set of inputs from a wider range of possibilities. Also, the changes that take place during the critical period will last for the entire life.

An interesting example of critical periods has been researched with songbirds. Their songs are passed along from one generation to the next through experience of hearing the song at the critical period. Songbirds that were placed in acoustic isolation during their critical period never learned the songs, even if they heard them later. Instead, these birds would sing what researchers call an "isolate song," a simpler, more general version of the song (Konishi 1985).

Critical and Sensitive Periods in Childhood

Childhood is a time when the brain is developing key skills that will be used through out life. Anyone who has had contact with young children knows how open and curious a toddler can be. In fact, a toddler's brain is two times more active than an adult's brain! Although the infant brain has most of the cells before birth, connections are forming and reforming all through childhood. The interplay of genetics (nature) and environmental influences (nurture) is an ongoing influence on how the child will develop.

Certain skills are attained during the critical periods of childhood. The window of opportunity is wide open for visual and auditory development during the first year of life. Infants learn to distinguish sounds and recognize shapes, colors, and objects. The ability to use both eyes together, binocular vision is a critical period developing between 3 and 8 months. The neurons involved continue to be in a sensitive period, where they are susceptible to damage until the child is three years old. Motor skills have a critical period during the first year of life, although there is quite a bit of variability on when the infant performs any of the motor skills such as sitting up, crawling, or walking.

Language development is ongoing for the first 10 years of life. But the critical period that sets the stage for language ability takes place during the first few years. Thus, when parents talk to their infants, they are providing the stimuli the infant's brain needs to be able to develop the language centers.

Critical Period for Primary Interpersonal Relationships

One of the crucial critical periods for healthy psychological development is forming attachments to primary caregivers. Our social interactions, especially with those who are closest to us early in life, have an enduring effect on the development of the brain. Parental imprinting takes place during the first 18 months and can continue to influence emotions, thoughts, and behavior ever after. If subjected to stress prenatally or separation from the caregiver during infancy, a child will have a tendency to increased anxiety later in life (Meaney 2001). However, a low stress pregnancy and nurturing early years with affectionate handling provides protection against later anxiety (Vallée et al. 1997). Family and couples therapies address attachment problems that may have occurred in early life by working on the original attachment to bring about corrective experiences.

Development of a strong relationship with a parent during critical phases of early development has a lasting effect on the individual's ability to form healthy relationships throughout life. The early attachment between mother and infant stimulates the neuronal connections that shape the brain's development in a healthy way for life. According to one of the founding fathers of this theory, John Bowlby (1907–1990), a healthy relationship with a significant other is a fundamental need that is wired into our very physiology. This primary relationship is called attachment. See Chaps. 18 and 19 for more about attachment theory and how to heal attachment problems in therapy.

Adolescence and Refining of the Brain

Adolescence is a time of impulsive behavior, increased risk-taking, and strong urges for social involvement with peers. New understanding of how the brain undergoes change during adolescence may help us understand just why adolescents typically act in these patterned ways.

The brain reaches its peak weight by around 10–12 years of age, but it will undergo a great deal of change before it reaches maturity. You may recall that during early development, the brain undergoes pruning, where synaptic connections and cells that are not being used die to make room for new cells and connections. A pruning process occurs during adolescence, where close to half of the synaptic connections are eliminated (Spear 2010). Typically, we think that neuroplasticity means adding new connections, and often it does. But some kinds of mental retardation are characterized by more synaptic connections than normal (Goldman-Rakic et al. 1982). Researchers believe that the refining of synaptic connections in adolescence makes space for mature patterns to form (Zehr et al. 2006). This sculpting process is influenced by experiences in the environment.

Another important brain change is that some of the axon pathways become myelinated which speeds the flow of information. The axons in the corpus callosum, which connects the two hemispheres, become increasingly myelinated during this period as well. The result of the decrease of synaptic connections at the dendrites (gray matter) and increase of axon myelination, (white matter) results in a higher ratio of white matter to gray matter than before adolescence. In general, the adolescent's brain becomes more efficient and streamlined, similar to how an adult brain is compared with childhood. Less energy is required for each region to function well, and blood flow declines (Spear 2010).

But we see an order in these changes, which might account for why adolescents are impulsive. The changes occur first in the posterior areas of the sensory and motor cortex. The result is that adolescents often develop sensory and motor skills, expressed in artistic talent and athletic ability. But anterior parts of the frontal cortex develop much later. Here is where our judgment and executive functions are performed.

The pruning, increased axon myelination, and enhanced brain efficiency correlates with quicker cognitive processing speed and efficiency. The working memory also improves during adolescence. Adolescents do show marked improvements in the complexity of their thinking as they become capable of tackling the more advanced academic subjects that middle and high school classes provide.

Despite the improvements in certain cognitive functions, adolescents are known for engaging in risk-taking behaviors and strong reliance on peer interactions. This may be accounted for by some of the other ways their brains are undergoing change besides the slower development in the frontal areas.

Adolescents have an increase in oxytocin, which correlates with the adolescent's intensification of social and romantic involvements. Oxytocin is sometimes called the prosocial hormone because it is increased by social situations (Taylor et al. 2006).

In addition, adolescents tend to feel their emotions more intensely than adults, and the shifts in hormones that occur during adolescence helps to account for this emotional intensity. They are exposed to higher levels of stress hormones, and then the stressful events in their lives such as academics and social pressures, increase activation of the HPA fear/stress pathway. Vagal tone, which is a measure of the how well people regulate their stress, increases through the adolescent years.

Adolescents show decreased activity in the PFC and ACC, areas of the brain that correlate with inhibitory control and decision making. They have lower connectivity between the frontal cortex and emotional brain centers. The connections between the frontal areas and the limbic system increase as the adolescent matures, leading to improved judgment and less risk-taking behaviors.

Perhaps, these are some of the key tasks of adolescence: to learn how to regulate emotions and make good decisions, even when emotionally aroused. Mature adults can successfully regulate stress and be more responsive to their nervous system signals. But for the adolescent, the process is an ongoing challenge, a time for experimentation and learning, that in the best cases, will lead to a fully functioning adult brain.

The Aging Brain

Developmental changes that occur in the mature brain are less dramatic than early development. But, some plasticity can occur at any point, and this exciting area, full of promise for therapeutic change, is covered in the chapter that follows. The next dramatic period for brain alteration occurs in the later phases of life.

Rumors abound about sharp declines and inevitable lose of cognitive functions. But actually, normal aging has a wide range of variations with many people functioning extremely well in their later years. For example, we can admire people like Grandma Moses (1860–1961) who began painting later in life. She could no longer do embroidery because of arthritis, and so turned to painting. She continued to be prolific during her long life; some of her works are in the Smithsonian. She painted her last work shortly before she died, at age 101. She did nostalgic scenes of Americana that portrayed happy people busily engaged in a simple farming life. She worked from her own memories, drawn from her childhood on a farm. Her life illustrates how people can remain optimistic and productive even into advanced age. She also exemplifies how we can learn new skills later in life, and even excel with them.

Most of us know people who continue to work well and live happily into old age with competence and wisdom. This may be helped by what is called crystallized intelligence; the accumulation of skills and knowledge built up over time and usually related to work or profession. This form of intelligence tends to remain stable with age (Kaufman, Reynolds, and McLean 1989).

Cognitive Changes

Unfortunately, many people undergo decline in memory and functioning as they age. A certain degree of decline is considered part of normal aging. The mild impairment of memory that accompanies aging is related to reduction in size of the hippocampus (Golomb et al. 1994). Generally, the type of cognitive function that seems to decline the most is fluid intelligence. Fluid intelligence means mental flexibility, processing speed, abstract and complex new problem solving, memory, and new learning. Two tests of fluid intelligence are Digit span and block design from the WAIS (Zillmer, Spiers, and Culbertson 2008).

There are two general theories used to explain why there is so much variability in cognitive function as people age. These theories are known as Cognitive Reserve and the Differential Preservation Hypothesis (Salthouse et al. 1990). The cognitive reserve theory is based on the finding that studies have shown a strong link between early intelligence and less cognitive decline. For example, a long-term project, known as the Nun Study (Snowdon et al. 1999) followed 678 Roman Catholic nuns who took regular cognitive and medical tests, and agreed to donate their brains following their death, for further research and analysis. The researchers found that those who had higher "idea density," which refers to complexity and sophistication of concepts when they first entered the convent, continued to have higher mental ability during their middle years in the convent than those who were less linguistically sophisticated. They also showed less cognitive decline in later years. Postmortem autopsies of the brains of some elderly nuns were found to have the plaques that are characteristic of Alzheimer's disease (AD), but they showed few symptoms in their functioning.

The differential preservation hypothesis (Salthouse et al. 1990) is that normally aging adults can maintain stable cognitive functioning by continuing to be mentally active. Just as we should keep physically fit, it is important to remain mentally fit. Challenging oneself mentally can keep mental functioning working well.

Although there are no definite answers as to which theory is more correct, and certainly no guarantees about aging, we have little to lose in exercising and expanding our mental faculties at every phase of life. Many of the exercises included in this book can be helpful for preserving and even extending brain function to help with healthy aging.

Brain Changes

As the variability in functions reveal, brain changes also vary greatly from person to person (Johansson 1991). But there are some structural trends. The size of the brain in terms of its weight and volume diminish with time. One myth is that as we age, large numbers of neurons die. But after analyzing results found by using the newer imaging methods, researchers believe that in normal aging the number of neurons remains relatively constant. What accounts for the smaller brain size is that individual cells are becoming smaller. Thus, cortical thinning could simply be due to the shrinking of cells, not cell loss (Haug 1985). An important area that seems to suffer with age is the prefrontal cortex, which may account for some of the loss in fluid intelligence (Esiri 1994).

Even though some parts of the brain alter with age, other parts remain robust. PET scans of elderly people showed that the cerebral metabolism stays relatively constant (Breedlove, Rosenzweig and Watson 2007).

Environmental factors can lead to more or less aging of the brain. Long-term stress has been shown to age the brain significantly (Epel et al. 2004) and shrink the hippocampus (Lupien et al. 1998). But exercise has been shown to enhance performance on cognitive tasks and increase brain density (Colcombe et al. 2003). Meditation has also been found to slow cortical thinning in the attentional areas as well as lower stress.

Disorders of Aging

Normal aging, although it has some of the symptoms and brain changes we have described, is not considered a disorder. A healthy lifestyle with exercise and mental challenge can maintain a normal quality of life. But aging has certain distinctive behavioral and brain disorders that exaggerate the declines of normal aging that can become debilitating.

Dementia is the term used to describe a group of behavioral symptoms that involve a wide range of impairments to memory, cognition including executive function and language, decline in social skills and work abilities, along with clouded consciousness.

Dementias are also classified in terms of the part of the brain that is affected. Two broad classifications are made: cortical and subcortical dementias. Knowing which classification can reveal a great deal about the nature of the behavioral, cognitive, and emotional changes will occur.

AD is considered a cortical dementia because it shrinks major areas of the cerebral cortex. As one would expect with loss to these brain areas, memory becomes severely impaired with a breakdown of semantic knowledge and long-term declarative memory. Interestingly, short-term memory is relatively spared along with language skills. AD also attacks some subcortical structures, such as the hippocampus and amygdala. Most of the subcortical damage occurs in the pathways connected to the association areas in the cortex. The subcortical damage is reflected in loss of visual–spatial functioning. AD patients often become lost or disoriented, even when traveling through familiar areas. AD also influences many of the neurotransmitter systems. Widely distributed neurotransmitters such as acetylcholine (ACh), catecholamines, glutamate, and neuropeptides are reduced in numbers. The brain losses, along with the neurotransmitter reductions, account for the general decline of intelligence that takes place. AD sufferers have difficulty with executive functioning, orienting attention, and regulating their moods. Often, exaggeration becomes characteristic of the personality style, either for better or worse.

Subcortical dementias include Parkinson's disease (PD) and Huntington's disease. PD is the result of a severe loss of dopamine in the substantia nigra of the basal ganglia. PD patients develop deficits in the functioning of the basal ganglia, especially in relation to the thalamus. Symptoms can become noticeable at any age when patients drop below a dopamine threshold of between 50 and 80 % loss of their dopamine levels.

Since the basal ganglia controls movement, PD sufferers have problems with motor movement. Patients develop a resting tremor, with uncontrolled small trembling movements that gradually worsen over time. They also develop involuntary rigidity in the muscles and joints.

PD brings problems with visual–spatial tasks as well, especially when motor movement is involved. These deficits differ from cortical dementias like AD. Mood problems that develop may be coming from the treatments and not the disease, although PD patients tend to suffer more depression than other chronic problems (Raskin et al. 1990).

Neuroscience has developed some new brain treatments for PD that show promise for the future. Transplanting cells that contain dopamine into the dopamine producing areas of the basal ganglia had limited success (Lindvall et al. 1994).

A recent innovation is deep brain stimulation, DBS. DBS uses a surgically implanted, battery-operated neurostimulator that delivers electrical stimulation to targeted areas in basal ganglia, blocking the signals that cause PD symptoms. DBS targets the subthalamic nucleus. It can bring a dramatic calming of the tremors and relaxation of muscle rigidities, and may improve mood as well (Deuschl 2009).

Conclusions

Clearly, brain development and environmental influences work together as reciprocal opposites. Change processes, recurrent and non-recurrent, are ongoing through the life cycle. Therapeutic input can help to promote growth where it is most needed to bring about these changes, to help move forward in an increasing cycle, or ease the journey through the declining phase. Correcting for deficits in development is one important way to promote growth. The next chapter, on neuroplasticity, will reveal even more possibilities.

Chapter 13
Neuroplasticity and Neurogenesis: Changing Moment-by-Moment

Neuroplasticity is the ability of the brain and nervous system to reorganize its neural pathways, connections, and functions. It occurs rapidly during brain development, from conception through the first few years of life. We have also described in Chap. 12 how neuroplasticity occurs at later developmental stages, such as during the teenage years and even with aging.

But neuroplasticity manifests in another way that is important for psychotherapy, when it occurs in response to experience. Experience-based neuroplasticity can happen all through life and in varied circumstances. It can be positive and expanding or negative and constricting. We see negative neuroplasticity associated with stress found in many psychological disorders including PTSD, major depression, and borderline personality disorder (Sala et al. 2004). People suffering from these problems show atrophy in their hippocampus. Psychotherapy can help to reverse this adverse effect. The hippocampus can also undergo a positive neuroplasticity, becoming larger than normal. An interesting study of London taxi cab drivers found that they had larger than normal hippocampus size as compared to bus drivers (Maguire et al. 2000). The explanation is that cab drivers spend more time navigating to new locations. One of the functions of the hippocampus is correlated with finding our location in space.

Other studies have found that hippocampal growth can be stimulated to enhance functioning. Seniors were trained to juggle three balls. Measurements on fMRI found increases in gray matter relating to skill acquisition as well as increases in the nucleus accumbens (part of the dopamine system) and the hippocampus (Boyke et al. 2008). Growth is often correlated with improved functioning, just as shrinking is associated with declines, and so whenever we can stimulate new growth, we are more likely to help our clients improve their functioning.

The discovery that the brain has this ability to change from what we experience presents a great hope for our ability to make significant changes, not just in thoughts and feelings, but also in the very structure and function of the brain.

C. A. Simpkins and A. M. Simpkins, *Neuroscience for Clinicians*,
DOI: 10.1007/978-1-4614-4842-6_13,
© Springer Science+Business Media New York 2013

We can make our lives, and our brains, better by what we do. And we can do so at any age!

> Indeed, it is now clear that constant remodeling is one of the brain's defining features. Just as important is the now abundant evidence that the relationship between changes in the brain and changes in behavior is bidirectional: experience also can alter neural structure. This plasticity is a feature of the nervous system that persists throughout life, from embryonic development until old age (Breedlove et al. 2007, p. 605).

These new understandings about how we can take a hand in fostering brain change makes neuroscience relevant for the helping profession. What we do with clients and patients can make a difference on many levels, including the physical. And so, we devote this chapter to the details of how neuroplasticity has been studied, what has been discovered, and ways that experience tends to bring it about. Later chapters will provide methods and techniques for encouraging brain change in positive ways.

How Neuroplasticity is Studied

Plasticity studies often involve three types of information. Some researchers have tried to get a picture of how the brain functions at a given time without any external input. This is often performed at the level of a single neuron, by observing data on interactions between two neurons at their synapses, often viewed as a bio-chemical or electrical interaction. Others have focused on what happens when experiences, as inputs to the brain, change the specific neuronal interaction. For example, how does it excite certain activities in the cells and how does this result in long-lasting alterations in the chemical channels, the synapses, and the neurons themselves? Finally, researchers ask, following the experiential input, how are the brain functions different from how they were before (Dudek and Traub 1989)?

The Microscopic View of Neuroplasticity

Neuroplasticity has been characterized as the brain's ability to undergo a change in the connections between synapses. Studies of learning and memory on the molecular and cellular level have revealed how the plasticity occurs at the synapses. Neurons seem to have the ability to modulate the strength and structure of their synaptic connections as a result of certain types of experience. This is why much of the research on plasticity is done at the neuronal level.

How does it happen? Neurons do not reproduce themselves. Rather, the brain has neurological stem cells that undergo neurogenesis. The old paradigm that we are born with all the brain cells we will ever have is not quite accurate. Although most neurogenesis takes place during development, it continues to occur to some

Fig. 13.1 Aplysia learning

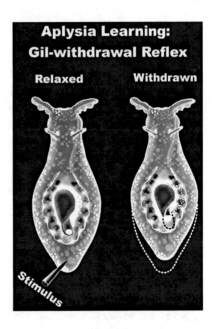

extent throughout life. And pruning occurs at every stage: the brain allows the unnecessary connections and cells to die off due to lack of use. Thus, we enhance neurogenesis at any point by what we do and how we do it.

One way to study what happens at the synapse is by looking at the simplest case (Fig. 13.1). Much has been learned from studying simple invertebrates such as the marine mollusk, Aplysia, because their neurons are relatively large in size and small in number. The Aplysia undergoes simple forms of learning with its gill-withdrawal reflex. As has been observed in evolutionary theory, nature has tended to utilize the same mechanisms in simple organisms and complex ones, such that many of the synaptic mechanisms observed in the Aplysia are also at work in the human brain. The difference is not so much in kind, but in number. From studying this simple brain, researchers have been able to perform controlled experiments to manipulate memory and learning and then observe synaptic plasticity in action.

Research shows that long-lasting change at the synapse requires gene expression. People used to think that the process took place mainly in the nucleus. At first, people believed that proteins were synthesized in the cell body and then transported to different parts of the neuron. But Steward and Levy (1982) opened the way for recognizing the possibility that the process also includes the synapses. Further research revealed that the cells, in a sense, think globally but act locally (Shen 2003). The mechanisms of plasticity involve local synthesis of proteins at the synapses. This means that the signal can be generated at the synapse and then travel to the nucleus. Once in the nucleus, the signal changes in gene expression. Then the signal travels back to the synapse with the new gene information to produce an enduring alterations in the strength of the synapse. Some of the mechanisms of synaptic plasticity in the Aplysia involve synthesis of messenger

RNA and proteins leading to growth of new synaptic connections (Martin et al. 2000). With plasticity occurring at the synapses, it becomes clearer why the possibilities for neurogenesis are much greater than had been previously believed (Martin and Zukin 2006).

This research can be a lesson to us all for keeping an open mind, even when the understandings are not immediately observable. Sometimes, taking the initiative to look somewhere else, as the researchers did in turning their attention to the synapses, can open a whole new world of potential.

History of the Discovery

Many people have been given the credit for coming up with the idea of neuroplasticity, such as Jerzy Konorski (1903–1973), a student of Pavlov who is credited with coining the term, along with Donald Hebb. But earlier psychologists, such as E. L. Thorndike and William James also referred to the idea. Thus, we can see that the concept has a rich network of proponents from many different backgrounds whose ideas now come together in this fertile ground of research that nourishes our understanding of the brain's potential for change.

William James believed that the habits that we engage in come about because of the plastic nature of the brain. The habits we form are the result of neuralplasticity. As he said:

> Plasticity, then, in the wide sense of the word, means the possession of a structure weak enough to yield to an influence, but strong enough not to yield all at once. Each relatively stable phase of equilibrium in such a structure is marked by what we may call a new set of habits. Organic matter, especially nervous tissue, seems endowed with a very extraordinary degree of plasticity (James 1896, p. 105).

James pre-visioned the modern direction of neuroscience. He recognized the interface between the actions we take and the brain, leading to the Hebbian position, as described in Chap. 3, that when neurons fire together, they wire together. But James's intuition would not be embraced right away. The acceptance of neuroplasticity has had a checkered career since James's optimistic statement more than 100 years ago.

Early Development

When Paul Broca (1824–1880) first located specific areas in the brain mapped for speech in 1861, researchers enthusiastically pursued the search for other brain areas that might be correlated with specific functions. In 1876, Carl Wernicke discovered an area close to Broca's area that was responsible for understanding language.

Korbinian Brodmann divided the brain into 52 distinct regions, known as Brodmann Areas, BA. Refer back to Chap. 1 for these early discoveries.

Scientists who were working on mapping out the areas of the brain turned to movement. Movement was a researchable area to explore because it could be observed. Early twentieth century researchers began to draw movement maps in the brains of animals by simulating specific parts of the brain and observing how the animal moved. What surprised them was how much variation they found between different individuals (Brown and Sherrington 1912). Sherrington hypothesized that the differences in the neural movement maps were a reflection of the animal's movement history. Repeated habitual movements left a physical trace in the animal's motor cortex. The brain, said Sherrington in his famous quote, is "an enchanted loom, where millions of flashing shuttles weave a dissolving pattern, always a meaningful pattern, though never an abiding one" (Sherrington 1906, p. 110).

Another neurologist, S. Ivory Franz (1874–1933), who compared movement maps in the motor cortices of macaques monkeys, also found that maps differed between individuals. He hypothesized that these differences were the result of varying motor habits between them (Bagley 2007).

Karl Lashley (1890–1958), at first set out to test whether the differences between individuals were innate, expecting that each brain should be fixed early in development. But his research revealed that the differences were, indeed, due to varying experiences, a plasticity in neural function.

Putting on the Brakes

Santiago Ramon y Cajal (1852–1934), the father of neuroscience and Nobel Prize winner, set the standards for neuroscience in general. He delimited what was and was not possible for neurogenesis. He affirmed that in early development, the brain is plastic and changing. But once the individual reaches a certain age, the potential for change is passed. "In the adult centers the nerve paths are something fixed, ended and immutable. Everything may die, nothing may be regenerated." (Zigova et al. 2003, p. 231). The birth of new neurons in the adult brain was considered to be was impossible.

Although psychologists such as Thorndike and Hebb (1904–1985) proposed the idea of plasticity at the synapses, neuroscientists continued on their course of building a clear picture of a largely unchanging brain organization with fixed functions. Thus, most of the early findings in plasticity were largely ignored.

As recently as the 1960s and 1970s, the brain was being viewed as similar to the hardware of a computer. Hubel and Wiesel (Hubel and Wiesel 1962) contributed to this idea by demonstrating that once critical development was over, the ocular dominance columns in V1, the lowest visual cortex area, were unchanging for the most part. In this climate, it was difficult to conceive of the possibility that neurons could form new connections beyond the early formative years of development.

Even as recently as a few decades ago, the consensus opinion was that the lower brain and neo-cortex were unchanging once the developmental period had passed.

Plasticity After Damage

But some researchers were unwilling to accept this limitation. Joseph Altman, an MIT neuroscientist, injected a tracer and then scanned the brains of rats and guinea pigs (Altman 1967). He detected new DNA, which indicated that neurogenesis did seem to be occurring.

Although movement studies had been the logical early choice for investigating plasticity, another type of plasticity studies involved somatosensory systems. Somatosensory links can be observed. A group of researchers (Jenkins et al. 1990) gave monkeys a new sensory experience to see if that would alter the brain areas that process information from the fingers. They positioned a four-inch grooved dick near the cage and trained the monkeys to reach out to touch the disk as it spun. The animals learned that if they applied just the right amount of pressure, the disc would spin and they would be rewarded with a banana pellet. Comparing pre and post maps of the cortex, they found that the map had changed: the areas receiving signals from the fingers were four times greater. This was evidence that behavior could bring about a significant, measurable change in the brain's structure.

Another technique, known as deafferentation, disconnected a nerve in part of the body of adult monkeys, such as the middle finger. At first, the researchers expected that once the nerve was cut, the cortex area would shut down. But in fact, they found that the cortical area devoted to the third finger did not stop functioning. Instead, the second and fourth fingers moved into the cortical area that had previously been used to respond to finger three. They also found that after extensive long-term deafferentations in adult primates, the cortical maps changed far more than was expected (Pons et al. 1991). The results of many types of experiments such as these, with small deafferentations in fingers and large ones with arms, showed the need to reevaluation the extent of cortical plasticity that could occur.

There are several ways to interpret the plasticity in adult somatosensory cortex. When small-scale shifts occurred, such as from one finger to another, the existing synaptic connections could be rearranging. Since the change was only over a small distance of around 2 to 3 millimeters, researchers speculated that perhaps no real new growth occurs. Instead, some synapses might be tuned up and others tuned down. But in large-scale deafferentations of a monkey's arm, the remapping moved over a much larger cortical area. These changes would more likely be the result of new axonal growth. Another possible explanation was that subcortical changes such as in the thalamus or even perhaps from the spinal cord to the subcortical areas, finally ended up in a change in cortex (Jain et al. 2000).

The findings from these studies have inspired work with human subjects suffering from strokes. In the past, stroke patients were forced to resign themselves to a permanent loss of function with very little hope for recovery, since it was assumed that the damaged brain area could not be repaired.

Based on the work done with deafferentation of animals and the significant plasticity in affected brain areas, researchers have developed methods for working with stroke patients suffering from paralysis in one arm using Constraint Therapy (CI). The stroke patients who participated in the study were 54–76 years old. Their treatment involved constraining the "good" arm for 3 h daily for 10 consecutive days, thereby forcing use of the afflicted arm. They combined this with a "transfer package" lasting 30 min each day to help the subjects transfer their gains to their daily activities. The study showed that not only did the patients improve the ability to use the impaired arm, but they also had measurable increase in gray matter in their sensory and motor areas (both primary and secondary cortices) for the affected arm as well as increases in both the right and left hippocampus (Gauthier et al. 2008).

Plasticity from Enrichment

Another approach to understanding plasticity comes from investigating ways that bring it about. Enriched environments can influence the developing brain in positive ways. William T. Greenough has shown in countless experiments that when animals were given a more enriched environment with ladders to climb on and wheels to spin and other rats in the cage, their cortices became thicker than animals living in impoverished environments. These differences were reflected in their behavior as well. They were able to perform better in learning tasks such as mazes than the deprived rats. The synaptic connections became denser and the dendrite branching became more complex. Furthermore, when animals were taken from the impoverished environment into the richer one, they underwent a surge in neurogenesis, especially in the dentate gyrus area of the hippocampus involved in learning and memory (Greenough et al. 1987; Briones et al. 2004). Researchers even found that mice suffering from a hereditary form of mental retardation were largely cured of the behavioral and neuronal abnormalities when they were placed in enriched environments (Restivo et al. 2005).

Based on Greenough's animal research, Fred Gage tested the effects of an enriched environment on adult mice. He placed adult mice raised in the typical impoverished laboratory environment, into an enriched environment for 45 days. Following this relatively brief period, the number of neurons in the hippocampus of these mice was far greater than the matched mice that remained in the typical laboratory cage (Kuhn et al. 1996). So, even the adult brain showed neurogenesis when given the right environmental stimuli.

Gage made another interesting discovery. He found that mice who engaged in voluntary running on their wheels seemed to show more intelligent behavior than those who were forced to exercise in a tank of water. He hypothesized that better results on learning tests wre dependent on physical activity *combined* with volition rather than just the physical activity alone (Bagley 2007).

Neuroplasticity in the Human Brain

Many people were still skeptical that the findings about plasticity in animals could apply to human brains. The rationale for the skepticism was that human neurons seemed too complex with the thousands of interconnections to undergo any type of cell division. Furthermore, some felt that if a brain structure such as the hippocampus underwent neurogenesis, it might destabilize memory ability and make it difficult to retain long-term memories. A more compelling explanation seemed to be that the brain was like a computer, a hardwired structure that could not change.

Thus, there was a need to test whether neurogenesis took place in human beings. Of course, one could not inject a person with the radioactive markers that were used on laboratory animals to observe change in neurons. But Gage realized that a marker, the molecule BrdU, was already being used with oncology patients to track the new growth of cancer cells. Since any cell that undergoes cell division incorporates the BrdU molecule, neurogenesis would be marked wherever it occurred.

Gage and Erikkson obtained the cooperation of terminal patients whose tumors were being monitored with BrdU. Upon their death, these patients agreed to donate fresh brain tissue to Gage's lab for observation. They found that neurogenesis did occur in the hippocampus of these patients, who ranged in age from the late 50s to 70s. The results showed that neurogenesis not only takes place in the brains of rats, mice, and monkeys, but also in human beings, and even into old age (Gage et al. 1998).

Remapping

Plasticity has also been studied by observing what happens when something goes wrong. One prominent researcher who has investigated the effects of damage on neuroplasticity is Vilayanur S. Ramachandran. He became curious why people seemed to suffer phantom limb pain, the pain from a limb that had been amputated. This mysterious sensation continued to bother people long after the limb was gone and the surgery had healed. Ramachandran suggested that the reason these individuals continued to feel sensations was due to the neural plasticity in the adult human brain.

The brain represents the body in the somatosensory area. Recall the homunculus (see Fig. 8.6) on the motor strip. The parietal lobe has a similar sensory strip that maps the body for sensory inputs. Input from the hand is mapped onto the brain next to input from the arm, as one might expect. But these inputs are then represented next to the face. Following the amputation, the patients experienced sensations in their phantom limb when the face was stimulated. This led Ramachandran to test his hypothesize that the brain had remapped the areas from the missing arm and hand onto the closest brain region, that representing the face. His research found that indeed, the arm and hand areas had invaded into the face area (Ramachandran et al. 1992).

Remapping also occurs with the sensory systems. People who have been deaf from birth do not lose functioning in the auditory cortex as one might expect. Although this area of the cortex does not respond to auditory stimuli, it does respond more strongly visual stimuli, indicating that the brain remapped to utilize this unused portion for something else. One study compared deaf subjects with normal subjects who were asked to look straight ahead while lights were flashed at the periphery of their vision. EEG for the normal subjects indicated response from the visual cortex, as one would expect. The deaf subjects had a four times larger response in the auditory cortex. The researcher concluded that the auditory cortex had been remapped for a more sensitive visual responsiveness (Bavelier et al. 2000). As an interesting note, the peripheral vision pathway is also involved with motion and location in the *where* pathway. Congenitally, deaf subjects were more sensitive to motion than nondeaf subjects. However, both sets of subjects responded similarly to visual stimuli presented at the center of the visual field. The signal travels along a different path when light falls at the center of the retina, in the *what* pathway. The *what* pathway is sensitive to form and color.

Novelty as a Stimulus for Neurogenesis

Ernest Rossi has been a strong spokesperson for the use of novelty and creativity to stimulate neuroplasticity through neurogenesis at the level of gene expression. Genes do not exist alone; they are in a dynamic, nonlinear relationship with the environmental inputs and the larger brain functions. The genes actively respond to inputs from life experience. These changes can be immediate, moment-to-moment, occur over hours to days, or they can take effect slowly, evolving over billions of years.

Thoughts, emotions, and behavior can influence gene expression According to Rossi, consciousness is an emergent property and as such, "conscious experience of novelty, environmental enrichment and voluntary physical exercise can modulate gene expressions to encode new learning and memory" (Rossi 2002, p. 12). This process, known as activity-dependent gene expression, can be harnessed in therapy to overcome problems and bring about new potentials.

Conclusion

The brain undergoes change in many ways, at numerous times, and with diverse timings. In fact, change is at the very core of the brain's nature. We can continue to introduce new experiences at every level to foster more optimal functioning. Therapeutic methods can facilitate these processes, as later chapters will reveal. And so, we can expand our expectations to include more, hope for better, and become our best.

Part V
Enhancing Brain Functions
for Treatment

Now that you have learned a great deal about the brain, its structures and functions, the ways they interact together in a network, and the kinds of changes that occur through time, clinicians will invariably want to know, how can I use this?

You can conceive of psychological problems as a brain-mind-body-environment network interacting together. Sometimes the unity manifests itself through its distinctive parts, but these parts are always within the larger network. Remember the jewels of Indra's net. Each jewel is a node in the network, interacting with each other and with the flow of the pathway. There are consequences from each individual approach. Keeping this perspective in mind can help in choosing the appropriate method and intuiting how a particular technique or exercise will affect the whole system of the problem, for healing at many levels.

Part V shows how to incorporate brain processes into treatments. Chapter 14 provides key principles to guide your integration of brain science into therapeutic techniques. Then the next chapters in this section give you ways to specifically develop client's attention (Chap. 15), emotions (Chap. 16), memory (Chap. 17), and interpersonal relationships (Chap. 18) to facilitate the therapeutic process. Attention, emotions, and memory all function within the context of the environment and other people. They form a network of brain processes, working together to help us pay attention, feel our response, remember, and relate with others. These processes can happen consciously or unconsciously, and we can use both ways in therapy. Each of these processes engages systems of brain activations and deactivations that come together when needed, as these chapters will describe. We offer sample techniques in the form of exercises, case examples to illustrate, and ideas about how to apply them. Creatively individualize these techniques to fit your approach, personality, and style as well as to fit your individual client's personal problem and needs. We hope these methods will inspire you to evolve your own work in innovative directions.

Chapter 14
Principles for Incorporating the Brain into Treatments

Neuroscience has brought many new discoveries that hold important ramifications for psychotherapy. As you come to understand the nervous system, and how the brain is always involved with the mind, body, and environment, you will be able to add these potentials from brain science into your treatment methods.

Certain principles emerge to help you get started in applying the new brain science to expand your methods. We gather an outline of key principles here as a quick reference, and develop them further in the treatment sections of the book. Use these principles to keep the brain in mind as you formulate your techniques to help meet the needs of each client and every problem.

Recognizing the Healing Tendency

1. Nature heals, *natura sanat*, and so does the nervous system.

The nervous system is exquisitely complex, acting within a highly organized network among the mind, brain, body, and environment. We have everything we need when things are functioning well. As the famous Zen master Lin Chi often told his students, "Nothing is missing". Many problems are best resolved by eliciting the natural abilities of the nervous system to heal itself.

Neurotransmitters are a good example of this principle. Clients often think that when they are taking a psychopharmacological medication for problems like depression or anxiety, they are swallowing a neurotransmitter their body lacks. In reality, the nervous system manufactures the neurotransmitters and sends it through pathways in the brain. Serotonin for emotions, dopamine for pleasure, and endorphins for pain reduction are already there in the brain. We have what we need. When these processes are interrupted or blocked medications can serve to stimulate the brain to manufacture more of the neurotransmitter in various ways, to help it flow as it should, such as by remaining longer in the synapse. The effect is to increase (or decrease if needed) the action of the neurotransmitter to help restore balance.

C. A. Simpkins and A. M. Simpkins, *Neuroscience for Clinicians*,
DOI: 10.1007/978-1-4614-4842-6_14,
© Springer Science+Business Media New York 2013

You can set a therapeutic process in motion by explaining to your client how the brain has innate wisdom to function well and find balance. The nervous system knows how to respond to threats and how to return to rest when the threat has passed. And it has a built in system to find pleasure in what is good for you, to enjoy good food, a restful night sleep, and the intimacy of a close relationship. When people have problems, they often lose faith in their natural abilities, and this negative expectancy acts against their change. You can foster positive expectancy based in the realistic abilities of the nervous system. You can give your client a tangible experience of being able to bring about a change by teaching her as simple a practice as a calming meditation or a relaxing hypnotic trance. The chapters that follow instruct on how to bring this about.

Working in Multi-dimensions

2. The mind, brain, and body function together as a network, one that is in more than two dimensions, so problems can be treated at different levels, not just the next point on a line of activity.

The nervous system, which includes the brain and all the nerves in the body, extends inside the body to the internal organs and endocrine system, and outward the extremities. These interconnections are also part of feeling, thoughts, and behavior. For example, when you feel afraid, the limbic system is activated, as well as the HPA pathway. When you feel pleasure, dopamine is released, activating the reward pathway in the brain. When you are thinking, the prefrontal cortex is more active. And when you are doing things, the motor cortex and basal ganglia are engaged. The nature of the connection between brain and mind is complex, as we discussed in Chap. 2. But clearly, there are interrelationships and you can use them for therapy.

Neuroscience research traces out the many interconnections between mind, brain, and body. Making a change in one area sends ripples through the system. For example, changing the breath can alter a chronic anger pattern. Or practicing tranquil, slow movements can raise energy in chronic depression. We can calm an over-activated limbic system with cognitive methods. Or a chronically over-activated sympathetic nervous system, which brings about stress to body and mind, can be altered and calmed by regular mindfulness practice.

With so many systems interacting together, you can approach a psychological problem at different levels. If you meet with resistance in one area, look for an alternative pathway. For example, when the client has trouble revising his or her thinking, a meditation to clear thoughts can help. Or if the client continues to have low affect, use energy-raising yoga postures. Since the mind, brain, and body are interconnected, there is always more than one way to resolve a problem.

Activating Neuroplasticity

3. Many structures and functions of the brain are plastic, with the potential for change, and so you can alter seemingly entrenched patterns by changing the brain.

One of the most exciting recent discoveries is the recognition that the brain can change far more than was previously believed possible. Part III showed how change could take place at any age and in many different systems of the nervous system. Because the billions of neurons that make up the nervous system are not actually touching each other, the connections between neurons are subject to change. Often, neurons learn to fire together by the experiences we have. So, depending on what you do, there is always the potential for changing the brain.

Therapy can provide experiences that foster neuronal firing in healing directions. For example, when someone has undergone a traumatic experience, they show shrinkage in the hippocampus, an area of the brain involved in forming new memories, learning, and navigating through space. When clients resolve a trauma in therapy, their hippocampus will actually re-grow. This process occurs by forming new neuronal bonds. You can foster these processes by providing some of the kinds of experiences that foster neuroplasticity. For example, give your client new and enriching experiences. Meditation has the benefit of awakening awareness, and so clients will discover new richness in the world around them without even having to leave home. It also helps to train attention so that clients will be better able to direct attention away from disturbing thoughts and toward positive ones.

Using What's There

4. The brain-mind-body network is neutral, with processes serving varied functions, so begin by working with the functions the client already has.

Typically, clients look at their problems as obstacles, something to throw away. Although their adjustment may not be working for them, the nervous system teaches us to look at problems more neutrally. You can begin by helping your client to use the processes they already have, and build from there. You can help them to see that their problem might actually hold the seeds of their potential.

The nervous system shows us how problems might actually become an opportunity to change for the better. For example, most people consider stress to be a problem. But if you think about the nervous system activity when people are feeling stressed, you recognize that it comes from the HPA pathway, a very adaptive and useful response system for dealing with danger. When we sense something threatening, the nervous system responds immediately to spur us to do something to meet the threat, either to fight, flee, or freeze, depending on which response is most protective to our safety and survival. Without this built in system, we could suffer injury or even death from the dangers we meet. And by using it well, we respond appropriately to the challenges of life.

When the reaction continues, perhaps under a sustained threat or a psychological problem, the nervous system remains activated, which can cause stress to the system. Thus, the very pathway that protects us can also be activated to stress us. But stress is not always a negative thing. There are times when remaining alerted is a blessing, not a curse. For example, for soldiers in battle, maintaining a high level of physiological arousal might make the difference between life and death.

Emotions and habits, even the unpleasant or troublesome ones, may serve an important purpose, often holding the seeds of healing. We are wired to feel a full range of emotions. Anger, sadness, happiness, and fear, are universal emotions, experienced by people in every culture (See Chap. 16). We are able to recognize emotions in other people's faces, perhaps because anger, sadness, fear, and happiness are built into our nervous system. When expressed and felt in the course of life, emotions serve to enrich our functioning. By becoming aware of feelings and accepting them, emotional reactions come into a healthier balance.

The nervous system teaches us that when we block a natural reaction, the response takes another path. Similarly, when we repress emotional responses, they come out in other ways. Often these alternative pathways are more troubling than the normal reaction itself. Begin with what is there and follow the thread back to the natural functioning to allow the mind and brain to find the best balance for that person in his or her situation.

Patricia was a young woman in her 20s. She wanted to attend college, but felt unable to take the steps she needed to register and go to classes. And so, she was working in her college town as a waitress, feeling uncomfortable. She came to see us because of troublesome nightmares. She dreamt of monsters chasing her, often waking up shaking with fear. Rather than rejecting the monsters, we invited her to accept them as a part of her trying to tell her something. She did some psychodrama techniques, taking on the role of the monsters and identifying with their motivations and needs. What she discovered was that these so-called monsters were actually expressing her own feelings of anger. Without realizing it, she had rejected her angry feelings. She told us that she was a sweet and kind person, and so anger didn't fit into her concept of herself. We encouraged her to include all her feelings in her life, that even anger had something to teach her. As she began to get in touch with her angry feelings, she realized there were some things that were bothering her. We encouraged her to express what she felt and talk about it. She brought this change into her relationship with her boyfriend. He told us in a session he attended with her that he appreciated when she expressed her anger, because now they could discuss things and work them out. And interestingly, as she got in touch with her angry feelings, they became less pressing. The nightmares stopped, and she came into a better balance in her personality.

Over time, Patricia felt motivated to start her college education and began to take the necessary steps to register and buy her books. Once she got started in college, she did well. By accepting all her feelings, she was able to reclaim her motivation and aggression, essential qualities for succeeding in life.

Begin by helping clients to accept that their mind, brain, and body interact in a system. Explain that there are built in pathways which use emotions such as anger or sadness for a purpose. Your client will discover personal strength by noticing and accepting uncomfortable feelings as holding the keys to helpful qualities, like aggression and motivation.

Enhancing Healing

5. Use feedback and feedforward to add weight to the bonds in healing pathways and lessen the weight given to unhealthy pathways.

Psychotherapy traditionally helps client to gain insight by experiencing and then coming to terms with uncomfortable emotions, to gain understandings that are helpful. But neuroscience offers an additional paradigm of feedback and feedforward. Since neurons that fire together wire together, the connections between neurons become more heavily weighted in terms of the likelihood that they will be activated when they are used more. Then the heavier weighted link probability is fed back into the system, to feedforward, so this connection becomes stronger. Applying this model to therapy, we can help our clients form healthy patterns by feeding back their strengths and resources so that these abilities can feedforward, and become stronger. Just focusing on the problems can sometimes reinforce those neuronal bonds. It is important to spend some time in each session facilitating positive abilities that the client has. Even the most troubled individual has some positive resources, although they may get little use. Work on these, either directly or indirectly, to foster neuronal firing. Work through the links, not the thing itself, because it is at the synapse where learning and change in learning takes place. Analyze links between events, objects, people, and things, to improve functioning.

Janice came to us because she wanted to lose weight. She felt compelled to overeat. Then, after binge eating, she felt guilty. She also complained that she was forgetful. She would lose her car keys and misplace her glasses. She complained that she often wasted many hours looking for things. She felt stuck in her patterns and unable to change.

We worked with her using hypnosis. We taught her to go deeply into trance. We talked about how sometimes a problem could become a useful ability. There are times when forgetting serves an important function. If people remembered every detail of every moment of their day, their mind would be cluttered. We suggested that her unconscious could find a way to make use of her ability to forget to help her with her eating problem.

After a number of weeks, she came to the session with a smile. She said that she went through a day and had been surprised to find herself forgetting to overeat! This happened another day, and then another. She just seemed to be engrossed in what she was doing, and forgot all about snacking. And at the same time, she told us that she had stopped forgetting where she placed her keys or her glasses.

In this example, we did not work directly on the problems. Instead, we worked on the connections between them. If you assume that people have what they need to change, you may be able to suggest a different set of interconnections, and then step back and let the client come up with the change, making them positive.

Eliciting Change

6. Make one small change to start a process.

The journey of 1,000 miles begins with one step. These wise words were uttered more than 2,000 years ago by the Taoist sage, Lao-tzu. The idea still rings true today, when we apply it to initiating change in the brain. Begin with a small change: a moment's meditation, a yoga posture, or an exercise in attention training. Any of the exercises in this book can initiate the process.

We had a client, Norma, who felt stressed. She was married, with two young boys, and worked at a grocery store. She could not control her eating and had gained 50 pounds in recent years. She had tried countless diets, only to find herself gaining the weight back. She felt that she was always in a hurry and never had enough time. She felt out of control in her life. She was argumentative at home and had difficulty getting along with people at work. She had difficulty sleeping but felt tired most of the time. She said, "I'm on a merry-go-round and I can't go off".

Norma responded defensively to any analytical work. She blamed everyone else and complained bitterly about her life. We could see that she needed something different. We asked her if she would be willing to do a learning task. She agreed to do it. We told her, "Make one small change".

Norma returned the next week looking visibly calmer. We asked her, "What did you change?" She replied that at first she could not imagine doing anything different, but that she had given us her word, so she had to try. She thought carefully, and decided she could change her hairstyle. She went to the beauty parlor and had her hair cut short. The results of this change surprised her. She said that with short hair, she could wash and blow-dry her hair very quickly. This gave her an extra 20 min. She no longer felt rushed, and was able to take the time she needed to prepare breakfast for her family and leave early enough to arrive at work on time. She said, "My whole day seemed to go more smoothly, because I wasn't running behind!" This feeling extended through her week, leaving her feeling more comfortable. Her stress levels began to ease and she met the responsibilities of her life better and better.

The small change she made seemed insignificant. But it allowed her to initiate a change that affected her both psychologically and physiologically. As her stress levels lowered, she felt better emotionally. Now she was able to address her problems.

The small change she made seemed insignificant. But it allowed her to initiate a change that affected her both psychologically and physiologically. As her stress levels lowered, she felt better emotionally. Now she was able to address her problems.

Conclusion

We have provided some of the key principles for applying neuroscience in your work with clients. We encourage you to discover the many others that are there, just waiting to be found and used by you to enhance every aspect of your therapeutic change.

Chapter 15
Working with Attention

Attention is one of our primary gateways to the world. How we use this built-in capacity makes a difference in the quality of life. Many clients have difficulties with their attention. They may feel unable to control it, or perhaps they overcontrol it. A number of psychological problems have been interpreted in terms of how attention is affected. Aging and Alzheimer's Disorder, ADHD, Autism, Borderline Personality Disorder, Neglect, and Schizophrenia all have distinct disruptions of normal attention. You can also see alterations in attention in typical psychological problems such as depression, where attention that is caught up in ruminating tends to be less aware of actual surroundings and events. Anxious people also experience alterations in their attention. They tend to be overly attentive to certain sensations, and underattentive or even avoidant of other concerns. And those with addictions have narrowed the focus of their attention to just one thing, their substance.

You probably ask your clients to turn their attention to feelings, thoughts, and behaviors, but those who suffer from these problems have difficulties doing so. Training attention will not only help clients with their psychological problems, it will also make it easier for them to engage fully in the therapy process itself.

There are additional benefits in working with attention. As clients begin to gain some control over their attention, their brain structures and functions change. The anterior cingulate cortex, which is involved in regulating conflicts, and the prefrontal cortex, which is correlated with executive functioning, are both enhanced through attention training (Posner and Fan 2004). Clearly, attention training can change the brain in ways that are specifically helpful for overcoming psychological disturbance.

Cognitive retraining can improve attention. Hypnotic methods depend on attention as part of the process. And the great meditation traditions work directly with attention. We also offer therapeutic techniques to stimulate the brain for improved attentional abilities. By making attention training part of your treatments,

C. A. Simpkins and A. M. Simpkins, *Neuroscience for Clinicians*,
DOI: 10.1007/978-1-4614-4842-6_15,
© Springer Science+Business Media New York 2013

you will find that therapy progresses more smoothly and your clients benefit in specific and nonspecific ways. This chapter provides current neuroscience models of the brain systems involved in attention.

Neuroscience Models of Attention

Most theorists agree that attention is a unified system that involves multiple behavioral and brain states with a number of subsets of cortical structures (Zillmer et al. 2008, p. 241). Neuroimaging has brought new evidence about the key brain areas and neural pathways involved. These findings tend to support psychology's earlier theories of attention, and take them further. With this information, you can base your interventions on helpful models to apply in therapy.

Arousal: A Key to Attention

Physiological arousal has been long recognized as the key to how the attentional system becomes activated. The reticular formation of the brainstem is central in producing arousal. The autonomic nervous system and endocrine system are both activated as well, bringing about increase in heart rate and blood pressure. Furthermore, the neurotransmitters dopamine, serotonin, acetylcholine, and norepinephrine are stimulated when we are aroused (Moruzzi and Magon 1959). This arousal helps us to become responsive to a stimulus. We find a certain level of physiological arousal that is best for optimal performance. You have undoubtedly noticed that some clients, especially those suffering from anxiety and trauma, are overly aroused. They react strongly to stimuli and startle easily, a sign that their nervous system is overactivated. At the other end of the spectrum are people who have low arousal, as in depression. Their tendency is to react slowly, minimally, or not even to react. They miss inputs occurring within and around them. Therapy helps clients to return to a more flexible and balanced responsiveness.

The Process of Attention

The sequence of mental processing in attention takes place in certain distinct pathways. Many different models have been created to delineate the steps on these pathways. But generally, for the purposes of psychotherapy, you can think of these pathways leading to attention: (1) Activating, (2) Selecting, (3) Orienting, (4) Maintaining, (5) Acting. Paradoxically, just as all roads lead to Rome, all pathways lead to attention.

The attentional system becomes activated from either an external or internal stimulus. Often when our clients respond to something in the environment, it triggers an internal concern. And so outer and inner become linked. Once activated, the attentional system selects, to focus on one thing and ignore others. This is because the conscious attentional capacities are limited. We can only consciously attend to a few items at once, (classically, seven plus or minus two items), so other things are ignored. We automatically pick the most salient, relevant inputs for allocating attention and orient there. This might happen spontaneously, as when you notice the sound of thunder during a storm. Or attention can be directed deliberately, when, for example, you turn your attention to read this chapter. Once alerted to something, attention will remain sustained there, often until some action is taken that will complete the process.

Attention Occurs Bottom-Up and Top-Down

Generally, the different components of attention involve a combination of top-down (conscious and deliberate) and bottom-up (unconscious and automatic) brain processes. In other words, some parts of attention are deliberate and conscious while other aspects occur automatically without quite noticing. Top-down attentional capacities are available to our conscious thinking, so they can be accessed intentionally. However, conscious attention is limited in how much information it can process at one time. But bottom-up attentional capacities are larger. We can register far more information unconsciously than we can possibly attend to consciously.

The capability to take in information consciously and unconsciously can be used in therapy. You can work with conscious attention directly, but unconscious attention often responds better to indirect work through hypnosis, meditation, and body therapies. Therapy should address both top-down and bottom-up processes to enhance your client's attention on many levels. The exercises later in this chapter and throughout the book provide techniques for utilizing both top-down and bottom-up aspects of attention.

Brain Areas Involved

From the perspective of the brain, a number of components are activated when we pay attention. If something draws our attention, we turn attention to it automatically, like a reflex, with bottom-up filtering. Attention responds to the properties of the object itself, such as a loud sound or a bright light. The brainstem is involved in orienting to the object in this bottom-up way. The parietal lobe activates as we focus on a specific relevant location, usually somewhere in visual space. This stage disengages attention from unimportant stimuli to move it to more relevant or pressing stimuli. When the stimulus is something visual, you automatically look. Visual orienting is activated from the superior colliculus in the visual pathway, which receives input from the eyes and from the thalamus, the gateway of the senses. Your eyes saccade, which means that they move all around the object, creating a spatial map of the object in the retinal field projected onto the visual cortex in the occipital lobe.

If the stimulus enters through a different sensory input, such as sound or touch, different areas of the brain are activated. For example, when the stimulus is a sound, your auditory cortex located in the temporal lobes registers the stimulus through the ears. Tactile stimuli come in through the sense of touch, through the thalamus, and are processed by the parietal lobe.

Thus, to summarize, different parts of the brain are activated depending upon the type of stimuli. The parietal lobe processes spatial and sensory input, the temporal lobe is engaged in auditory and some visual input, and the occipital lobe becomes activated with visual stimuli. The processing occurs outside of conscious awareness, as a bottom-up process that is immediate and often automatic. Therapeutic methods can work with these bottom-up processes as well as working top-down.

Once the attentional system is activated, it maintains an alert or vigilant state by involving top-down with bottom-up processing. Stabilization of the attentional system comes from the rostral midbrain structures and brainstem and involves flexibility and consistency in attentional efforts. Sustaining of attention engages the motor system, which helps you to keep your attention alerted and placed on the chosen stimulus. This process relates to executive control since you are deliberately keeping your attention focused where you want it. The basal ganglia involved in motor action works with the thalamus to guide deliberate, strategic placement of the alerted attention. This alertness engages the right frontal area of the parietal lobe along with several parts in the front of the brain: the anterior cingulate and the lateral and orbitofrontal prefrontal cortex. Now, attention can be deliberately responsive to goals and executive control. If a shift in attention is needed, the prefrontal cortex is activated to help redirecting the attentional system. The hippocampus and amygdala also become engaged in the encoding, which is being attended to, into working memory. All these brain regions interact together when the attentional system is activated.

Unconscious Attention

As the models of attention describe, we are continually taking in much more information than can be registered by conscious attention. Prolific neuroscience researchers, Larry Squire and Eric Kandel recognize that even when sensory input is not attended to, it has none-the-less been perceived (Squire and Kandel 2000). This unattended information, even though it remains outside of conscious awareness, can also influence behavior (Posner 1978). All of the ways we take in and use information, even if we attend unconsciously, offer exciting potentials for therapeutic work, especially when people are struggling with resistance or troubled

Fig. 15.1 Tichener circles

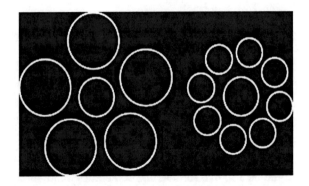

histories that present obstacles to change. Skillful therapy that works with both conscious and unconscious attention can activate the pathways, which facilitate real brain changes for new responses and potentials.

How Do We Attend Unconsciously?

When we pay attention to something we see, we typically think that the visual processing is fully conscious. But the visual system contains two streams—one conscious and the other unconscious. The ventral stream, for processing information for what something is, is a conscious stream. We can know the object consciously. The dorsal stream, for processing where something is, its location in space, is an unconscious stream. It is "just there". Information processed through the ventral stream is consciously attended to and understood explicitly, whereas information processed through the dorsal stream processes is unconsciously attended to, without being known explicitly, or understood.

Research has shown that even though dorsal stream attention is unconscious, subjects can respond with information gained from dorsal stream attention without understanding or knowing how or why (Squire and Kandel 2000). An interesting experiment will illustrate this point (Fig. 15.1). To experience the illusion, look at the two sets of dots surrounding the center dot. Does one of the centers look larger than the other? Most people will see the center dot surrounded by smaller dots as larger. The surrounding context makes a difference in how the dots are perceived. In reality, the two centers are identical.

An interesting experiment was performed to expand on this illusion (Milner and Goodale 1995). Subjects were asked to pick up the center circle they thought was larger. The researchers recorded how the subjects reported seeing the size of the centers. Then, they measured the distance between the fingers as the subjects reached to pick up the dot. When the researchers made a slightly smaller center dot, the subjects adjusted their fingers to match the true size even though they still saw the illusion that the dot was larger. The unconscious dorsal stream had assessed the true size more accurately than the conscious ventral stream!

This experiment is an excellent illustration that the unconscious can know things accurately and correctly without knowing why or how. This capacity is helpful in therapy, and in everyday life. A fine cabinetmaker, for example, can often estimate sizes of wooden objects quickly and accurately without actually measuring them with a tape measure. After years of careful measurement and integration of designs with construction, the sizes and shapes of things that are, are easily discerned. Inaccuracies are clear. Ask the cabinetmaker how he or she knows, and the reply will often be, "I just know."

Milton H. Erickson (1901–1980) gave us a therapeutic application for using unconscious attention. Erickson believed the unconscious mind is intelligent, and could be a source for healing; sometimes, more directly and effectively than what we consciously attend to (Rossi et al. 2006). Erickson evolved and explored creative ways to activate unconscious attention along with other unconscious processes to help his patients' bypass conscious limits and make real therapeutic changes (Erickson and Rossi 2006).

Working with Conscious and Unconscious Attention Therapeutically

You can work with both top-down and bottom-up attention in therapy. CBT, for example, directs attention consciously to problems and their faulty interpretations. Sometimes, false beliefs are held outside of conscious awareness and so cognitive therapy helps to bring unconscious thoughts and ideas to consciousness. Bottom-up attention is stimulated more indirectly, when we work with hypnosis, for example, which elicits automatic responses that lead clients to spontaneously attend to what they need to notice for healing. The exercises that follow offer methods to work with attention first consciously and then unconsciously.

Developing Attentional Skills Consciously

We all find it easier to pay attention to something of interest. You can engage the emotional limbic system to help intensify and improve attention. Most people find that something they like is interesting, and therefore, easy to pay attention to. Of course, tastes vary widely. Some may choose something beautiful, such as from art or nature. But others might be more involved in fantasy figures from movies, books, or cartoons. Some people may think of playful variations, which we often encourage. Instead of making focusing attention into a chore, it can become an opportunity to express creativity and humor, as well as to have experiences that satisfy personal tastes.

Narrowing Attention

People often have difficulty focusing on something because their attention becomes distracted. You can cultivate the ability to focus by the practice of deliberately withdrawing attention from outer stimuli. This exercise, drawn from the great yoga tradition of pratyahara, turns attention away from the outer world to focus it inward.

Sit comfortably and quietly for a few moments. Then, notice everything around you: the sounds around you, the temperature of the air, any aromas, and the quality of the light in the room where you are sitting. Focus on everything surrounding you for several minutes.

Next, turn your attention to whatever is close to you such as the feeling of your clothing, smooth or textured, loose or tight. Notice the chair you are sitting on, whether it feels hard or soft, warm or cool. How do your feet meet the floor? Are they resting lightly on the floor or pushing against it? Keep your attention focused on what is close to you for several minutes.

Next, notice inner body sensations. Scan through your body to notice your muscles. Are you holding them tight or are your muscles relaxed? Is there any unnecessary tightening? If possible, relax any tensions that can be relaxed. Turn your attention to your breathing. Allow it to be natural as you notice how the breath goes in through your nose and then out again. Perhaps, you can sense your pulse or heartbeat.

Next, withdraw attention from your body sensations to focus on your thoughts. Notice the stream of thoughts, like sitting on the bank of a river and watching each thought flow by. If your attention wanders back to sensations, to the stimuli close to you, or to the outer environment, gently bring it back to your thoughts as soon as you notice it. Maintain attention to thinking for several minutes. The ability to withdraw your attention from distractions responds to practice, so repeat this exercise in different rooms of the house or even outdoors.

Selective Attention

You can select one area of focus and then move on to another. Using your body as the object of focus can be easier for many people, because it is concrete and provides you with experiential feedback.

Begin at your head. Turn your attention to any sensations you notice in your face and neck area. Perhaps, you are holding your mouth tight or loose. Maybe you feel tension in your forehead. Or perhaps you feel tightness in your neck. Notice whatever sensations you can for a moment or two. Then do not attend to the face and neck and instead, direct attention down toward the shoulders. Pay close attention to them. Mentally trace out the distance between the shoulders. Notice any tension in the shoulder muscles and let go if possible.

Continue down through your body, first by withdrawing attention from the previous area and then by paying close attention to the next area. If your attention wanders back to an earlier area, let yourself notice the sensation for a minute and then gently bring it back to the area you are focusing on now. Do not force your attention. Instead, simply notice where your attention goes, encourage it to return to the area of focus, and wait for it to do so. Move all the way down through your body, attending to your torso, both front and back, and then to your extremities, arms, and legs.

Focusing Attention

Correctly directing the focus of attention can be trained to remain steady on one thing. Deliberately cultivating the ability to focus helps gain control of attention. Then, you can direct your attention at will. The focus developed by this practice becomes a great resource.

Method acting is a highly regarded classical application of intense focus of attention. The founder, Constantin Stanislavski (1936, 1984) believed that when actors learn to concentrate their attention on the circle of the stage they occupy, awareness of the audience would diminish and the action on the stage would intensify. Then, the method actor could delve deeply into the unconscious and feel intensely memories and experiences, from which a character and role would creatively emerge and take form for the stage without being distracted by the audience. This same skill is helpful for therapy, where clients may need to be able to temporarily disengage from surrounding problems and people in order to explore deeply their inner experiences.

Begin by focusing on one thing. Pick any object, picture, or piece-of-art that is personally interesting or meaningful. Place it in clear view. Sit upright cross-legged on the floor, on a small pillow, or on a straight-backed chair and look at the object. Keep your attention focused on it and notice as many aspects of the picture that can be thought of: color, shade, texture, shape, size, function, and meaning.

Now, focus on the quality of your attention itself. Is your attention focused and fully engaged? Or do you find that your attention is vague and undifferentiated? For example, if you are attending to a picture, do you find yourself looking at what is in the painting? Or do your thoughts associate to something else? If your attention wanders away, where does it go? If possible, retrace the links back to the object of focus. Otherwise, gently bring your attention back to the object originally selected, noticing the process of doing so. Begin with just 2 or 3 min of focusing. Gradually, increase the time as you become able to maintain focus. Skills in focusing attention improve with practice.

Focusing on an Inner Image

After focusing on an object with your eyes open, close your eyes and picture the object. People who are naturally able to form visual pictures will see a vivid image of the object. For others, the picture may be vague. Keep your attention focused on the image even if the imagined image is vague. Notice all the details that were noticed when looking at the object such as texture, color, patterns, etc. If you find that you have forgotten some of the details, open your eyes to look again. Notice more about the object, and then close your eyes again and visualize the object as an inner image. Keep switching between eyes opened to look and eyes closed to visualize for several minutes until you feel that you are able to hold a clear sense of the object within.

Focus on One Thing

Correct focus is the ability to direct attention to a single point and keep it there. This one-pointed, selective attention can be trained with practice. Deliberately training of attention becomes a valuable skill for therapy when you need to keep attention focused on inner concerns and experiences, even if they are uncomfortable. This skill exercises the executive control of the brain, an important and helpful ability to develop.

An easy place to begin is by focusing attention on a color. Pick your favorite color. Once the choice has been made, close your eyes and think about the color. Some may be able to see the color fill the horizon of their mind's image, while others may need to think of something specific that is that color, such as a blue sky, a large blackboard, or an enormous yellow sun. Keep your attention directed to this color and only this color. If another thought emerges, gently let it go and return to focus on the color. Begin with as little as 30 s for those with attention problems or young children and gradually increase the time to between 2 and 5 min. This skill improves dramatically with practice. You will enjoy the experience of control and calm that results.

Training Attention to be Flexible and Open

Attention can be focused, but it can also be flexible, open, and resilient. People who suffer from psychological problems are often stuck in redundant patterns of thoughts and emotions. Attention is directed in a habitual pattern of focus. Eventually, people will tend to notice only those things that fit into the expected patterns, reinforcing the aberrant system. LTP of neuronal response may perpetuate the patterns. Training in flexible attention can begin a change process that ripples through the entire system, forming new neuronal connections to help elicit a change. The exercises that follow can help to develop a more open, flowing, and flexible use of attention.

Focus on your favorite color for a minute or two, as in the previous exercise. Once you have pictured your color for several minutes, imagine that it becomes smaller and smaller, until it shrinks down to a single concentrated dot. Then, let it become larger and larger again. For a variation, let the shape change from a square to a circle or to a triangle. Be playful with the shifting. You can have fun with this exercise, allowing their imagination to alter the image in many creative ways.

Opening Attention

Another way to broaden attention is to let your attention be open. Notice a time with no immediate responsibilities or obligations coupled with less spontaneous mental activity. Look for such a moment, perhaps at night, just before sleep, during a lunch break, a time alone with nothing that has to be done. Another possibility is to find a time when attention wants to drift or the mind feels blank. At moments like these, you might try to force yourself to do a chore or task. But instead, such a moment can be used as an opportunity to develop open attention.

Spend a few minutes permitting your mind to be blank for a moment and explore how expansive that blankness can be. Do not try to discern what it is exactly, but allow this spontaneous tendency to do nothing. This state of mind may happen sitting, standing, or even when waiting in line for a long time. The important thing is to notice the moment of opportunity and allow the experience to take place when circumstances permit. Let your thoughts drift. Do not do anything and do not think about anything in particular. Simply sit quietly, allowing this experience to develop. After allowing the naturally occurring blankness, even if only for a brief time, you may find that you can deliberately access this open attention as a helpful alternative to focus.

Working with Free-Flowing Attention

When focusing attention takes people away from what might be most important, you can work with unconscious attention more effectively. Sometimes, the things that are on the periphery of awareness are significant and need to be brought to light. This exercise develops free-flowing attention to broaden the scope to include subtle cues.

Widen your attention to include what you are not focusing on, outside of the spotlight of your attention. If attention is like a searchlight in the dark directed upon an object in the dark, free-flowing attention includes attention to things that are outside that focus of light. Listen to yourself as you speak. Instead of simply staying focused on what you think you intend to say and mean, let your attention be broader. As you speak, listen to any other areas that you are not expressing that are beyond what you are consciously aware of. If some inkling occurs to you, take note of it and see if you can capture and express it. So, let yourself include even subtle cues, seemingly unrelated thoughts, fleeting associations, or whatever is outside the single beam of attention's light.

Ways to Work with Unconscious Attention Therapeutically

Unconscious attention is responsive, continually reacting to stimuli just outside of awareness. The unconscious attends to stimuli in a free flow of natural, active, creative processing often without intervention of conscious purpose. Unconscious attention can be worked with and directed therapeutically.

Incorporating Hypnosis

Hypnosis is a method that focuses attention so deeply that people experience a deeper absorption that leads to a shift in their consciousness. It tends to deactivate the executive functions, allowing more activation of unconscious processes. In fact, when people are in hypnosis, they are more sensitive to these kinds of experiencing that are occurring outside of our usual consciousness. Therefore, hypnosis is a premier method for working with unconscious attention.

   We offer some exercises for activating unconscious processes to induce trance. You can readily integrate hypnosis into your work. We suggest further reading our most recent book on hypnosis, *Neuro-Hypnosis: Using Self-Hypnosis to Activate the Brain for Change* (Simpkins and Simpkins 2004, 2005, 2010). We also recommend Milton Erickson's collected works (Rossi et al. 2006), books by Rossi (2012, 2002), Jeffrey Zeig's works (2006), Michael Yapko's books on this topic (Yapko 2001, 2003, 2006), and John Lentz (2011), for methods to use hypnosis indirectly. If you decide to go further, you can also get certification from training through the Milton H. Erickson Foundation.

Begin by Exploring Unconscious Attention

The first two exercises introduce how to explore unconscious attention as it manifests itself in everyday life.

   Begin by turning your attention to your hands. You probably were not thinking about your hands, but now that we mention them, you become aware of whatever sensations you are having. Perhaps your hands feel cold, or tingly or maybe light. You cannot accurately guess what you will experience without simply paying attention and waiting for the response. The experience occurs in its own way and in its own time. Sometimes, it is interesting to place one hand on each knee and pay attention to the weight of each hand. You might find that one hand feels immediately lighter or heavier than the other, or that at first they seem the same, but as you pay attention one becomes heavier than the other. You may be surprised by your unconscious response. While you are waiting for one hand to become lighter you might discover unexpectedly that one hand becomes cooler, or maybe you have a new experience of your hand feeling very far away, or growing larger. You will respond in your own unique way. Your conscious mind does not know how this will be. As you learn to allow and be attentive to your spontaneous responses, you will become acquainted with your natural unconscious which you can then learn to develop.

Exploring Peripheral Associations

This time try to recall how you felt after you completed the previous exercise. Picture yourself sitting or lying comfortably and remember how your hands felt. Wait until you feel ready to try the exercise. As you focus on this, you will probably find your body beginning to relax a little. While thinking about the previous experience, peripheral thoughts probably flicker in the back of your mind. Shift your attention to a peripheral thought or experience. For example, as dinner-hour approaches, notice if a vague thought or image is present about food. Perhaps, you realize you are thinking about a pleasant moment. These less obvious thoughts are present peripherally, but you usually do not bring them into consciousness. In this exercise, try to mentally reach for those flickering thoughts as they appear briefly in the stream of awareness.

To work with this, let your thoughts drift for a moment. If you notice a flicker into awareness that you cannot quite recognize, wait and allow any related feelings, images, or a thoughts to emerge. As you become more at ease with your unconscious attention, you will be surprised to discover that unconscious thoughts and images will emerge more readily.

Learning Self-Hypnosis

Unconscious attention is more accessible and easier to work with in hypnosis. If you decide to use hypnosis in your practice, we strongly advise you to practice self-hypnosis yourself. Here is a simple way to induce self-hypnosis. You can adapt this induction to use with clients. Other exercises in later chapters provide ways to work with unconscious processes that can also be adapted to hypnosis.

Entering Self-Hypnosis

Find a quiet and comfortable place where you can do this without pressure or interruption for at least 15 min. Sit or lie down and relax for a few moments. As you relax, let your thoughts drift and your attention roam wherever it likes. Try not to get lost in any one thought-path; simply notice associations and let them go. Do this until you notice some settling or calming.

Knowing that you will be trying hypnosis for the first time, you may feel excited or nervous. You might wish to pay attention to your feelings about hypnosis, and perhaps listen to the inner dialog of your thoughts: What are your reactions? Glance inwardly now to note any attitudes you might have about doing self-hypnosis. Sometimes, people have superstitions about the powers of hypnosis from television and movies, about how it can take control of the mind. According to research (Kroger 1977, p. 104), no one has ever been harmed by hypnosis itself. Hypnosis allows you to be in touch with inner needs and motivations. You will not do or experience anything that is inconsistent with your true nature, including your ethics and morals. It is reassuring to realize that personality remains constant. Your nature does not change; you are just hypnotized.

Imagine for a moment what you expect trance to feel like. People sometimes expect to relax, to feel calm, to have their body become cool or warm, to become light or perhaps tingly. Picture yourself in trance. Would you look relaxed? Are your eyes open or closed? Or perhaps it is easier for you to think about yourself in hypnosis, or sense your way in. Notice your response. Does it surprise you or is it consistent with your expectations? If you truly feel surprised, you have probably had a genuine unconscious response. As you imagine yourself in trance, do your eyelids feel heavy? Suggest that your eyelids grow heavier and heavier and want to close. Wait for your eyes to feel ready to close. Then, allow your eyelids to close. If they do not want to after a long wait, close your eyes anyway. Relax your eyelids and allow your entire body to relax very deeply.

Chapter 22 on Substance Abuse offers a method for discerning your typical perceptual mode, the way in which you tend to notice things. Most people are visual, but many perceive more through thinking first, emotions, or sensing. Turn to this chapter if you have difficulty imagining now, what it would be like for you to go into hypnosis, and then come back to this exercise and try again.

Allowing Unconscious Attention in Hypnosis

Now, let your body relax even more. It is possible to visualize color in formless, abstract, or symbolic form. At first, the color may appear as just one shade. Gradually, it could alter in its shade, depth, or even change colors. Sometimes, people see a kaleidoscope of color. Other times it is simply white or black afterimages, lights, or streaks. Experiment with thinking about, imagining, or sensing a color you would like to experience. Wait for your response. Watch it evolve. And as you imagine, sense, or think about your color, breathing can become comfortable and your body can relax even more deeply. Allow yourself to drift comfortably for several minutes.

Coming Out of Hypnosis

When you are ready, you can begin coming back to full consciousness. At first, you can help yourself in comeing out of hypnosis by counting backwards from five to one. With each number, as you approach one, you will become more alert, all your sensations returning to normal. If you finish but continue to feel unusual sensations, wait for a few minutes. Then if you need to, close your eyes again and go back into trance for a minute or two. Suggest that your sensations will now return to normal and again count backwards from five to one. Transitions in and out of trance become smoother and easier with practice. You may evolve your own way.

Ratification of Hypnosis

The hypnotic experience is not always easy to recognize at first. You may have noticed that you were relaxed and calm but may feel that this is not anything unusual or different from ordinary waking. Ratification, that hypnosis is happening, helps to intensify the experience and leads to an increase in hypnotic abilities. Ideomotor signaling is useful for this.

Unconscious Sensitizing to Hypnosis: Sit or lie down so that your hands rest either on your legs or by your side. Experiment with one of the previous exercises where you felt responsive. Invite yourself to become even more relaxed than before. Once you feel comfortable, focus on your hands. Consider how frequently people move their hands in conversation without thinking about it. Sometimes, the gesture is even more meaningful than the words. Do you talk with your hands?

Now, ask your unconscious a "yes" or "no" question. Designate one hand as "yes" and the other hand as "no." Choose a question for which you do not have the answer, such as: Would my unconscious like my legs to relax? Could I feel tingling in my fingertips? Can I have a pleasant memory? Could I see colors when I close my eyes? Now, wait and pay close attention to your hands. Do not try to move them, simply notice. Sometimes people feel the answer as tingling, lightness, or heaviness in one hand or the other. Sometimes, a person will notice warmth or coolness. Still others will feel a finger raise in one hand or the other or maybe a feeling jumps from one hand to the other and back again. After a little while, you will know what your response has been, and in which hand it occurs. If you felt something in your hand, you unconsciously answered the question.

Your first experience of hypnosis may not be what you expected or exactly as you predicted. These mysteries make communicating with your unconscious interesting. Your conscious, rational thinking does not know what your unconscious already knows. In his seminars, our teacher Milton H. Erickson often said, "Your unconscious mind knows a lot more than you."

Conclusion

Attention facilitates contact with the environment and is a key element in therapy. People can enhance their abilities to pay attention, both consciously and unconsciously. Work with the exercises in this chapter, to narrow and broaden focus, and to be able to sustain attention when needed. They do respond to practice. You can also learn to allow attention to flow freely and attend unconsciously to subtle cues, a useful skill for enhancing therapeutic process.

Chapter 16
Regulating Emotions

Psychology has given us many time-honored theories of emotion. William James taught us that emotion is not just a response to a stimulus, but it is the perception of feeling our bodily changes that occur in response to the stimulus (James 1896). Walter Cannon related emotions to homeostasis. When the body receives a stimulus through the autonomic nervous system, a physiological change takes place that disturbs equilibrium, giving an emotional feeling and the needed information for restoring balance (Canon 1927). Later theories included cognition. Stanley Schacter and James Singer stated that emotion involves two components: physical arousal and cognitive labeling of the arousal (Schacter and Singer 1962). Three decades later, Richard S. Lazarus extended the theory further, claiming that cognitive appraisal and interpretation of a situation is primary in emotion (Lazarus 1991). Most of our modern psychotherapies assume this, as part of the rationale for creating techniques to help people regulate their emotions and find a more comfortable balance.

Meanwhile, at first, neuroscience de-emphasized the study of emotion. For years, many people believed that study of the brain was about cognition. They considered affect to simply be a special case of cognition. But debate through the years and many research projects later showed that although emotion is certainly intertwined deeply with cognition, emotion and cognition is not just one thing (Cacioppo and Berntson 2007). As Joseph LeDoux, one of the prominent researchers in fear conditioning, points out, "A pure cognitive approach, one that omits consideration of emotions, motivations, and the like, paints an artificial, highly unrealistic view of real minds. Minds are not either cognitive or emotional, they are both, and more" (LeDoux 2000 p. 157).

Today, a number of neuroscientists have dedicated their work to the study of emotion (LeDoux 1996, 2000, 2003; Damasio 2010; Ekman 1992, 2003; and Levenson 2003 to name a few), and these studies have moved the study of emotion forward. We now have a new science, known as affective neuroscience, specializing in understanding emotion and the brain. It combines the findings from

C. A. Simpkins and A. M. Simpkins, *Neuroscience for Clinicians*,
DOI: 10.1007/978-1-4614-4842-6_16,
© Springer Science+Business Media New York 2012

multiple disciplines, including psychological, and cognitive models along with brain research. You can gain a better understanding of the brain areas and processes involved in emotions, which will provide additional tools to use to help clients alter their emotions in helpful directions.

Evolutionary Development of Emotions

The emotional system is considered an older part of the brain in an evolutionary sense. Emotional responses are innate, selected through evolution to help us adapt and flourish. We are motivated to live and perpetuate ourselves, and emotions are tied into these fundamental drives. Thus, we feel discomfort with what threatens our survival and feel pleasure in those things that perpetuate it.

Specific pathways that evoke emotions are built into the nervous system to help us keep ourselves save and pursue what is healthy. For example, the HPA pathway warns us of danger with feelings of fear that help us to take action by fighting, fleeing, or freezing. The pain pathway prompts us to protect ourselves from further injury when we feel pain. And the reward pathway brings pleasurable feelings that reinforce us to pursue behaviors that foster life.

Emotions do more than simply promote our safety and survival. They are also a source for regulating ourselves through life. They serve as a source of information that helps us process our decisions and take action. The intensity and strength of an emotional response can tell us how important the stimulus situation is to us. This relationship is known as *salience*. Emotions also serve the purpose of motivating. For example, if something is frightening to you, you might get motivated to become stronger in order to cope with it. Emotions serve us well for regulating our life. Thus, emotions are a key focus during psychotherapy, because successful self-regulation is a primary goal.

Emotions are Embodied

When we have an emotion, corresponding changes occur in our body. They invoke an internal set of sensations, known as interoception. And they involve movements, the way we express them, such as facial expressions of smiling when happy, or stomping a foot when angry. They also involve characteristic changes in the nervous system, such as a quickening of the pulse and shortening of the breath when angry.

So, after a stimulus is presented, your attention is turned to your body sensation, because you notice that you are feeling something. Meanwhile, your body responds with certain facial expressions such as a frown or a smile, along with alterations in your autonomic nervous system, such as a faster heartbeat, a knotted feeling in your stomach, or a flushed face.

These emotional body feelings give us information on the salience of the stimulus (how important it is to us), which then leads to taking appropriate action. This link between emotion and action explains why getting in touch with emotions, which we help clients do during therapy, helps them to truly know themselves better, all the way down to their neurobiological reactions.

Components of Emotional Experience

Emotions have several components, involving a felt-body sense, actions, and cognition. They are triggered by a stimulus of some sort, either external in the environment or internal, from a body sensation, thought, or memory. All of these components have correlates in the brain, which become active when we are having an emotion. Thus, you can see that the process of emotion engages a network of interacting brain systems involved in body sensations, and motor movement and attention.

How Emotions are Evoked

Emotions arise through and in the interconnections in response to stimuli, from the body, the environment, and from our own cognitive processes. Many different areas all around the brain are involved in this combination of responses, which helps to explain why emotions are so important to us. The unity of all these areas functioning together results in this dynamic pattern we call emotion.

Emotions can be triggered by stimuli occurring in the actual environment, such as facing a menacing robber, playing an exciting game, or interacting with a loved one. Or emotions can be evoked by memories, images, and thoughts from the past that you recall in the present.

Once the process is initiated, a cascade of events takes place. In the case of fear provoking stimuli, the amygdala sends signals to the hypothalamus and brainstem, altering the autonomic nervous system to increase blood pressure, breathing rates, and contract the stomach and blood vessels in the skin. The facial muscles react with the look of fear and cortisol is secreted, to give the organism an extra boost in energy that may be needed to respond to the threat: to fight, flee, or freeze. Attention and cognition are adjusted, so that full attention can be directed to dealing with the threat. You probably won't be thinking about your credit card payment while dodging a car speeding toward you as you cross the street! The cerebellum also plays a role in modulating our expression of fear, which is why things like military service and psychotherapy can alter how people respond to threat. For example, military training makes a soldier's reaction to a threatening battle very different from the reaction of civilians caught in the crossfire.

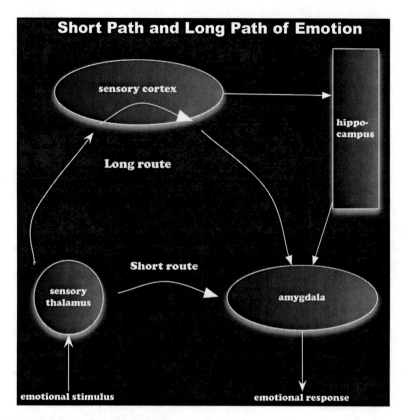

Fig. 16.1 The short path and the long path

Timing of Emotional Reactions through the Short Path and the Long Path

As explained earlier, lower brain areas in the brainstem and cerebellum are interconnected with the cortex, which makes it possible for emotional response to either follow a short path or a long path. These two emotional timings engage somewhat different pathways in the brain. Both the bottom-up and the top-down processes can become tools for therapeutic interventions, and we will describe this in more detail and provide techniques for using these pathways later in this chapter (Fig. 16.1).

The Short Path: The short, bottom-up path of emotional response tends to occur automatically and unconsciously. Thus, we can feel like an emotion just happens to us suddenly, without much prior awareness until we find ourselves having it.

A fast emotional system of response travels directly from the thalamus to the amygdala, bypassing the cortex. This shorter pathway accounts for the automatic, pre-conscious, or even unconscious quality of many emotional responses.

Evidence for this theory comes from the extensive research on fear (LeDoux 1996), which has revealed a long path and a short path depending upon how fear is processed. For example, if we see a poisonous snake on the path where we walk, we feel afraid. The short path has been activated to make our heart rate increases, palms become sweaty, and our face to flush. We are alert and ready to run from danger. But then, if we stop and take a more careful look and then perhaps, realize that the snake is really just a piece of rope, the long path is activated. We think about how we feel, recognize what we see, and realize that there is no real threat. The physiological reaction abates, and we might even feel amused and laugh at the reaction as silly.

Sometimes emotions are triggered as an automatic response seeming to occur instantaneously, as with fear from the speeding car example or from a traumatic event, and also from a spontaneously positive experience, such as the happy emotion you feel when smelling fresh-baked bread in the oven. These kinds of emotional reactions occur very quickly, traveling bottomup through the brain, within a few hundred milliseconds.

The Long Path: By contrast, the top-down long path engages higher brain functions in the cortex, our thinking brain, and takes slightly longer for the brain to process. When a sensory stimulus is processed, the information travels from the senses through the thalamus where it is relayed to the cortex, and to the subcortical limbic systems. We recognize consciously that we are having an emotion and think about it. This pathway, the long path, sends signals through the sensory cortex in the parietal lobe, thereby making emotions consciously recognized. We know that when we deliberately think about a sad event, for example, we will feel sadness. Thus, emotions can be triggered top down through the brain, traveling a slightly longer path that is influenced by our thoughts about situations that elicit an emotion more slowly.

Primary emotions such as happiness, sadness, anger, and fear may travel the short or the long path. Often emotions begin through the automatic systems in the body, arising from sensory experience that is processed by the limbic system. If the emotional response travels the short route, directly from sensation to the limbic system, it arises before interpretation and is unconscious, outside of deliberate conscious control. If, instead, we think about the response, the emotion that occurs is consciously recognized, so different parts of the brain are engaged for this.

These two paths of emotional response help to account for the quick reactions we may have, usually occurring unconsciously, as well as the slower responses that have a conscious element involved. Sometimes we may switch to a different path when necessary, for example when our quick, first reaction is not the best or the most mature response for the situation when considered in context. Therapy can offer methods for working either with the long path or short path to facilitate change.

Fig. 16.2 Papez circuit

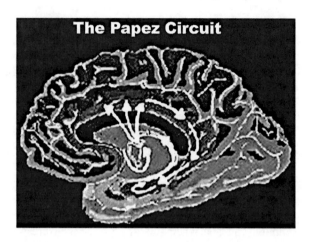

Fig. 16.2 Papez circuit

Emotion Areas in the Brain

So, what brain area is the center for the cognitive, physiological, and action components of emotion? Instead of one single area, we find a network of patterns and pathways linking together many parts of the brain with the body and with cognitive processing. Structure and function interacts (Fig. 16.2).

Back in 1937, James Papez (1883–1958) an American neuroscientist specified a group of organs in the brain that he believed to be activated when feeling emotions. The Papez circuit, as it has come to be known, forms a neural pathway to process emotion including the hippocampus, the hypothalamus, the anterior thalamic nuclei, and the cingulate gyrus. As more was learned about emotions, other organs, and pathways were added, now known as the limbic system (See Fig. 8.3). Today it is believed that the prefrontal cortex, the amygdala, the insula, and brainstem also play key roles in emotion (Ramachandran et al. 1999).

Amygdala

The amygdala is an almond-shaped organ located toward the center, one for each hemisphere of the brain, functioning as a gateway to emotional processing. Stimuli enter through the senses and travel to the amygdala. The amygdala helps to monitor how aversive or dangerous a situation might be, signaling the nervous system for a quick response. If a stimulus is salient, the amygdala passes this information through the limbic system. Thus, the amygdala is a key player in the short path.

The emotional qualities of memory are also processed in the amygdala, similar to how explicit memories are processed in the hippocampus. For example, when you reconnect with a good friend, the memory system in the hippocampus helps in recalling the details of when you last saw that person. Meanwhile, the amygdala and its neural pathways process the unconscious, implicit emotional memories associated with your past experiences of that person. This emotional aspect is experienced directly as changes in your body, such as a smile or a rise in heart rate. You have a conditioned association, an emotional learning, between the past situation, and the happy emotion you feel in the present. This process of emotional learning can explain how traumatic experiences from the past continue to have an influence on emotional experience in the present.

A large body of research shows that the amygdala plays a primary role in fear conditioning, influencing not just the unconscious, autonomic, and hormonal changes, but also inducing changes in behavior and attention (LeDoux 1996). The amygdala helps in recognizing the emotions of others; expressed in facial expressions, especially fear, which can help to make appropriate responses and decisions (Adolphs 1999). This is how the amygdala helps protect us from threatening situations, by alerting us to the danger by means of an emotional response.

The amygdala registers both positive and negative emotions. Negative emotions, especially fear conditioning, have been studied widely (LeDoux 2003). But more recently, researchers are finding the amygdala is also involved in processing positive emotions (Davis and Whalen 2001).

Amygdala responses are unconscious, going directly through the short pathway. But people can learn to modulate this short path emotional response. Researchers showed that therapy shifts the balance in the relationship between the amygdala and prefrontal cortex, helping to regulate emotional reactions (Banks, Eddy, Angstadt, Nathan, and Phan 2007). This regulation of emotions changes an automatic short path emotional response into a more well thought out long path reaction.

The Insula

The insula was first understood as being involved with taste. Early studies mapping disgust reactions indicated the central role of the insula for this feeling. But the insula plays an important role in our embodied sense of ourselves and in decision making, two crucial components of emotional experiencing.

The insula is located in a deep fissure under the lateral part of the frontal lobes, with a number of different sections. It receives information from sensory pathways through the thalamus and sends outputs to limbic structures as part of the short path. It makes connections to key limbic structures: The hippocampus, amygdala, and prefrontal cortex. The insula is also involved in movement through its connections to the motor cortex and the basal ganglia. Thus, the insula's location and interconnections show its importance in sensing, processing, and responding to emotions.

The insula is the central structure involved in interoception, our internal body sense. By providing a map of the internal viscera, much like the motor cortex for movement and the parietal cortex for sensations, the insula map correlates with a felt sense of ourselves. These perceived internal body feelings influence our emotions, mood, and sense of well-being.

Through this map, the insula helps form the links between body sensations and emotional feeling. In therapy, we can help our clients to get in touch with their emotions by developing their body sense. This map also helps to explain how we might have a gut feeling, or intuitive sense about a decision, since the insula is also linked to decision making and risk taking (Paulus, Rogalsky, Simmons, Feinstein, and Stein 2003; Krawitz, Fukunaga, and Brown 2010). We offer techniques for developing interoception sensing later in this chapter.

Cingulate Gyrus

Researchers first learned about the role of the cingulate gyrus in emotions from studies originally done back in the 1950s with people who had lesions to this brain area. These patients suffered symptoms including apathy, inattention, and emotional instability (Tow and Whitty 1953).

The cingulate gyrus has separate areas devoted to processing emotion and cognition. The dorsal parts process cognitive functions and the rostral-ventral sections process emotion (Bush, Luu and Posner 2000). Thus, the cingulate gyrus may help in shifting between the short and long pathways.

Recent fMRI and PET studies have helped in the understanding of the role of the cingulate gyrus in cognitive and emotional processing. Both the emotional and cognitive systems operate in a network with many other brain areas.

The emotional side is highly interconnected with many of the limbic system structures including the amygdala, nucleus accumbens, hypothalamus, anterior insula, hippocampus, and orbitofrontal cortex, as well as interacting with the endocrine, autonomic, and visceromotor systems (Bush, Luu and Posner 2000). The emotional part is crucial for maternal instincts and stability of emotions.

The cognitive area is part of an attentional network involved with the lateral prefrontal cortex, parietal cortex, and motor cortices. It participates in such functions as motivation, novelty, error detection, working memory, and anticipation of cognitively demanding tasks (Carter 1999).

The emotional and cognitive parts of the cingulate gyrus operate separately, but they modulate each other. The emotional system can deactivate the cognitive one; and the cognitive system has a deactivating influence on the emotions (Drevets and Raichle 1998). Either system can predominate as a pathway, through the switching action of the cingulate gyrus (Posner, Rothbart, Sheese, Tang 2007). Thus, therapies that work with emotions cognitively and therapies that work with emotions experientially can both be effective. Often, they are complimentary, as the techniques later in the chapter will show.

Patterns of Emotions

Primary emotions are limited in number and wired into the nervous system. These primary emotions tend to be shared by all ages, cultures, and species. The primary emotions give rise to the many varieties of secondary emotions that are mixtures and shades of the primary emotions.

Brain activations for happiness, sadness, anger, and fear have significantly different patterns all over the body. These primary emotions engage some of the emotion structures more than others in distinctive and unique patterns. The neural patterns found for each emotion seem to constitute a multidimensional map that forms a basis for what is experienced as an emotion (Damasio et al. 2000). Thus, each emotion is actually a conglomerate of responses and reactions that come together in patterns that form a map leading to the feeling of the emotion.

Information from the pathways in the lower brain areas such as the brainstem and hypothalamus tends to be unconscious, moving through the short path. Activations in the insula and cingulate cortex can be accessible to consciousness, activating the long path. The interaction of these brain pathways: conscious, long path, and unconscious, short path, help to account for how sometimes people are aware of their emotions while at other times they are completely unaware of them.

Each emotion has unique patterns of activation and deactivation. Happiness correlates with increased activation in the right posterior cingulate cortex but sadness is accompanied by decreased activation in that area. The dorsal pons, a lower brain area, is active in anger and fear, both emotions that often occur automatically without conscious control (Damasio et al. 2000). Thus, the subjective feeling of emotions is grounded in dynamic neural systems for each emotion, some conscious, and others unconscious. Therefore, therapists need to work with emotions using both conscious and unconscious methods depending on where the activation is situated.

Working with Emotions

One of the byproducts of successful psychotherapy is learning how to attune to emotions as they occur, sometimes consciously and other times unconsciously, and then to be able to regulate them appropriately. Recognizing that emotions involve a sensory, motor, and cognitive brain component, suggests varied ways for working on emotions by using one or several of these components. These exercises introduce methods for working with each component. The interpersonal aspect of emotion will be discussed in Chap. 18 on the social brain. We encourage you to incorporate your own innovations based in the therapeutic approaches you already use.

Working with Interoception Sensing

One level of emotional experiencing comes from feeling a physiological response to bodily changes, interoception. From this perspective, the body can become a means of regulating emotions. Every cell is alive and active, with a kind of intelligence. All the parts work together in vibrant harmony. The nervous system allows us to interface from body to brain, to thoughts and feelings, setting in motion a healing process.

Awareness of the body involves attuning to subtle cues. For psychotherapy, body awareness methods offer powerful methods to get in touch with unconsciously generated emotions. Often people with psychological problems lose touch with their bodies, or else they develop an exaggerated concern for one body sensation, often ignoring others. They form inaccurate concepts about their sensations that lead them into problems and away from solutions. Bodywork on internal sensations can attune the client to the moment. Clients may be able to bypass defenses as they get in touch with what is happening within. Through the process, they can take the first steps toward a more realistic sense of themselves.

Many people ignore their internal sensations. And yet, these sensations offer a direct link to emotional experiencing. When worked with sensitively, they can become a source for healing and change.

Begin with Breathing

Breathing is one of the most accessible internal sensations involved in emotions. Often, breathing reveals a great deal of information about how you are feeling. This exercise can be used to help you become more in touch with what you are feeling. To start learning the skill, try this exercise when you are feeling comfortable. Once you are successful with doing this when calm, perform the exercise while having a particular emotion. To compare and contrast, try it when feeling one of the primary emotions: happy, sad, angry, or afraid. Paying attention to breathing while in the midst of many different feelings will be instructive, adding a new felt-sense about what is happening. Furthermore, you may find an uncomfortable feeling begins to lessen somewhat or even dissolve completely.

Now, sit quietly and begin by deliberately turning your attention to breathing. Notice the rhythm of your breath. Is it fast or slow; and what is its quality: is it pushed, labored, shallow, or perhaps deep? Sit quietly and feel any other sensations that accompany breathing with this feeling such as heat or cold, tingling or numbness. Also, note what thoughts are racing by or moving slowly.

Later, repeat this exercise when you are feeling another emotion. Compare and contrast the quality of breathing with different emotions. Getting to know the quality of breathing that accompanies different emotional states may be helpful for bringing about change.

Follow Internal Sensations

Turn your attention to your breathing as you did in the previous exercise. Notice the sensation of breathing in and out for several minutes. Don't alter your breathing. Simply attend to the sensations. Can you feel the air as it flows through your nose and down into your lungs? Do you feel coolness or warmth?

Next, turn your attention to your heartbeat. If you have trouble sensing it, begin by placing your palm over your heart and notice the sensations of the pulsing beat. Another way to perform this exercise is to place your finger on your pulse at the wrist. Allow yourself to attune calmly, quietly sensing what you notice. As you perceive, notice how each moment is a little different. Perhaps as you sit quietly, your heartbeat slows or quickens, gets harder or softer. Allow yourself to be as you are, just noticing.

Now turn your attention to your stomach area. Pay attention to the sensations there. Do you sense movement, warmth or coolness, tingling, or any other interesting sensations?

As you attend to all these internal sensations now, what do you feel emotionally? Breathing, heartbeat, digestion, and emotions are all linked together as part of your emotional experiencing. As you attend to these sensations, you may begin to notice your emotions in a different way.

Attending to the Motor Response in Emotion: Tension and Relaxation

Often when people experience strong emotions, especially disturbing ones, there is a corresponding tension in the muscles of the face and the body, to instinctually prepare for action. This exercise series can help to become aware of such patterns and then initiate a process to mediate your responses if needed.

Begin with Awareness

Scan through your body and notice the general tone of your muscles. Then, starting at your head, notice the muscles in your face. Are you holding any part tightly, such as your mouth, forehead, or eyes while other areas are relaxed? Each emotional experience has accompanying muscle patterns of tension and relaxation. Using your attention, carefully trace out the patterns in your muscles now. So, if you feel a pulling sensation between your eyes and forehead, notice each part of the pattern. Do you feel discomfort? Or perhaps you feel nothing. Sometimes such patterns have a habitual sense of familiarity, and so they are hard to notice. Keep paying attention, and you will begin to perceive subtle distinctions between muscles that are relaxed and those that are tense. Do not changes anything; simply notice.

Then move down to your neck and shoulders. Sense whether you find tightness in your neck and shoulders. One clue to whether your neck and shoulders are tight is found by noticing how these areas meet the floor, bed, or couch you are lying on. Notice how these areas meet that surface. If they are tight, you will probably feel that they are either lifted slightly away from the floor or bed or pressing down hard. Once again, notice any accompanying sensations without changing anything. Continue down until you have attended to your entire body, just noticing tension areas and relaxed ones in this way. When finished, move on to the next exercise.

Easing Tensions

Start again at your head by paying attention to your face. If you discovered tension in a particular area such as in your eyes, forehead, or lips, turns your attention there. First, carefully trace out how you are holding these muscles as you did in the last exercise, then, if possible, allow these muscles to relax. Letting go involves an opening, releasing feeling. Muscles will feel softer, lighter, larger, or smoother. Breathe comfortably and let the tension go with each exhale of breath.

Then direct your attention down towards your shoulders. Pay close attention to them. Notice whether you are holding the muscles tight and let go if possible. If you noticed that your shoulders were held away from the floor, can you let them sink down and take support from the surface? Continue down through your body, first paying close attention and then trying to relax any extra tension. You may be surprised to notice areas that are held tight, but don't need to be. If attention wanders away from to outer concerns, bring it back. Do not force yourself to relax. Simply notice where you can or cannot relax, and gently keep trying to let go of unnecessary tension. When you feel that you have relaxed some of your muscles, notice whether you have a change in your emotional experience. Often people find a direct link.

Noticing Emotional Sensory and Motor Responses in Different Situations

Using the skills from the previous exercises, turn your attention to your sensations when you have a spontaneous emotional reaction. Therapy offers opportunities to do so under guidance. The opportunity might also arise when you are watching a movie or television show or when you are reading a book. Notice how your breathing might alter, skin temperature might change, or heartbeat might increase. If appropriate, allow an easing of any unnecessary tensions. Does the emotional tone change?

As you are able to notice your emotions when watching a movie or reading a book, turn your attention to sensations when having emotional responses in situations during everyday life. Directing your attention in this way often has a paradoxical effect. You are now adding awareness to what is usually unconscious, switching the track of emotional reactions from the short path to the long path. Often people learn about their reactions and may have other feelings arise that can be worked on in therapy. Over time, you may find that emotional reactions become easier to sense and are more moderate.

Thinking About Emotions

The long path of emotion engages higher brain areas of cognition. The influence of cognition can be helpful for moderating overly strong emotional reactions, and exercises will follow. By contrast, cognition in the long path can sometimes interfere with healthy emotional reactions, as when people suffering from depression ruminate negatively. Cognitive therapies work with these problems by fostering better cognitions, leading reactions into a positive direction. Exercises that bring people back to their senses through meditation and hypnosis may help to reverse these patterns. We offer exercises to add to your repertoire as well.

Recognizing Emotions

Emotions can be our signals of what to do and how to act. But people who have difficulty managing emotions often do not recognize what they feel until the level of emotion is too high to control. Learning to notice the signs of an emotional response as it begins to build is a helpful first step.

Ron was an angry man, but he did not think of himself that way. He told us that his problem was other people. He described himself, "I'm pretty laid back most of the time." He did not really feel too much emotionally and prided himself on letting negative things roll off. He had come to therapy because his girlfriend recommended it. She was bothered by his angry outburst, but he did not quite notice them. In fact, he told us, the only time he recognized that he was angry was if he found himself in the middle of a fistfight. His therapy involved some of the exercises in this chapter, to help him get in touch with what he was sensing within, and also to learn how to think about his emotions and then act more appropriately.

Recognizing Early Warning Signs

Meditative calming can be helpful, but modifying cognitions that increase emotion is also important for resolving the problem. Becoming aware of the early signs of troublesome emotions can begin a process to change an emotional pattern.

You can use cognition to alter a pattern of unawareness. Although you may not notice what you are feeling in a situation, you probably understand when people typically have emotional reactions. You have likely observed your friends and family reacting emotionally at times or seen people having strong feelings in movies or novels. Take note of the kinds of situations where people typically have emotions. Then, observe your own situation. Perhaps you have tried to do something and been unable to accomplish it. Or maybe there was a circumstance at work is going badly. Or maybe someone you care about is yelling at you, crying, or refusing to talk. Pay attention to your external situation, noticing that it is happening. Now, stop and deliberately turn attention to your sensory and motor experiencing as in the earlier section. First, notice your breathing. Use the attention to your body using the previous body awareness exercise. Notice your breathing, heartbeat, patterns of muscles tensions, and body sensations such as in your stomach, chest, or head.

Now, pay attention to the thoughts that come to your mind. Observe the thoughts as they occur, but mentally step back, as an observer. Have you had these thoughts before? And have you had this emotional state before. Often when people have recurring emotions, especially problematic ones, they have corresponding recurring thoughts that go with the emotions.

Take note. These body and mind experiences are your signs of having an emotional reaction. Get to know these signs well. Sensitize yourself to them by becoming consciously aware of them. This will help you in the process of learning to recognize your emotions. In doing so, you will be able to moderate your emotional reactions, as the next exercises guide in doing.

Accepting Emotions Mindfully

When people are able to notice emotional reactions as they begin to occur, they may feel uncomfortable. Usually people avoid their feelings for certain reasons. Therapy can help to unravel the thread, to help clients discover the sources of their discomfort. One source is from attitudes and beliefs, often judgmental in nature, such as "I hate this feeling," or "Feelings are for the weak minded," or defensive thoughts such as, "Who cares—not me!"

Mindfulness meditation can help people to become aware of and accept whatever they sense, feel, or think, without judgment. The practice is helpful for therapy and has been widely accepted now as an efficacious method. Kabat-Zinn's mindfulness based stress reduction research (Kabat-Zinn 2003) showed that mindfulness meditation can lower reaction to stress. And research has found that meditation moderates strong emotional reactions and enhances emotional regulation (Aftanas and Golosheykin 2005). This mindfulness series will teach clients how to become aware of their emotional responses, and to notice how the accompanying thoughts and judgments are creating reactions.

Taking a Non-judgmental Attitude

Mindfulness begins with you in your own experience, here and now. Work with the exercises to develop your skills. Use your body, feelings, thoughts, and objects of your thought. In time, your awareness will spread into every moment.

Mindfulness gives you the opportunity to get to know about all your life, actions, thoughts, and feelings as well as others. But as you learn to observe more deeply at first, you may not always like what you see. And as a result, you might be tempted to pass judgment on yourself or others before you fully understand. Moralizing will not help you on your path. In fact, it may interfere.

Mindful awareness should be non-judgmental. Like a scientist who is gathering data, don't jump to conclusions or use the new information you gain from being mindful to form biased opinions. Wait until you have a fuller picture. Trust the process and cultivate an open mind.

If you notice something about yourself that you don't like, take note of it. You may decide that it is a quality to change in the future, but do so without criticizing yourself. There is an important difference between simply observing something that may need changing and moralizing about it. Whenever you can simply observe, you will find that your inner mind opens up to you. As you begin to experiment with mindfulness exercises, try to observe without making judgmental pronouncements: just become aware of yourself in the situation.

Fostering Mindful Acceptance

You can practice refraining from judgment as you notice details of your experience, your actions, and their effects. Begin by becoming aware of what is there, but keep your observations clear and descriptive. Learn to accept your experiencing, without making comparisons or criticisms. Then you will be able to appreciate your qualities, just as they are.

To apply this nonjudgmental attitude to mindfulness, survey yourself from head to toe and recognize all your different parts. Describe each part to yourself. Notice, for example, your hair, its color, texture, style, your eyes, etc. But stay factual. For example, observe that your hair is, for example long, dark brown, and curly. But don't add an evaluation, such as unattractive.

This exercise might be easy to do if you are handsome or beautiful, with a perfect body and no faults, but it can also be done even if you are filled with faults. Mindfulness can help you to accept yourself, even if you believe that you have serious deficiencies. Embrace what you are and what you feel, without belittling yourself. Some problems are made worse because of self-doubt. With a mindful approach to your experiencing, you have everything you need to be happy.

Mindful of Emotions

Emotions are an important component of living, and so mindfulness must include attention to feelings. Mindfulness deals with emotions in a way that will overcome suffering from uncomfortable feelings and maximize fulfillment from positive ones.

Feelings can be categorized as pleasant, unpleasant, or neutral. People tend to cling to pleasant feelings and reject unpleasant ones. But this clinging and rejecting sets in motion a secondary set of reactions that interfere with awareness and causes suffering. You will be able to drop the secondary reaction as you become more aware of the feelings themselves, leading to more comfortable, aware, enlightened reactions.

The Impermanence of Feelings: Think about how feelings are impermanent. As we understand the findings of neuroscience, we know that emotions, like every other aspect of human experience, correlate with a pattern of activations and deactivations in the brain. Clinging to a pleasant feeling will inevitably lead to frustration because the brain pattern always ends. Conversely, trying to avoid an unpleasant feeling will also bring suffering, since you cannot escape from your moment-to-moment ongoing experience in the brain.

Identify Feelings: Mindfulness of feelings begins when you can identify the emotion you are having. To start the process, sit down for a moment and close your eyes. Turn your attention inwards. Try to put a name to your emotion or mood. Then match the description of your internal sensations, facial expression, and muscle tension pattern. If it is not quite right, modify your label until you feel satisfied.

Next, identify whether the feeling is pleasant, unpleasant, or neutral. As you identify it, try to remain calm. Be like a benevolent mother who watches over the children as they play in the backyard. When the children begin fighting, she does not become angry with them. Instead, she tries to calmly settle the differences and attend to each child's needs. Benevolently observe all your different feelings, even the ones you have labeled as unpleasant. By eliminating the secondary aversion reaction to a negative feeling, you will significantly lessen your discomfort with it.

Observe Mindfully: Sit quietly and observe what you are feeling now. In each moment, notice any experience you are having. Remain non-judgmental, simply noticing any thoughts, sensations, or movements as they occur. If your mind wanders, gently bring it back to your emotion in this moment. Observe how each moment is a little different, as your ongoing moment-by-moment experiencing occurs. Don't judge what you notice as good or bad; simply accept each experience, just as it is, without trying to change it deliberately. You will notice that things change of themselves, as you simply stay attuned here and now.

Mindful of Emotions in Action: The most challenging and at the same time rewarding application of the previous exercises is to perform them in the midst of the situation. Allow yourself to be mindful as you have an emotional reaction in your real-life situation. Stay aware of what you are feeling and doing as you do it. If the situation becomes too difficult to handle, step away, stop, breathe, and relax. Give yourself the space you need to handle the situation maturely.

Beyond Cognition

Emotions are often fueled by thoughts. The flow of uncontrolled thoughts can give a strong push to emotions. Thoughts get in the way of direct, clear perception. People add complexity with worry and rumination that fuel emotions in negative directions. Learning how to let go and be present can interrupt these harmful patterns and provide space for a different emotional response. Then, from a mind clear of negative thoughts, emotions can settle, allowing the natural mind-brain system to return to balance. In the quiet, open moment, new possibilities emerge. These exercises help to clear away the patterns of thoughts that tend to deregulate healthy emotional responses. In addition to helping to clear away negative thinking, the research on meditation shows that brain activity become synchronized together for greater awareness, sensitivity, and more highly attuned attention. (Hankey 2006) These skills will have broad, nonspecific effects that can be used at many points during therapy.

Clearing the Water Meditation: Begin with this visualization and you will find your distracting thoughts clear away naturally. Sit quietly with your eyes closed. Imagine that you are sitting on the shore of a pond. The pond is alive with activity. Frogs croak; crickets sing; birds fly overhead; a fish jumps out of the water, feeding on insects, splashes back, and jumps again after a bit, in another spot. Wind whips over the water, stirring up the muddy bottom. All is movement. Then gradually as the day passes, the conditions begin to shift. The wind dies down. The frogs settle in for a nap, the crickets become silent, birds perch in the trees, and the fish stops jumping and waits. The pond is quiet. The murky rippled surface calms as the mud sinks to the bottom, and the water is crystal-clear, reflecting the natural surroundings. All is stillness. Then the frog jumps into the pond. Splash! The moment when you can vividly see the pond reflect the surroundings and hear the frog splash into the water, you tap your clear open mind. Imagine this scene vividly. Stay with the moment.

Opening and Clearing: Classic Zazen: Zazen, sitting meditation, is the classic exercise for letting go to the moment in Zen Buddhism. Zen monks spend many hours meditating in this way, seeking to bring about an open, undifferentiated state of consciousness that continues to unfold. Follow the instructions carefully, and with time and practice you will experience a special calm and alert awareness. Attention is focused in the present moment, free of distractions while also being open and receptive. Brain patterns are both alert and calm together, a unique state with positive long-term effects on the cortex.

Sit upright, cross-legged, with your hands palms up and the backs resting on your thighs. Let your body be upright and straight–but not rigid–without leaning either left or right. Your head should be held straight with your ears and shoulders parallel to each other. Hold your tongue loosely against your palate and keep your lips closed and teeth together. Eyes should be closed or half open. Breathe calmly and regularly. As you begin to meditate, clear your mind of all thought. When a thought does arise, notice it and then dismiss it, returning to your calm, clear mind. By continuing to do this over time, you will eventually find that thoughts intrude less and less and that your concentration becomes natural and profound.

Conclusion

Emotions are embodied in a pattern of interaction involving the mind, brain, and body. Emotional pathways in the brain involve a number of structures functioning together, from the lower brain stem areas all the way up through the center of the brain through the amygdala and hippocampus, and all the way up into the cortex. We can intervene at any level: sensory experiencing, motor activity, and cognitive thinking. Sometimes begin by altering the short path of emotion, by altering the nervous system balance nonverbally and experientially to bring about a spontaneous change. Other times deliberate cognitive interventions can help bring a client back to a healthier center. Emotions are broadly intertwined with the mind, brain, and body, and so multiple levels of therapeutic techniques will help to get a change process happening.

Chapter 17
Reconsolidating Memory

Introduction

Recent brain research, combined with the earlier psychological models, has brought great progress in understanding of memory. We have long known that there are different kinds of memory. Short-term working memory is temporary, but once the memory is consolidated, it goes into long-term storage. Many of the memory processes take different forms, conscious or unconscious, as this chapter will describe. Clients often carry traumatic memories. What is most hopeful for therapy is that you can approach memory work using many different tools. The consolidation process can be altered, reprocessed, and then reconsolidated through therapy. This chapter shows you how to help clients work with and reconsolidate painful memories to bring about change.

Basic Forms of Memory: Declarative and Non-Declarative

Today, memory is generally considered as two separate but interacting systems, each with its own neural counterpart: declarative and non-declarative. These systems have their own unique logic, conscious recall for declarative and unconscious performance for non-declarative.

In general, a memory moves from short-term memory to long-term memory when certain genes and proteins are activated along a shared pathway between short and long-term systems. This process involves new growth of axons and dendrites at the synapse (Squire and Kandel 2003). Through this process, short-term memories are converted into long-term storage. These processes take place in many brain areas.

C. A. Simpkins and A. M. Simpkins, *Neuroscience for Clinicians*,
DOI: 10.1007/978-1-4614-4842-6_17,
© Springer Science+Business Media New York 2013

The Microscopic Picture of Memory and Learning

Learning and memory begin at the synapse with long-term potentiation: LTP. Studies of the neurons in the hippocampus (Bliss and Lemo 1973) revealed a process where axons from a presynaptic neuron bombard dendrites at a synapse for a brief but very rapid series of pulses, 100 per second for 1–4 s. The result is that some of the synapses become potentiated, which means they are more responsive to similar types of inputs and thereby become strengthened. Another way to think about it is that following the rapid stimulus, the postsynaptic neuron adjusts its weight of association through a different chemical balance. Long-term depression (LTD), is an opposite process that occurs in the hippocampus (Kerr, Huggertt, and Abraham 1994). LTD happens when axons are active at a low frequency, firing only one to four times per second. As a result, there is a prolonged decrease in response at the synapses. This pattern of potentiation and depression can last for minutes, days, or even weeks.

The neurotransmitter involved in LTP is glutamate, the excitatory neurotransmitter found all through the brain. Glutamate depends on the polarization of the cell across its membrane. When the cell is at rest, glutamate attaches to one of its receptors called NMDA, which blocks the channel with a positively charged magnesium ion, Mg^{++}. But when the cell becomes depolarized by stimulation, glutamate attaches to one of its other receptors, AMPA, and the magnesium ion moves out of the way, opening the channel to allow calcium ions, to flow in. The influx of calcium ions is important for LTP and initiates a molecular biochemical cascade that changes the transcription of proteins. This process causes more glutamate at the synapse, more glutamate receptors to be produced (both AMPA and NMDA receptors), and more dendrite branches to form. This whole process affects future responsiveness to glutamate. All of this takes place in the postsynaptic neuron, which releases a retrograde transmitter, usually nitric oxide, NO, to travel back to the presynaptic cell making it more likely to produce action potentials (Ganguly et al. 2000) and extending the axons. Thus, the process is perpetuated from both pre and postsynaptic sides.

LTP has become an appealing explanation for learning and memory at the cellular level because of three qualities it has: specificity, cooperation, and association. First, LTP tends to be specific to those synapses that have become highly active. Only these synapses become strengthened. Second, axons that are located close together tend to work together cooperatively to bring about an even stronger LTP effect than would occur from one axon alone. Third, when a weak input is paired with a strong one, the weaker response is enhanced.

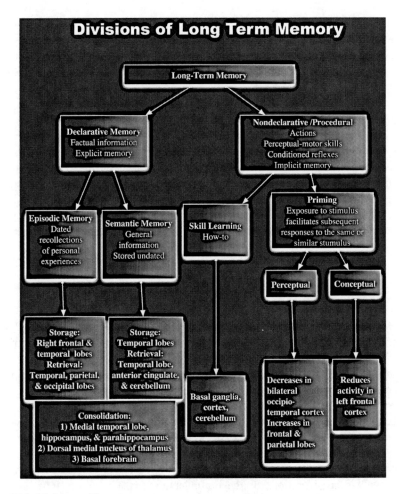

Fig. 17.1 Divisions of long-term memory

The Big Picture of Memory and Learning Processes

LTP describes the activity at a single synapse, and so it has provided a useful cellular model. But it does not give a complete description of learning and memory. Research with individuals who have suffered memory loss through lesions or damage has helped to extend the model into larger groups of brain areas and pathways that may be involved. Thanks to the findings, the earlier psychological theories have become grounded in neural structures and processes (Sternberg 1996) (Fig. 17.1).

Memory involves a number of separate and distinct structures. First, information comes into the sensory store and is processed as it is held in short-term memory. If damage occurs, these short-term memories are lost. But if the memory is consolidated, it is successfully sent on to the area of cortex where it was first processed for storage (Squire 1987). For example, experiences with a strong visual component tend to involve the visual areas of the cortex for working memory, whereas experiences of odors activate the olfactory areas (Zola-Morgan and Squire 1990).

Spatial memories for where things are and how to navigate in space involve cells known as place cells, located in the hippocampus. Thus, smelling a fresh baked pie, hearing an old song on the radio, or returning to the place where we grew up might stimulate a memory of an earlier experience. Place cells are also involved in traumatic memories, which can be evoked by the place where the negative experience occurred. Even going somewhere that shares some similarities to the original setting can set off a traumatic memory. For example, soldiers suffering from post-traumatic stress disorder (PTSD) may react to the sound of a car backfiring that brings back the traumatic experience of gunshots during battle.

Long-Term Memory Areas in Declarative Memory

Long-term memory has many structures and different brain areas associated with each. Declarative memory is responsible for encoding, consolidation, and retrieval of episodic and semantic memories.

Encoding of episodic memories tends to involve specific areas of the dorsolateral area of the pre-frontal cortex in one of the hemispheres: Verbal episodic memory occurs on the left side. Non-verbal episodic memory involves the right side. Semantic encoding occurs in the left prefrontal area for both verbal and non-verbal information (Cabeza and Nyberg 2000). Retrieval tends to involve more areas. For episodic memories, many areas of the brain are involved, such as the temporal, parietal, occipital lobes as well as the cerebellum. Retrieval for semantic memory also involves large portions of the brain including the temporal lobe, anterior cingulate, and cerebellum.

Consolidation of declarative memories involves three main areas (Zillmer et al. 2008). First, we have the areas involved in moving the information to long-term memory. H. M. is a good example of a person who suffered damage to these areas and could retain short-term memories briefly but was never able to consolidate the information from short-term memory to long-term memory. The areas that handle this transfer are the medial temporal lobes, the hippocampus and the parahippo-campus. Higher processing tends to be stored in the later evolutionary areas of cortex and the hippocampus. The hippocampus and surrounding areas are involved in encoding declarative information, complex learning, and in encoding learning consolidation for long-term store, but memories are not stored there (Cabeza and Nyberg 2000; Zola-Morgan and Squire 1990). Thus, once memories are consolidated, they become independent of the medial temporal–thalamus–basal forebrain areas.

The second area that helps in memory consolidation is a section of the thalamus, the dorsal medial nucleus. People with Korsakoff's syndrome, a frequent consequence of chronic alcoholism, suffer from both retrograde (the loss of old memories before an injury or illness) and anterograde (loss of memories for events after trauma or disease) amnesia and show damage to this area of the thalamus.

The basal forebrain is the third area involved in long-term memory consolidation. This area includes clusters of large cells that are cholinergic called the basal nucleus of Meynert. It also includes portions of the basal ganglia, nucleus accumbens, and amygdala. The basal forebrain structures actively output the neurotransmitter acetylcholine to the cortex and this cholinergic system is important for making memory structures function properly. Alzheimer patients, who suffer from a degenerative memory loss, show a marked reduction of acetylcholine in the frontal lobes, and this was correlated with a loss of cholinergic cells in the basal forebrain (Hendelman 2000).

The basal forebrain area also has strong links to the limbic system and is even considered a part of the limbic system by some researchers (Crosson 1992). Thus, we see the important link between memory and emotion reflected in the brain systems involved. The amygdala is primary for emotionally charged memories. When the amygdala was lesioned in rats, their memories involving affect were impaired. But rats showed no impairment for emotional memories when the hippocampus or the cortex dorsal to the hippocampus was cut (Kesner and Williams 1995). Humans who have damage to their amygdala cannot remember emotionally charged words (Richardson et al. 2004).

Amnesias are one side of memory. But, what about problems such as PTSD in which memories are overly vivid, continually recalled and relived with stressful, disturbing consequences? Clients who suffer from traumatic memories become caught in a feedback loop to re-experience the episode as a flashback or intrusive memory that stimulates the stress response, which then activates certain adrenal receptors that adds reinforcement to the memory (Pitman 1989). CBT, meditation, and hypnosis have all been used successfully for treatment of PTSD. Each of these forms of treatment offers unique approaches to problems. What they share in common is that they lower the stress level while fostering a calmer response.

Long-Term Memory Areas in Non-Declarative Memory

There are many varieties of non-declarative memory and learning. Each does different things. Procedural know-how is a separate process from knowing something implicitly, perhaps from past experience but not consciously being aware of it, or forming a habit. But these types of learning do share certain common qualities of not being directly, explicitly monitored by rational deliberate thought and also being different from declarative systems. Patients who had severe long-term amnesias in the declarative systems could still improve on various kinds of non-declarative learning tasks, which indicated that a different kind of memory was involved.

The brain areas involved are not completely understood, but researchers believe that these areas involve evolutionarily older brain areas such as the cerebellum and basal ganglion, both part of subcortical motor processing. For example, the circuitry for perceptual motor learning begins as information flows from the appropriate sensory cortex through the basal ganglia to the thalamus and up to the premotor cortex (Petri and Mishkin 1994).

Another distinction has been made between habit learning and declarative learning. Habit learning involves unchanging tasks that are repeated day after day. Research with rats searching for food in a radial maze compared the two types of learning (Bear et al. 1996). In the declarative learning situation, finding food involved remembering where it had been located the day before. This form of learning was disrupted by damage to the hippocampus. Thus, the hippocampus is thought to be involved in declarative learning and memory.

To test habit learning, the food was located in the same place in the maze and never changed. Damage to the hippocampus made no difference in how well the mice found the food. However, damage to the caudate nucleus did disrupt this learning. Here we see evidence for two distinct locations in the brain and two different types of learning and memory: habitual activities that we do the same way over and over versus declarative learning that occurs as we gain new information.

Emotions have a strong influence on non-declarative memory. Our evaluations of inputs, whether we like or dislike something, or whether an input is associated with a traumatic or uncomfortable past experience, will have a profound effect on how we learn and remember. Emotional learning is highly linked to the amygdala, and the response is often unconscious. One study tested the unconscious response in the amygdala to fearful and happy facial expressions by using a condition of binocular suppression. People were not consciously aware of what they saw, but the amygdala was aroused anyway, indicating an unconscious emotional response (Williams et al. 2004).

Having an emotional response can increase or decrease the strength of learning. Researchers have shown that unconscious emotions influenced how well people learned. Students were exposed to certain shapes flashed too fast for conscious recognition. Later when given a memory test, they performed better with shapes they had "seen" unconsciously. They also felt more positive about these shapes, even though they did not know why (Squire and Kandel 2000). Psychotherapy can draw upon these emotional factors to facilitate therapeutic learning. Patients can be primed to learn.

Recent research shows that non-declarative memory can be updated, literally forming new synaptic connections in the memory circuit. This process is known as reconsolidation (Ecker 2008). Therapy helps people to revise traumatic memories. In a sense, we do not just look forward to a brighter future, but we can also look backward to a brighter past. Clients can work therapeutically with a painful memory and then reconsolidate it, so that they literally form a different neural pattern in long-term storage. In this way, real change occurs. The client literally feels differently about past experiences and is able to move forward unhampered.

Reconsolidating: Updating Unconsciously

The unconscious responds through a number of naturally occurring mechanisms: priming, association, and suggestion. You can use these mechanisms therapeutically to help the client elicit change that reaches deep into the recesses of even repressed, unconsciously held memories.

Over a century ago, William James carefully defined and described associative principles. James believed that when two brain processes have been active at the same time or in immediate succession, one tends to excite the other. This is basis for the law of association (James 1896).

Association often occurs outside of deliberate attention, arising by similarity from the free flow of thoughts. Similar ideas become linked forming compounds that join to other ideas. Seemingly dissimilar ideas can end up mentally connected.

Some associative processes are influenced by learning and memory. For example, if we mention the word swallow, ornithologists will think of birds, throat specialists will think of throat diseases, and thirsty people might realize how much they want a drink of water. Associations are also influenced by how recent, vivid, or congruent the ideas are. The process of association is complex and multifaceted, and influences the direction that attention might flow.

Despite its complexities, the flow of unconscious associations is not random, but evolves out of a person's individuality. Patterns of association reflect our past, including likes and dislikes, conflicts and agreements, needs, actualities, and expectancies.

Suggestion can be defined as the natural ability to respond to an idea. A suggestion directs the attention to something and then allows the response to take place. This ability is especially helpful for therapy, and in fact, helps to explain the effectiveness of many therapeutic interventions.

Sometimes suggestions happen deliberately, as when you read the book your friend suggested. But often you respond to suggestions automatically, such as when you hear an old song on the radio and it suggests a memory of when you heard that song long ago.

Neuroscience research has shown that measurable brain changes occur when people carry out a suggestion. One study compared responding to suggestions of seeing colors on a grayscale pattern, imagining seeing a colored pattern, and actually looking at a real colored version of the pattern. When people imagine seeing color, only the right hemisphere visual areas are activated (Howard et al. 1998). But when seeing color that is suggested during hypnosis, both the left and right hemisphere color areas of the brain were activated, similar to seeing a real colored pattern. (Kosslyn et al. 2000). The brain responds to suggestions by literally seeing, hearing, or feeling as if the stimulus is actually present. Thus, working with suggestion can stimulate real responses in the brain and body.

Suggestions can be direct or indirect. They use qualities such as our tendencies to make associations, and the effects of priming. For the purposes of working with unconscious processes, indirect suggestions are often more helpful. The next two exercises will illustrate the difference between direct and indirect suggestion.

Direct Suggestion Experiment

Clasp your hands together with fingers intertwined. Sit for a moment and pay close attention to your fingers intertwined. Then tell yourself that your fingers will begin to feel stuck together. Let the stuck together feeling increase. Now try to slowly separate your fingers. Do you feel some resistance as you move your fingers apart? If so, you have responded to a direct suggestion.

Indirect Suggestion Experiment

Clasp your hands together again. Pay attention to the sensations of your hands intertwined. As you sit comfortably noticing your hands, you might enjoy feeling uncertain of which finger goes to what hand. You could wait for any interesting sensations that occur naturally such as a stuck together feeling, warmth or coolness, tightness or looseness or anything else that you spontaneously feel. When you are ready to separate your hands, let them come apart as slowly as they would like. What do you experience? Some people will feel like their hands become stuck together and that separating them is difficult. But this is just one of many possibilities. If you had any kind of experience in your hands, you have felt a response to indirect suggestion.

Reconsolidation of Declarative Memories

You can use association, priming, and suggestion to help clients work with and overcome disturbing memories. Once you have built some trust in the therapeutic relationships, clients will be able to disclose a traumatic experience from their past. At first, they might describe the painful memory rationally, keeping it at a distance and showing little emotion. Simply airing the memory can be helpful in a limited way, but more work is needed for reconsolidation and lasting change to happen. Experiencing the memory at a sensory, level can begin deeper work.

A mariner returned from battle suffering from PTSD. He was physically fit, respectful, and well mannered. He had been through a great deal of training and took pride in being part of the marines. He told us that he came to therapy because he was having nightmares and was easily startled by sounds, even a door closing. He was not sure if therapy could help him, but he wished to return to active duty if he could just get past these problems. He told us that he had always been able to withstand harsh conditions and was surprised that he was facing something that he could not overcome with self-discipline. During the first session, he described some of the battle situations he had experienced. He spoke in a crisp voice with little emotion, as if he was describing the plays of a football game.

We began by utilizing his ability to be disciplined, and taught him some yoga attention focusing exercises. It made sense to him to practice one-pointed awareness and keep his attention fixed on a single point. He told us it reminded him of target practice. We also taught him some of the meditative relaxation methods presented in this book. He practiced between sessions and seemed more relaxed and trusting in as therapy progressed. We began to explore some of the sensations he experienced as he recalled the battles he fought. Certain character-istic sensations emerged, as he got in touch with his body experiencing. He felt knots in his stomach, a strong beating in his chest, and tension in his muscles. As he followed these body sensations, he suddenly had a flood of memories come over him, emotionally recalling the deaths of comrades, the stifling smell of burning bodies and buildings, and the deafening sounds. We offered him support along with this suggestion: that something positive could be found in the battle experience. We encouraged him to allow these memories as he let himself relax enough to handle them. Then one day, as he was going over his battle memories, he recalled something else which surprised him. He remembered something that he had completely forgotten. He felt a surge of energy as he recalled when one young soldier started screaming. He vividly recalled walking over to him and placing his hand on the young man's shoulder. The gesture calmed the other soldier down, and he was able to carry out his part in the battle. He remembered feeling a flickering moment of satisfaction from having been able to help this man. Other examples of humanity amidst horror emerged, as he recollected small acts of kindness and compassion that he had either witnessed or participated in. Gradually, he began to reconsolidate his memories of the battle in a new way. He felt renewed faith in humanity as he realized that even the worst experience sometimes elicits the best in people. His nightmares subsided and he was no longer startled by sounds.

Even traumatic memories can be re-evaluated so that something new can be learned. You can suggest other possibilities, which will allow your client to rewire his neural memory pathways to include something new that helps to move forward.

When working with traumatic memories, provide a safe environment for exploration within the supportive therapeutic relationship. Gently guide the client to turn attention to body sensations and notice them as just sensations while holding the memory in mind. Encourage opening new perspectives, calling upon other brain areas that are part of the memory, perhaps from a more secure position now, or maybe, as with our client, recalling something else that modifies the memory somehow.

A one-sided perspective may be driving memories that are filled with resentment and anger. A different point of view can bring a change and even calm an over-activated limbic system reaction by eliciting a broader, more rational perspective that correlates with greater prefrontal cortex activation.

Childhood memories are encoded from the immature perspective of youth. The brain is not fully formed and memories may be biased by an earlier developmental stage. Clients become stuck in these redundant memory patterns from the past, rehashing the same experience as a child. The adult brain has more resources.

Invite your clients to revisit a troubling memory from a new, more mature perspective. Encourage reframing and opening up other alternatives. Help the client enrich the memory to include new information and the more that will allow them to outgrow the redundant pattern and form a new, more mature perspective as an adult.

Working with Non-declarative (Implicit) Memory: Priming and Seeding

The unconscious continually absorbs much more information than consciousness perceives at any given moment, as we discussed in Chap. 15 in the section about unconscious attention. More information is received through the senses and stored unconsciously, outside of awareness. Even when we do not pay attention, an input can be perceived. These processes usually do not involve intervention from conscious attention. In certain situations, we receive unattended information without registering it, especially when our attention is directed towards something else.

Priming influences non-declarative implicit memory. This process takes place unconsciously. Often, stimuli that we do not deliberately, consciously pay attention to are nonetheless unconsciously attended to. And from this unconscious attention, learning and memory can be altered, often enhanced. Therefore, you can use priming to help clients therapeutically.

Priming occurs unconsciously, without deliberate attention. For example, the subject is given an ambiguous word such as "organ" and then primed before either with the word piano or donor. Subjects performed better when their test sentence was congruent with their primed word such as "Kevin played hymns on the organ" given to those primed with "piano" and worse when the primed word was "donor." (Zeelenberg et al. 2003). This process occurs without conscious attention or even recognition that it has occurred.

There are different kinds of priming involving perceptual, conceptual, and semantic areas. Sensory-based priming involves the visual-perceptual systems. Another type of priming, called category-examplar, involves category and meaning-based processes (Roediger and McDermitt 1993).

PET images of individuals doing a task following priming revealed a reduction in activity in the visual cortex, specifically the lingual gyrus of the posterior occipital lobe (Squire and Kandel 2000). This lowered activity could be interpreted as meaning that when priming occurs, the visual system has already processed part of the work of learning; so, less of the higher processing is required. Another experiment studied the timing of lower and higher level processing. They found that priming took place faster than higher cortical processing. The researchers believe that the priming effect for visual stimuli is processed earlier in the visual pathway, utilizing the same cortical areas that process visual information (Badgaiyan and Posner 1997). Single cell recordings along with EEG showed that priming leads to more efficient processing of sensory stimuli, and that cortex activation was lowered (Schacter et al. 2007). Recent findings with visual priming are consistent. They show reduction in gamma frequency oscillations even though perception and behavior are improved. The researchers explain the effects using a network model, where priming leads to less activation needed in the lower visual areas and less resistance in the higher cortex areas for making a decision (Moldakarimov et al. 2010).

If we do not interfere consciously, we can remember even a single presentation of an item that has been received unconsciously, even after many years have passed. In 1988, Mitchell and Brown showed that normal subjects retained priming effects for one week. Eighteen years later, Mitchell and Brown performed a long-term follow-up of some of their priming subjects. They found that amazingly, these subjects retained priming 17 years later, with a significant difference between non-primed materials. These subjects also showed a significant improvement as compared to controls (Mitchell 2006).

Priming can be helpful, but it can also have a negative influence. We see this phenomenon at play when clients are having psychological problems. They often have taken in experiences unconsciously and stored them in implicit memory for many years, without having consciously noticed.

Using the mechanism of priming therapeutically, is known as seeding (Zeig 2006). The principle works similarly, that by introducing a stimulus uncon-sciously, later experience is changed.

Milton Erickson used seeding to help a client reconsolidate his memories for a better future. His client had a terrible past. He grew up with no supportive care-givers, enduring a traumatic and disturbed childhood. Using hypnosis, Erickson suggested the man recall his past, but with a difference. Erickson seeded a stimuli: that he, Erickson, was there in the client's past as a benevolent uncle. The man vividly re-experienced his past anew. Following this work, the client was trans-formed. No longer did he feel completely alienated, since now he remembered his benevolent uncle who had been there to help him through difficult times.

Indirect work elicits a response. It builds on the natural capacity to respond. Rather than trying to reason with his client that the past is over, or perhaps it was not as bad as he believed, Erickson elicited the natural response by seeding the idea that if client had a benevolent person in his life, he would feel differently. This kind of work must be individualized to fit the needs of the client in order to be most helpful.

A lonely, angry young woman recalled repeatedly how unreasonable her mother had been when she was growing up. Her mother would break into a rage with little provocation and then lock our client in her room. She suffered many long hours alone and felt enraged whenever she thought about her mother. Early in the therapy, we talked about the Dalai Lama, how he had suffered from losing his country. We discussed how he had developed a perspective of compassion for angry people, because they are suffering themselves. We can learn from them, what not to do. Then we moved on to other topics, with no further mention of the Dalai Lama. We had seeded a new way of construing her situation, but she did not make any conscious connection. Later in the therapy, she began to feel her way into her mother's perspective. She recalled for the first time how much her mother was suffering. She knew that her mother had a hard life, but she had never really felt the impact. As she empathized, she revised her memories to include some compassion for her mother. This reconsolidation of her difficult past allowed her to stop feeling so resentful. Eventually, she formed a satisfying relationship, which brought her happiness.

Changing Memories Indirectly

Memories are always recalled through the window of the present, and so, there is always a new opportunity for change. A client who recalls a stormy past might be inspired by a metaphor about the eye of a tornado, or the calm after a storm. Someone who was abused may grow from hearing a story about a real world hero who used his insights from suffering to help others who suffer. You can use metaphors, stories, and indirect suggestions that imply change. Plant seeds for a broader perspective, a different way to construe things. This kind of indirection invites the client to think about how one thing leads to another, and allow clients to make their own connections.

Positive Forgetting

There are two sides of memory as well: remembering and forgetting. We are often concerned with wanting to fill the mind, wishing to have a better memory of things and events, but we must not overlook the great value of emptying the mind, forgetting. Many who suffer from problems remember too much and too well. What is needed is positive forgetting. Our thoughts reach out to the world around us. We think about our plans, our wishes, or our worries. In order to overcome problems, forgetting may be just as important as remembering. Both are a natural and important part of the process. Each person has an optimal balance between remembering and forgetting. This exercise invites positive forgetting, a helpful resource to develop.

Close your eyes as you sit in a quiet, comfortable place, without any distur-
bances. Imagine that you are going on a journey. It could be a place you have
actually visited and enjoyed or a fantasy vacation you would like to take. Pick a
spot that is quiet and peaceful, preferably close to nature. As you allow yourself to
imagine being there, feel all the concerns back home slipping away for now. You
have plenty of time to relax and enjoy the beauty around you. You do not lack
anything or desire anything. You simply delight in the experience. Maintain the
image for several minutes. Return to this calm, desire less state whenever you need
a break.

Remembering and Forgetting

Sometimes before you can change a memory that you have to forget. We have all
had the experience of first meeting a person and having an initial impression and
then later, after knowing that person for a long time, experiencing her or him very
differently.

Think about someone you know well. Think back to yesterday, when you were
with this person. Let yourself sense how you experience her or him. Now recall
your feelings about this person last week, last month, a year ago, then several
years until you recall the first time you met this person. Let yourself re-experience
your first meeting as vividly as you would like. Now compare that first impression
with how you experience the person now. Have you taken on some limited attitudes
toward this person? Next time you are together, can you experience this person
openly, as if for the first time? Or perhaps can you reformulate your concept of this
person that look for some of the better qualities you may be ignoring? You can use
this method to step away from fixed perceptions about people or situations,
opening the way for a better experience in the future.

Building Tolerance and Acceptance

Memories of trauma and loss lead to chronic stress and an over-activated nervous
system. However, no matter how negative or traumatic the past has been, there is
always the possibility for something new in the future. Recall the Markov chain
and the random walk described in Chap. 6. The client may know where she has
been, but the next step is not entirely determined. The future can be open if we
know how to free ourselves from the past. Mindfulness meditation provides a
viable way to stay in the present moment and accept what is. Through the practice,
people learn to tolerate what has been before and accept it as it is now, freeing the
next moment to be open.

Mindful Memory Exercise: Sit comfortably and turn your attention to a memory.
Describe the memory from the present moment. Now I remember this….Now I am

recalling that… Stay with the description of the memory, but do not leave the present moment perspective. You are recalling the memory now, and so it takes on a new life in the present moment. You may find that the memory begins to alter as you express it in the context of here and now.

Conclusion

Memory, like many of the brain functions, does not have to be as fixed as we might think. You can work with memories to allow them to change in helpful ways. We have presented a few possibilities. By recognizing the potential for rewiring the memory circuits of the brain, you will undoubtedly discover methods that work for the individual client's situation and specific problem. We encourage you to use the many methods given in a later chapter of this book to help your client discover a future freed from the shackles of painful memories. For more information on working indirectly, see our book, *Neuro-hypnosis* (Simpkins and Simpkins 2010).

Chapter 18
Maximizing the Social Brain

We do not live alone. We are in continual interaction with others in the world. These interactions with others influence and are influenced by our nervous system. In fact, the nervous system is literally wired to be responsive to other people in a number of different ways. This chapter describes the built-in wiring we have that is engaged in human interaction. When you understand how these different systems function, you will be able to add the brain into your treatment methods when working with interpersonal problems, to facilitate their resolution and enhance the enjoyment of living, working, and loving together.

Hard-Wired for Relationships

We have been discussing how we are embodied beings, with our feelings, thoughts, and actions always in an intimate interaction with our brain–body reactions. But our embodied reactions expand out into the world. Our embodied responses extend to include other people as well. The idea that we are completely separate from others does not account for the wiring in the nervous system that is intimately reactive and responsive to others. Perhaps the philosophy of Martin Heidegger (1962) comes closer when he envisioned being-in-the-world as inescapable. We could not be alive without a world to be alive in. Existence is always in a world.

Recognizing Faces

Darwin (1872) first recognized that our emotional state reflects the responsiveness of our nervous system to the facial expressions of others. When we see someone

C. A. Simpkins and A. M. Simpkins, *Neuroscience for Clinicians*,
DOI: 10.1007/978-1-4614-4842-6_18,
© Springer Science+Business Media New York 2012

Fig. 18.1 Recognizing
emotions in simple line
drawings of faces

laughing, we often find ourselves laughing too. Or when we perceive a sad face, we often feel sad too. This is partly because the brain is wired to respond to facial expressions. Researchers have discovered that we have a separate area in the visual system for recognizing faces (Sergent, Ohta, MacDonald, 1992; Kanwisher, McDermott, and Chun 1997) located in the temporal lobe, known as the fusiform face area (FFA). The FFA is part of the ventral stream (the *what* pathway) of the visual system. In addition, people from different cultures all over the world can recognize a happy, sad, angry, or fearful face, and will respond to pictures of different facial expressions automatically (Ekman 2003). We can even recognize a few random lines as a happy, sad, angry, or fearful face, despite the lack of real information. This ability is built into the nervous system (Fig. 18.1).

Responding to Language

In addition, our nervous system reacts directly to a variety of social cues. Language is one of the key methods we have for communicating interpersonally. Although words convey meanings, processed by the cortex, the autonomic nervous system is also keyed to language, just from hearing speech sounds. For example, if someone is expressing anger in a foreign language, we can recognize the emotion, even if we do not speak that language. We can identify expressions of all the different emotions, happiness, sadness, and even love without having to understand the words. The nervous system reacts directly, sending feedback signals that help us to discern the emotions of others. You hear angry voice tones and the nervous system is triggered by the fear/stress pathway (See Chap. 9) to fight, flee, or freeze.

You recognize that a threat is there and feel a little unsafe and wary. Or, when a close friend looks you in the eye and speaks to you in a warm and friendly tone, your nervous system reacts, and you feel comfortable and safe.

Keyed to the Environment

The environment can also influence how your nervous system responds. A noisy setting or unknown surroundings may trigger a threat response, felt perhaps as a vague discomfort. But then, if a friend walks over to you and smiles, you find yourself feeling at ease. These are automatic responses wired into the nervous system to help us assess whether we feel safe or threatened.

Thus, many of our social experiences engage automatic body reactions. The reactions our clients are having correlate with the way that their nervous systems are responding to experiences of safety or threat. Often they are living in a world they perceive as perpetually threatening. Their fear/stress pathway is activated, keeping their sympathetic nervous system reaction constantly on alert.

The Polyvagal Theory

This model of sympathetic–parasympathetic nervous system reactions signaled by activation of the fear-stress pathway has been expanded to include the tenth cranial nerve, known as the vagal nerve. This theory helps to explain how the heart, lungs, and stomach are all interconnected in our autonomic nervous system responses. Since all of these organs are linked to the brain, therapeutic methods such meditative breathing or hypnosis that alter the heart rate and change the breathing can have a direct effect on lowering the experience of threat.

The vagal nerve has multiple parts, and so the model is called the polyvagal theory (Porges 2011). The vagus nerve plays an important part in the autonomic nervous system's responses to threat and safety. For example, the vagus nerve is involved in lowering the heart rate, as part of the parasympathetic nervous system's response that everything is safe. Or it can raise the blood pressure from the sympathetic nervous system in response to an experienced threat. The vagus nerve extends from the brainstem and then forms two branches that regulate many of our organs including the heart, lungs, and stomach. It has outputs to a number of different organs in the body and conveys sensory information about the state of these organs back to the central nervous system, known as neuroception. Thus, clients can learn to *neuroceive*, i.e., perceive their inner state and alter it.

This polyvagal system includes diverse dimensions that involve the heart, the speech centers, the breathing passages, and the movement of the face and head. We receive and send social cues through facial expressions, eye gaze, vocalizing from the speech centers, and head positioning. Through these facial and verbal centers, our social communications take place. When we look at someone's facial expression or hearing what they are saying, we get information as to whether to come closer to others or move away. This takes place through signals from the polyvagal system.

This system of reactions evolved over three evolutionary stages, which can be represented as the triune brain (See Chap. 11 on evolution). First is the oldest evolutionary phase of reflex reactions. Next is the limbic system's emotional level of responding. And the cortex, with its deliberate conscious thinking, is more recent in the evolutionary timeframe. We use all three when responding to other people and to the environment, from the thinking and feeling brain to fundamental functions such as breathing and heart rate.

Based in the vagal nerve and the autonomic nervous system, our mind–brain–body system engages in continual feedback–feedforward interaction between our internal reactions, other people, and our environment. Our nervous system, heart, and breathing are part of a network with our emotions, thoughts, and social feelings. Our treatments can target any or all of these levels, from the automatic unconscious to the deliberate, cognitive.

Reading the Signs of Problems

Psychological problems can be looked at through the lens of this mind–brain network, leading to enhanced diagnosis and treatment. People suffering from psychological problems will exhibit deficits in their nervous system responses. As a clinician, you can discern signs of these nervous system deficits just by carefully observing certain qualities in your client.

Observe how the client looks at you, at objects, and at other people. Can she look you right in the eye? Or does she tend to look down, away, or not at anything at all. Poor eye gaze is one of the signs of a nervous system disturbance.

Observe the client's level of emotion. Does he lack affect, expressing himself in a flat, non-emotional way? Or perhaps the client expresses overly strong emotion with a harsh tone of voice or elevated volume. Another clue to the nervous system network is whether the client has difficulties with his verbal communications. Does he speak quickly or does he answer your questions in monosyllables. Look beyond what the client says about what she feels to how she is expressing herself emotionally. All of these signs can indicate whether the client is feeling safe or threatened in the world. As you notice these qualities, you will have data to help guide your treatments.

Sara was a successful professional who pushed herself to succeed. But she was plagued by panic attacks, which often interfered with her work. She had done many different forms of therapy including CBT and dynamic therapy to try to overcome her anxiety, but her anxiety kept returning. She came to us for hypnosis, something she had never tried before.

She saw herself as organized and hardworking. We observed that she spoke quickly, looking past us. She had a pleasant expression, but showed very little change in her face, even as she talked about problems. She controlled her feelings by being rational and keeping her feelings at a distance. She had a great deal of insight from all her different therapies, but the insight was not helping her to overcome her anxiety.

We taught her to go into hypnosis, and she was able to feel comfortable in the office. We told her of our belief that all hypnosis is ultimately self-hypnosis since it takes place within the person experiencing it. We taught her to do hypnosis on her own, and she practiced self-hypnosis between sessions. As she became skilled in hypnosis, she began to gain confidence in her inner ability to develop her own inner calm.

In time, she had an insight. Her early years until she was 12 years old, had been spent in a communal group with a great deal of interdependence and sharing. Her memories of those times were positive. But suddenly, she realized that even though there were many benefits growing up in this protected community, she had subtly been taught to mistrust people in the outside world. As an adult living outside the community, she continually felt threatened.

Sometimes the best first step is to provide an experience of safety for your client. As Sara felt safe in hypnosis, first in the office and then at home, she was able to calm her nervous system. Then by contrast, she was able to recognize that she felt unsafe in the world, and then came the insight about her feelings of insecurity in the world. She continued to develop her own ability to regulate herself using hypnosis, and gradually the life-long anxiety pattern diminished and then went away.

Treatments such as hypnosis and meditation can calm an over-activated nervous system directly by providing an experience of safety and wellbeing. In time, the nervous system reaction alters, as the client no longer feels threatened, opening the way for better receptivity to the cognitive and dynamic methods you use for resolving problems and developing resources.

Rewiring Attachment Networks

People who have had early attachment problems often lack the ability to handle stress and problems. Instead of modeling patience, tolerance, and self-regulation, they often observed You can help them, you can put their attachment experience into the context of a larger network, including what was then with what is not, and potentials for their future. You can include their talents and capacities, those they have developed and those that are potential. Thus, in working on changing relationship patterns, you not only look forward to a brighter future, but also look back to a brighter past. The failings of caregivers, the traumas, and suffering, when viewed from a more centered present, grounded in a calmer nervous system and more self acceptance, take on a different future.

As you become sensitive to the kind of relationship problems your client may be experiencing, you will be able to offer appropriate corrective experiences. Their reactions are usually unconscious and automatic, so the best ways to open the door for overcoming them is often through bottom-up experiences. Underlying insecure or avoidant patterns is a nervous system reaction of flight or freeze. And an aggressive, angry client is showing you a fight response. People often feel their inability to successfully regulate their discomfort and feelings of threat, and on some level, they know that they are lacking something. Most likely, they may never have been taught these skills. You can begin by providing a safe and nurturing environment in the session, and then teach skills to help clients regulate their nervous system responses, opening the way for the healing process to begin.

Developing the Ability to Relax

These exercises develop skills in relaxation, first of the body, then of the mind. The first few exercises are designed to help relax the tensions that facilitate nervous system regulation and healthier attachment styles. Take frequent short breaks to practice these exercises. Meditate intermittently every day for a stronger effect.

Lie on your back, with knees drawn up and feet flat on the floor. Some may want to add a small, thin cushion under the lower back to relieve tension. This position offers comfort to the back muscles, which often become tense when the body is stressed. A variation of this position is to lie on your side on the floor or on a soft rug, with your head on a pillow, legs bent, and knees pulled in. Again, a thin pillow under the hips or shoulders may help align the body comfortably. Close your eyes and allow your breathing to be relaxed. Imagine, with every breath out, that tensions in your muscles ease. If any muscle groups seem tighter, try to let them go using this breathing. Rest comfortably for 5 min or so, breathing out tightness. When ready, sit up and stretch.

Mind Relaxation

Sometimes a brief moment of relief from stress can help to ease the symptoms and allow the nervous system activations to lower. Training makes such an interlude possible. Skilled meditators can create a peaceful moment no matter what mood they had before meditating (Kohr 1977).

Evoke the moment of relief by filling your mind with a calming symbol from nature. We offer this picture, (Fig. 18.2) but you may find a flower, a tree, your pet, or even the clouds in the sky to inspire you. Focus your attention on this and think about it. You might sense beauty, playfulness, or other qualities. Let your thoughts roam around this symbol, enjoying it for a moment. Do this regularly for a few minutes at different times spaced out through your day.

Fig. 18.2 Nature

Experiencing Attachment Styles Bottom-Up

Neuroscience shows us how we are not separate from the world; we are always part of it, interacting and responding. People become embroiled in relationship struggles, for dominance, for rightness, for winning, for escaping. The kinds of disputes they find themselves in often reflect the way they form attachments. Through the back and forth arguing, a healthy, respectful, and sensitive communication is lost. When therapy gets stuck trying to help people resolve their disputes, real or imagined, you may have more success addressing the deeper attachment issue. You can work bottom-up, to help your client become grounded in the ongoing experience of relating itself.

The next two exercises are drawn from an interactive style of martial arts known as Wing Chun, originally founded by a woman named Yim Wing Chun. Chi-sao, as this practice is called, offers an alternative way to experience relationship by attuning to the other's movements. This exercise gives a definite experience of staying in tune with an ongoing relationship as it is happening. It can be a metaphor for sensitively interacting with others. Through chi-sao practice, you can develop a way of perceiving as if listening with the skin. Each person senses the other in a new way. Partners often find that they can stay attuned without contending. This experience opens a new potential for harmonious

Fig. 18.3 Chi Sao

interaction. Following carefully, without changing the course of things, permits healthy relating in a bottom-up, automatic way. The effects can ripple through to higher brain functioning as therapy helps the client to make sense of his or her reactions.

Chi-Sao Exercise

Therapist and client can perform this exercise, if the rules of your practice permit, or alternatively, while observing, with the client and his or her significant other. Take turns guiding and sensitively following (Fig. 18.3).

Stand a few feet apart, facing each other. The partner raises one arm, bent at the elbow, extending the hand forward. Raise and place your hand lightly over the wrist of the other and close the eyes. The partner moves his or her semi-relaxed arm around slowly, extending back and forth, up and down. Stay with the partner's hand, lightly moving in unison. Sense the force of movement without adding any force of your own.

As you follow your partner, notice your instinctive reactions. Do you tend to want to push back? Do you try to take the lead? Or do you find yourself wanting to withdraw from contact? These natural tendencies can teach you about your own attachment style. Observe, but try to return to doing the exercise, just following and staying with the other sensitively.

After several minutes, switch roles. You lead, and the partner follows. When awareness has been correctly focused, the arms may feel tingling from the raised chi energy in that area Repeat the exercise with the other hand as well. Both people should remain as relaxed as possible.

Cognitive Chi-Sao

Sensing the other person's communications, much like in the movement chi-sao just practiced, listen carefully to what the other person is saying. Stay with his or her logic, thoughts, and feelings. But also maintain self-awareness of feelings and thoughts without letting your own perspective engulf you. Shift back to the other.

Sense the other's point of view and infer the likely perceptions; as real to the others as yours are to you. Often when the other person is truly experienced, not just listened to, he or she will be more open as well. Communication can flow, back and forth in harmonious chi-sao.

You can learn about your own attachment style by noticing your tendencies. Do you find yourself wanting to interrupt and contend? Do you feel compelled to pull away and say nothing? Notice your tendencies, but keep trying to stay attuned to listening and then changing roles to be the one who expresses and is listened to. Differences may resolve naturally, by working together to restore balance for both.

Fostering Loving Connections

Living with relationship problems may interfere with feelings of love and benevolence. But as we have seen, love is a wired-in need and a vital component for a healthy life. Changing relationship patterns involves the ability to develop respect and valuing of oneself. Then people can more easily reach out to another to seek, accept, and give love. We can foster our loving nature to encourage the brain to wire the neurons as nature intended. Attune to built-in motivations toward a natural and healthy life. Foster the healthy, natural instincts and let nature take its course. This meditation can stimulate the natural wiring for love and enhance the potentials for forming loving bonds.

Sit quietly. Then, think of someone you love, your spouse, parent, child, or friend. For those who do not have a specific person come to mind, allow a general feeling of benevolence to develop. Allow your positive feelings of love, warmth, kindness, or caring for that person fill your thoughts. Extend that feeling out to people you know and like. Then reach further out, to people of your community, and then to all humanity. Feelings of loving kindness directed outward to the world also reach inward through the whole being.

The Mirror Neuron System

There is another system located in the brain that is also wired to respond to others.

The brain contains a system of neurons known as mirror neurons that respond directly to the intentional actions and emotions of others. The mirror neuron system makes it possible for us to know the actions of others just as we know our own actions. Traditionally, the human ability to understand other people has been thought of as a conceptual understanding. We believe that other people think and feel, as we do. As explained in Chap. 3, we all have a Theory of Mind (ToM) and believe others do too. Our understanding of others engages our thinking brain, the higher level processing in the cortex. These processes are top-down: So, we think our way into empathizing with someone else by imagining what it would be like to walk in their shoes. But then, we are left with this dilemma: How do we know that our ideas and concepts about another person's mind are true for that other person? For example, how do we know that two people see the same blue when they both look at the sky? How can we be sure that what we imagine someone is feeling is really what he or she feels?

Mirror neurons provide a possible resolution for this problem (Gallese 2009). The sense of another person does not start from an intellectual understanding that assumes another person is feeling or acting as we would. Our neurons fire as if we were literally doing what we observe the other person doing. We resonate within, with the firing of our mirror neurons, giving us a literal felt sense within of what the other person is doing. So, in this way our feelings and thoughts about others are embodied, just as our own feelings and thoughts are embodied. We begin with neuronal activation in our own brain that imitates or mirrors the action of another. Embodied cognition begins with our own brain responding.

According to mirror neuron models, this system of neurons in our own brain is the basis for our understanding of others in many different ways (Gallese, Keysers, and Rizzolatti 2004). Through the activation of the mirror neuron system, we have the capacity to literally feel what and how others think, feel, and experience. Mirror neurons may even be the basis for language. And there is mounting evidence that mirror neurons form a basis for empathic experiencing and social cognition in general.

How Mirror Neurons were Discovered: Monkey Studies

As with many of the world's great discoveries, mirror neurons were first discovered by accident. Researchers in Italy were studying the premotor cortex of macaque monkeys (Gallese, Fadiga, Fogassi, and Rizzolatti 1996). The experiment was originally designed to monitor the monkey's planned and carried out movements by attaching a sensor to a single neuron in the monkey's ventral premotor cortex. While the researchers were on lunch break, a graduate student returned to the lab eating an ice cream cone. The monkey's neuron began firing just as it did when the monkey moved food to its own mouth. Even though the monkey was simply watching the student eat the ice cream cone, his brain was reacting as if he were eating! This exciting finding inspired a flurry of studies to explore this strange new phenomenon.

The fundamental understanding that came out of this discovery was that a certain group of neurons found in the monkey's ventral premotor cortex activate either when a monkey makes a certain movement or when it observes another monkey or even a human making a similar movement. Thus, observing an action that others perform leads to the same neurons firing as when executing a similar type of action oneself (Gallese, Fadiga, Fogassi, and Rizzolatti 1996). The mirror neurons do not just respond to any movement, they respond to the intentional actions of others. So, even if the monkey is interested in a moving object, the

Fig. 18.4 Human mirror
neuron system

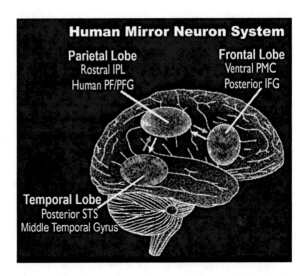

mirror neurons will not fire. They will only activate from observing human or animal movement. This distinction adds strong evidence that the mirror system has an interpersonal element to it.

Further mirror neuron studies in monkeys found neuron systems in areas for vision, sound, and touch as well as motor functions (Gallese, Keysers, and Rizzolatti 2004). The early work stimulated an entire new field of study that has the promise of answering some old questions about our social being.

Human Mirror Neuron Systems

Human beings also have mirror neuron systems. The human mirror neurons are found in the frontal, parietal, and temporal lobes. In a broad sense, mirror neurons are involved in seeing, sensing, and participating in certain types of motion with a purpose.

Mirror neurons have been found in humans that correspond to those found in monkeys, F5 for motor cortex and PF for parietal area (Buccino et al. 2001). Just like in the monkeys, what distinguishes mirror neurons from other neurons in humans is that mirror neurons fire both when executing a movement oneself and when observing that motion in another (Fig. 18.4).

The frontal area involved in mirroring is found in the premotor cortex. We would expect to find a motor area activated, since mirroring involves observing or participating in a movement. The second mirror area that is consistently found corresponds to the monkey area PF. In humans, this area is located in the inferior parietal lobe (IPL). The parietal lobe is involved in sensory processing, and thus it is activated during hand grasping, specifically in two sub-areas of the inferior parietal cortex, PF and PFG. Many of the first mirror neuron studies involved grasping of an object, and so the activation of this area is not surprising.

The mirror neuron system in humans is more widespread than in monkeys. A third area is also activated, found in two sections located close together: the middle temporal gyrus and the superior temporal sulcus (STS). These areas take part in motion perception and vision.

How We Measure Mirror Neurons in Humans

We can measure a single neuron in a monkey's brain by penetrating the skull to insert an electrode. Such invasive measurements are not possible with human beings. Therefore, human mirror neurons are measured indirectly, using instruments that capture brain wave patterns such as EEG, or using imaging methods such as fMRI, MEG, and TMS.

EEG reveals two types of rhythms when the person is resting: alpha rhythms (8–13 cycles per second c/s) and mu rhythms, also 8–13 c/s. These two sets of rhythms differ in how they react to stimuli: alpha rhythms disappear when the visual and sensory systems are activated by sensory stimuli. The mu rhythms disappear when the individual makes a motor movement or observes one in another person (Chatrian 1976). Forerunners in this field found that when humans observed an action made by another person, mu rhythms desynchronized, meaning they disappeared (Gastaut and Bert 1954; Cohen-Seat, Gasaut, Faure, and Heuyer 1954). These findings were replicated in later studies (Cochin, Barthelemy, Lejeune, Roux, and Maritneau 1998). Cochin also tried presenting non-human movement, such as waterfalls, for example. He found that the mu rhythms did not desynchronize when people watched nonhuman movements. Thus, similar to the monkeys, human mirror neurons seemed to have an interpersonal focus of activation (Cochin et al. 1999).

Other forms of measurement were used, showing a correspondence between an observed action and an executed action with MEG (Salenius et al. 1997) and TMS (Fadiga, Craighero, Buccino, and Rixxolatti 2002).

More recently, fMRI imaging has been used to locate specific brain areas that became active when an individual observes the actions made by another person. Researchers found that when subjects observed object-related actions in another, an activation that was somatotopical organized, similar to the classical motor cortex homunculus. Thus, when people watched an action in someone else, an inner replica of that action was generated in their premotor cortex. When the action was related to using an object, the action was reflected further in the parietal lobe, just like when the subjects themselves used these objects (Buccino et al 2001). Combined together, many studies give compelling evidence for the existence of mirror neurons located in the motor cortex related to hand movements and located in the parietal lobes when people observe object-directed actions (Fogassi et al. 2005).

New studies found that mouth-related activity, such as watching someone eat, hand movements such as reaching for an object, and foot movements such as kicking a ball each have separate mirror neuron locations. There are a large number of frontal and parietal areas with mirror properties (Rizzolatti and Craighero, and Fadiga 2002). Simply hearing the sounds that are typically part of a particular goal-oriented or intentional action led to firing of the mirror neuron system in monkeys. The audiovisual mirror neurons respond whether the actions are performed, heard, or seen (Kohler et al 2002). Thus, humans may have a more diversified mirror neuron system than was previously thought (Rizzolatti, Craighero, and Fadiga 2002).

Mirror Neurons Respond to Intention

Researchers found that the mirror neurons in monkeys fired more vigorously when the action of the experimenter had a clear goal or intention, such as grasping food to eat it as opposed to simply grasping a neutral object. This extra activity in the monkey's parietal lobe could help to explain a basic aspect of intention (Fogassi et al. 2005).

The premotor cortex mirror system in humans is also involved in more than just the recognition of actions in general; it is also involved with understanding another person's intention for the action. Humans' mirror neurons display the ability to distinguish between intentional and non-intentional actions. Mirror neuron areas of humans' fire more strongly in situations with intention, adding evidence that mirror neurons seem to be capable of understanding the intention of other people. In context-related cases with intention, fMRI indicates significantly more activity in the posterior part of the inferior frontal gyrus, as well as the closest section of the ventral premotor cortex (Iacoboni et al. 2005). Thus, action and understanding seemed to be linked with each other when observing a meaningful goal-directed action of another. Studies suggest that the way we reflectively understand complex cognitive functions such as intention may be grounded in the mirror system (Wohlschlager and Bekkering 2002).

At the neuronal level, we see even more fine-grained distinctions being made. Mirror neurons are sensitive to both the action itself and to the goal or intention of the action. Some mirror neurons are broadly congruent, which means they fire when observing and performing a goal directed motion no matter how the goal is achieved, such as reaching for a cup by moving straight in or around to grasp it. Other mirror neurons are strictly congruent: they only respond to the exact same goal directed movements such as just reaching straight in (Gallese, Fadiga, Fagassi, and Rizzolati 1996).

Mirror neurons seem to be even capable of inferring intention. For example, the mirror neurons of monkey's fire even when an object is obscured from view (Umilta et al. 2001) indicating that the intention can be understood even if it is not directly seen.

These many studies point toward how we come to understand other people, beginning with personal firing of our own neurons. Our understanding of others is not abstract or a conceptual grasp. The grasp begins from within, in direct resonance between self and other.

From Embodied Cognition to Social Understanding

Embodied cognition assumes that we know what we feel because of the sensations and brain states in our own body. High level processing always involves reactivation of interconnected states of the body that are intimately involved in our symbol making process. Embodied cognition theories begin at the level of the body (Winkelman, Niedenthal, and Oberman 2009). This view works with the older symbolic interactionist view (Blumer 1969) that higher level processing interprets sensing, and represents it as feelings and thoughts. Because of embodied cognition, social functioning arises from our ability to construct an embodied simulation of others, an empathic felt-sense. The two approaches work together. Intellectual construction is part of symbolic interaction, giving form to the interpretive processes.

Mirror neurons may be the foundational building block of social cognition. Pineda proposes that mirroring is involved at every level of social functioning, from the simplest priming to a sophisticated experience of ToM involving cognitive and emotional understanding. Mirroring processes engage a re-representation of information in our own brain that feeds back into other systems, from one modality to another, in a complex, interrelated matrix across all the different neural systems. In this way, mirroring is fundamental to every level of processing, and especially important for social cognition (Pineda et al. 2009).

Empathy and Emotions

The mirror neuron theory has been expanded more recently to include emotions. For example, observing the faces of people feeling a particular emotion has been researched. Mirror neuron areas along with the anterior insula and also (to a lesser degree) areas of the anterior cingulate cortex were activated when people viewed faces showing disgust. (Wicker et al. 2003). Mirror neuron systems have been found to be involved when subjects performed and observed emotional facial expressions for other emotions as well. When we perceive faces exhibiting emotion, the inferior frontal cortex, superior temporal cortex, insula and amygdala are all activated. Some of these areas overlap with mirror neuron areas, (Carr, Iacoboni, Dubeau, Mazziotta, and Lenzi 2003). And so, mirror neurons may be involved when we interpret other people's facial expressions of emotions. (Oberman and Ramachandran 2009).

Even the emotional centers of the brain have mirror-like qualities. We seem to be able to empathize with specific emotions as well as pain experience of others, especially when someone we love feels pain (Singer et al. 2004). Emotional responses of others may actually be *felt* in our own parallel brain area, which is associated with that feeling. So, when the therapist empathizes with the client, it is possible that the same brain areas of the therapist are being activated, thereby allowing the therapist to be able to understand the client fully and deeply.

Thus, we can see why mirror neurons have been of great interest to the helping professions as holding a key to the neurological component of empathy. Edward Tichener (1867–1927) who firmly believed in introspection is often credited with coining the word *empathy*. Empathy comes from the German word *einfuhlung*, which translates as *feeling into*. Tichener (1909) believed that we could not understand other people by simply using our reasoning powers. Empathy requires putting ourselves in the other person's shoes, to personally feel our way into what the other might be experiencing. Now, with mirror neurons lending evidence that we do just that when we observe other people's actions, we can answer the question of whether other people feel as we do. Mirror neurons tell us that as others feel, we feel. Empathy as Tichener first proposed it may now have a scientific basis.

Mirror Neurons and Language

The implications of mirror neurons move us from the realm of meaningful actions to meaning and language itself. Mirror systems are located close to Broca's area, the part of the brain involved when in speaking language located at the base of the primary motor cortex in the left hemisphere. Rizzolatti proposed early on in his work with mirror neurons that because there is a mirror neuron system for gesture recognition in humans, and also since it includes Broca's area, we may have a bridge from moving to communicating (Rizzolatti and Arbib 1998).

The motor theory of speech perception (Liberman and Mattingly 1985) proposes that words and sounds are literally gestures and actions of the mouth that act as signals. Language is embodied in the movement of our mouth and tongue to form the sounds. Thus, in a sense, language is an intentional action, movement of the mouth with the intention to communicate meanings.

This idea has inspired others to investigate ways that the mirror neuron system could form a neuronal basis for language functions, working memory capacity, and higher cognitive functions. Mirror neurons fire more when people listen to the production of sounds that require complex movements of the tongue. But they do not fire when we hear sounds where less tongue movement is needed (Fadiga, Craighero, Buccino, and Rizzolatti 2002).

Physical action is involved in both producing speech and perceiving speech. A motor movement of the mouth is part of speech production; since similar mirror neuron areas fire when both seeing and hearing other people speak (Calvert and

Campbell 2003). Thus, mirror neurons, which activate when we hear and observe speech, may add to our understanding of what other people say (Oberman and Ramachandran 2009). Ramachandran also proposed evolutionary stages that led to the language skills we have. These stages begin with basic mirror neuron understanding of grasping actions seen in primates, and evolve through imitation, elaboration, simple sign-making, to conventional gestures, and finally result in the cultural evolution of language itself (Ramachandran 2000).

Autism as a Deficit in Mirror Neuron Firing

Autism may involve a deficit in the ability to empathize in the embodied way through the mirror neuron system. The exact point of disturbance, whether it is at the level of mirroring, imitation, or social understanding is being explored (Oberman, Ramachandran, and Pineda 2008).

One research group has found evidence that people with autism spectrum disorder (ASD) have trouble spontaneously imitating, whether they have an emotional or non-emotional stimuli (Hamilton, Brindley, and Firth 2007). Some have proposed that what may be causing the problem in imitation is a mirror neuron dysfunction. A recent study used fMRI with children who were diagnosed with autism and compared them to matched normal children of similar age and IQ. They imitated and observed facial expressions of a variety of emotions including anger, fear, happiness, sadness, and neutrality while in the fMRI machine. The control group exhibited similar mirror neuron activity to normal adults, but the autistic children had little to no activation in typical mirror neuron areas (Dappreto et al. 2006). These and other studies have led to a Mirror Neuron Theory of Autism. This theory proposes that autistic individuals do not have the typical observation and execution matching system that mirror neurons produce. As a result, they do not get the immediate, direct experiencing of others. This leads to deficits in empathy, ToM development, and the ability to imitate others. As a result, autistic children do not develop normal social skills (Bernier and Dawson 2009).

Suppression of mu rhythms correlates with mirror neuron activity. Pineda and his research group used neurofeedback to teach autistic children to have more normal mu rhythm activity (Oberman, Ramachandran, and Pineda 2008). Pineda explained the hypothesis: "This idea implies that retraining mirror neurons to respond appropriately to stimuli and integrate normally into wider circuits may reduce the social symptoms of autism" (APA 2007). This preliminary research has been able to produce temporary changes. The results are promising and offer new directions for treatments.

Encouraging Mirror Neurons to Fire

Mirror neurons fire as a response to the intentional actions of others. Learning to attune to intentional action can enhance mirror neuron activation. Sometimes we are aware of our intentions and at other times intentions are hidden. Noticing the small intentions involved in an everyday task can enhance awareness of the intentional nature of actions. By deliberately practicing these skills, you can enhance your understanding of your actions and the actions of others, a key component in empathy. The exercises that follow can be a non-threatening way to help couples or families who are angry and out of touch to begin developing empathic understanding.

Enhancing Awareness of Personal Intention

Pick a small task that needs to be done and follow each intention while doing it. This exercise could be considered as to an algorithm that maps out every step in a computer program. So, perhaps you have chosen to follow your intention as you get up and go to work in the morning. The algorithm might go as follows: I hear the alarm and register that I intend to get up to go to work, but I feel a second intention pull on me to get a few more minutes of sleep. And so, I close my eyes as I reach over to the alarm and press the snooze button so that I can fulfill both intentions of getting more sleep while still awakening soon. I hear the alarm again and this time I follow my intention to awaken by raising my body. I intend to get out of bed, place my feet on the floor, and stand up. Now I feel the urge to use the bathroom and my intention to go there. Follow each detail of intentional action through washing, dressing, eating breakfast, or whatever you do. This is just one everyday example. Please pick one that you find interesting to follow.

Inferring Another Person's Intentions

This exercise is similar to the one above; only follow the intentions of another person. Enlist the help of someone to perform a small task, and verbalize each intention as you observe it. So, for example, if your partner is getting up in the morning, verbalize: "You intend to get up but want to sleep, you intend to wake up as well, and so you press the snooze button on the alarm," etc. Following the exercise, get feedback from the other as to whether you captured the intentions accurately. For angry couples, this exercise should first be performed during a session so that the therapist can make sure the exchanges are accurate and objective. The describer must be descriptive and not comment on the observed intention, nor personalize it in any way not intended.

Shared Intention: Meditating as a Couple or a Family

People will enhance their empathy by participating in experiences together. With shared intentions, mirror neurons resonate together. Meditation can be an excellent way to break down barriers and build positive mutual experiences. We have often had families and couples take time each day to share in meditation for several minutes or longer. People find the experience interesting and enjoyable. It initiates a shared intention to be conflict free, even if only for a few minutes. It also forms a basis for better times together, something to look forward to, and solid basis for a positive future.

Pick a meditation method for the family or couple to perform together. For those who have never meditated, use one of the early meditations such as, think of a color or attention to breathing. For more advanced meditators, use clearing the mind or mindful in the moment, or focus on a symbol or sound. Ask everyone to find a comfortable position, either on the couch, chairs, or floor. Sit together and invite the group to perform the meditation for several minutes. When finished, ask them to stretch a bit if they would like. People often smile and feel refreshed. Share what the color was or what the experience was like for each participant.

Conclusion: Implications for Treatments

Usually, psychotherapy works from the top-down, helping people to conceptualize how the other might be feeling, and thereby extending emotional understanding. But the neuroscience implications of interpersonal relationships points to bottom-up approaches as being helpful. We can work directly with the autonomic nervous system and vagus nerve. And we can enhance empathy and even enhance personal emotional understanding by activating the mirror neurons. The treatment chapters provide methods for working directly with these neurobiological reactions, thereby altering experiencing on a deep level, to provide the basis for successful therapeutic work.

Part VI
Shifting the Nervous System in Common Disorders

Today there is mounting evidence that psychotherapy is not only effective, but that it literally alters the brain and nervous system. We have long known that therapy changes attitudes, thoughts, and behaviors. Now we also know therapy influences the brain systems, activating some areas and deactivating others, which may be similar to or different from how pharmacological treatments work. Clinical acumen would predict that different types of therapy tend to bring about different effects. Certain methods seem to work better than others for specific problems, and sometimes one person will respond better to a particular treatment than another person. One might wonder, why does this occur? The neuroscience answer is that unique patterns of brain activation are involved.

Each form of therapy has distinctive effects on the brain. The many approaches to psychotherapy offer different options for brain changes, as Chap. 19 will show. All of these changes in the brain elicited from different treatments demonstrate the relationship between structures, functional systems of pathways, and experiences. As you learn to work with these different effects as needed, your clients will be able to make changes more quickly and easily, with the sense that change takes place naturally, of itself. We have often had clients tell us that they are surprised to find the problem is no longer bothering them and that they feel that they have inner resources they never knew were there.

Therapists are usually pragmatic in their approach to helping. Although they have trained in particular methods, most are open to new techniques to help. Activating the brain deliberately in targeted ways offers great potential for change. Since different forms of therapy affect the brain in specific ways, an educated application of methods to activate the brain can have lasting benefits.

Chapters 20–22 will describe how the brain is altered from different psychological problems and how to stimulate brain changes that will help clients overcome their problems. Chapter 20 gives methods for changing clients who suffer from problems with anxiety and stress, using this nervous system paradigm. Chapter 21 offers ways to work with depression and bipolar disorder. Chapter 22 helps substance-abusing clients to free their pathways that have become compromised by frequent drug use, to make it easier for them to live a happy, drug free life.

Psychological problems are not in a simple one-to-one relationship with the brain. The interactions are more multidimensional. So, here you will find methods that initiate brain change, sometimes working top down (consciously), and other times working bottom-up (unconsciously). In addition, non-specific factors and individual differences should not be ignored or deleted. Sensitive application of techniques that take these and other factors into consideration are also included to deepen the methods. And case examples are given to show how these techniques have been applied successfully. Using both conscious and unconscious brain functions, specific and non-specific pathways, along with global effects and individual differences, your methods will bring the most powerful and lasting results for brain and mind change.

Chapter 19
How Psychotherapy Changes the Brain

We have long known that psychotherapy alters how people think and feel. But it is only more recently that we have come to realize that therapy also has measurable effects on the brain. From the interventions we make with our clients, some activations and deactivations are altered, shifting the balance in structures, functions, and pathways.

Each approach to psychotherapy offers different options for brain change. Thus, by applying several of these different options, problems can be approached from more than one angle. Combining different therapeutic methods together may be most effective. In order to best utilize a variety of therapeutic methods, you will find it helpful to know which brain structures and functions are changed by each approach.

This chapter will discuss some of the frequently used forms of therapy in terms of their unique brain changes. We also provide a few approaches that may be less common, but have a particularly powerful effect on the brain. You probably already use some of these methods, and so you will gain insight for you to choose the right method to stimulate brain change that best meets the needs of each individual client and problem. And we hope that you will consider adding treatments such as meditation, hypnosis, and body work. We cover traditional methods of cognitive therapy, behavior therapy, and psychodynamic therapy and methods that are more recently being applied to therapy: meditation, yoga, and hypnosis.

Cognitive Therapies (CBT and RET)

New evidence is emerging to show that clients who undergo successful cognitive therapy will have measurable changes in various brain structures and functions. Many studies have also observed how cognitive therapy effects the brain activations and patterns that help correct brain imbalances found in people with different

C. A. Simpkins and A. M. Simpkins, *Neuroscience for Clinicians,*
DOI: 10.1007/978-1-4614-4842-6_19,
© Springer Science+Business Media New York 2013

Fig. 19.1 Cognitive
behavioral therapy

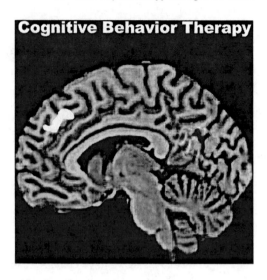

psychologic problems. We offer a few exemplary studies, but there are many others (Fig. 19.1).

Increase of Gray Matter

Cognitive therapies such as CBT (cognitive behavior therapy) and RET (rational emotive therapy) assumes that the way people think about themselves and their problems will have a profound influence on what they feel and do. But cognitive therapy does more than just alter cognitions, emotions, and behavior. It also changes the brain.

A recent study found that in addition to the experiential changes, cognitive techniques literally increased the volume of key areas in the brain in 22 chronic fatigue syndrome (CFS) sufferers. Subjects were given cognitive therapy along with a rehabilitative approach of gradually increased physical activity. CBT addressed their faulty thoughts and beliefs about CFS, a difficult problem involving persistent fatigue and a decrease in cortical gray matter volume. The subjects not only felt better, but they also showed a significant increase in gray matter volume localized in the lateral prefrontal cortex. This brain area is important for the speed of cognitive processing (de Lange et al. 2008). This study reveals how a psychologic intervention can bring about a measurable change in the brain's anatomy.

Balancing the Limbic System with Frontal Lobes

We all know that cognitive therapies help people to make reappraisals and reinterpretations. The methods involve challenging irrational assumptions and reframing the situation in more reasonable, neutral terms. This process of reappraisal leads to a rebalancing in the activation and deactivation in the brain. The activations in the emotional limbic areas are decreased while the activation of the executive frontal areas is increased. This change in balance between the frontal lobes and limbic system correlates with better regulation of emotions.

The neurobiology of reappraisal involves the way that the frontal cortical regions modulate how the amygdala reacts. One study measured people performing reappraisal tasks and found that activity in specific areas in the anterior cingulate and frontal lobes increased while activity in the amygdala decreased (Banks et al. 2007). This study shows how emotional areas of the brain interact with cognitive brain areas: The two are intimately linked.

A recent study compared people diagnosed with major depressive disorder to people who had never been depressed. They found that the more severe the depression, the less capable the subjects were in regulating their emotions when viewing emotionally arousing pictures. fMRI data was gathered using whole-brain analysis and connectivity measures. What they observed was that the depressed subjects were not able to lower the activation in their amygdala. They also showed diminished activation in the prefrontal cortex along with lower coupling between these frontal areas and the limbic areas as compared to the non-depressed controls. In addition, people with more severe depression showed less emotional regulation, and an exaggeration of the higher activations in the limbic region and lower activation in the frontal areas (Erik et al. 2010).

Lowering the activations in the limbic system using CBT helps people with phobia. This deactivation allows them to literally feel less fear as they adjust their cognitions to help them overcome their fear (Linden 2006).

CBT Alters the Basal Ganglia in OCD

Cognitive therapy also influences the motor areas of the brain in the basal ganglia. People who suffer from obsessive–compulsive disorder (OCD) have over-activation in a part of the basal ganglia, the caudate nucleus. Recall that the basal ganglia are located deep inside the brain, regulate motor movement, and are part of the dopamine pathway. OCD sufferers have hyperactive neurotransmitter circuits between the cortex, basal ganglia, and thalamus, which may explain the obsessive worrying and difficulty stopping habitual behaviors (See Chap. 20). Linden (2006) measured changes that occurred in OCD patients before and after cognitive therapy. He found that following treatment, obsessive–compulsive subjects had decreased metabolism in the right caudate nucleus.

Another study of the brain activations of OCD patients following a cognitive therapy that used exposure and response prevention found a significant decrease in the thalamic metabolism along with a significant increase in the right side of the anterior cingulate cortex. The anterior cingulate is involved in reappraisal. It also helps in the regulation of negative emotions and is more active when people are feeling optimistic (Saxena et al. 2009). As the balance shifted to more activation in the anterior cingulate cortex, these patients felt improvement in their OCD symptoms.

All of these different studies show that CBT does not just correct a problem cognitively, it can also alter the structure of the brain. So, you can begin devising strategies for calming the limbic system or activating the areas that help to regulate affect in the cingulate gyrus. Or you might need to help ease activation in the basal ganglia to help clients discontinue harmful habits and rituals.

Behavior Therapy (BT)

Most therapists are familiar with the classical learning theories that form the basis for behavior therapies. Classical Conditioning of Pavlov (1856–1936), the Law of Effect of E. L. Thorndike (1874–1949), and Operant Conditioning of B. F. Skinner (1904–1990) all highlight the interesting ways that behaviors can be learned, maintained, and influenced. Neuroscience helps to explain how associative pairing and reinforcement can have such a strong influence on behavior.

Associative Pairing

Associative pairing helps us understand how a fear is acquired, as a learned pairing of stimuli. Pavlov observed that presentation of food to dogs, which is an unconditioned stimulus (US), would lead to an automatic unconditioned response (UR) of salivation. When he paired the presentation of food (US) with an unrelated stimulus, the sound of a bell (CS), the two stimuli became associated, for the dog. After many repeated pairings of the US and CS, the dog would salivate soon after it heard the sound of the bell alone. These experiments led to Watson and Rayner's (1929) famous experiment in which they conditioned fear in the infant Little Albert. They were able to show that fear could be learned by mildly scaring the infant with a loud sound while playing with a small white rat. The infant's learned response was to exhibit fear and to cry when the rat was presented without any sound. The response soon generalized to other animals, furry coats, and even Watson with a Santa Claus mask on (Fig. 19.2).

The link is made in the amygdala LeDoux (2003). A conditioned stimulus (CS) (for example when a soldier hears gunshots and explosions in the midst of a battle) comes in through the thalamus and then goes directly to the amygdala where it gets

Fig. 19.2 Pavolian
Conditioning

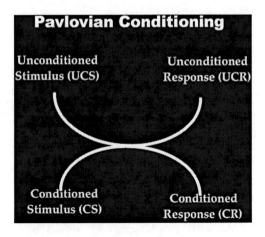

stored, as does the unconditioned stimulus (US) of fear and/or high emotional arousal. After pairing of the US and CS, the cells get triggered to respond to the CS from the amygdala. The conditioned response (CR) might be triggered when the soldier has returned home, from the benign sound of a car backfiring, for example. The CR is the result of a learning process that depends on the amygdala. Evidence suggests that synaptic plasticity within the lateral nucleus of the amygdala is responsible for storing memories of the CS–US association. Once triggered, a signal goes directly to the hypothalamus, activating the fear/stress pathway that raises blood pressure and hormone levels (Medina et al. 2002). Thus, the stress reaction is paired with the traumatic event as well. We often see the process work in reverse when an unrelated stressful event in the client's life can trigger the traumatic memory.

Both implicit and explicit memory are involved. Recall that implicit memory tends to be nonverbal, automatic, and unconscious, whereas explicit memory includes conscious memories that can be accessed and talked about. So, when clients say that they do not remember the specific details of the traumatic incident, the memory is more likely to have been retained by the implicit memory system and stored in body sensations and emotions. Thus, associative pairing and memory processing occur unconsciously during learning, taking the short path that bypasses processing from the cortex. This suggests an explanation for how fears are often experienced as compelling. Since they are learned outside of conscious awareness, they are not subject to conscious monitoring and thus are hard to control.

The memory of the aversive event lives on in the amygdala. Neuroscientists have learned that extinction involves the amygdala neurons, which signal the hippocampus and the prefrontal cortex. If any of these areas are cut, extinction does not occur (Maren and Quirk 2004). Thus, our psychotherapy methods help to extinguish the aversive associative pairings. Through therapy, the amygdala remaps the connections with the hippocampus and prefrontal cortex.

One of the challenges for therapists is to help the client generalize extinction, which may be learned during the therapy session, to other contexts. Through therapy, the client builds a separate extinction memory through the medial pre-frontal cortex and the hippocampus that suppresses the fear memory. Thus, when you help your client to substitute experiences that are more positive, you are facilitating a new set of memory connections. In this way, the fear is literally replaced in the brain.

The Neuroscience of Reward

Later theories looked at how such associations are perpetuated. Thorndike was one of the early psychologists to address this issue in the Law of Effect. In an experiment he performed with cats learning to escape a puzzle box, he observed that successful responses were "stamped in" by the satisfactions that came with success leading to these successful behaviors being performed more frequently, while the unsuccessful responses that produced annoying consequences were "stamped out" and subsequently occurred less frequently. The consequences strengthened or weakened the responses.

Later, Skinner formulated a conditioning paradigm that recast Thorndike's concepts of satisfiers and annoyers as the concept of Reinforcement in Operant or Instrumental Conditioning. Positive reinforcement occurred when the animal received a reward for performing a certain behavior. Negative reinforcement was the removal of an aversive stimulus when the animal gave the desired response, which causes an increase in the behavior.

In reinforcement learning, dopamine produced in the midbrain areas, the ventral tegmental area, (VTA) and the substantia nigra, form the reward pathway (See Fig. 9.5), that produces dopamine and sends it out into other areas of the brain. When there are deficits in the reward system, as seen in drug addiction and schizophrenia, the neurons do not react to rewarding environmental stimuli in the way that the brain of a person without such problems would respond.

The dopamine system is involved in the ongoing prediction of whether or not a reward is expected. Dopamine is delivered to target areas throughout the brain to signal structures that influence the process of choosing actions that maximize reward. Neurons in the target brain structures such as the striatum (part of the basal ganglia), orbitofrontal cortex, and amygdala code the quality, quantity, and preference of rewards (Schultz 2001).

The process takes place over time. There is a time delay for receiving the reward. Experimental evidence shows that animals keep track of the time elapsed from the presentation of a CS and make predictions accordingly. The process proceeds as an ordered set of events: First, the potential to receive a reward is detected. Next, an expectation that reward is coming is maintained in working memory to keep motivation high. These reward expectations guide behavior. If the reward received is similar to what was expected, the firing of neurons becomes

Fig. 19.3 Psychodynamic
therapy

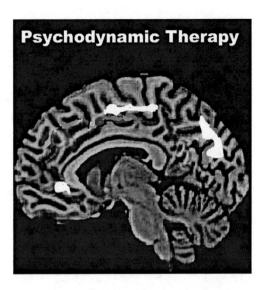

Fig. 19.3 Psychodynamic therapy

stronger. But if the reward does not meet expectation, the firing stops. Thus, if the prediction proves to be in error and the reward does not come, behavior eventually alters (Schultz et al. 1997). Thus, reward expectations are learned and can be unlearned if the client is guided correctly in therapy, or else uses relevant life experiences therapeutically.

Behavior therapy methods such as contingency management programs, systematic desensitization, exposure therapy, and modeling are based in the fundamental principles of association, conditioning, reinforcement, and their counterparts: dissociation, deconditioning, reconditioning, and punishment. The new neuroscience findings about the role of the amygdala, hippocampus, and stress pathway, along with the dopamine reward pathway, offer a rationale as to how such methods appeal to some of the natural neural mechanisms and instincts built into our nature. By altering associative pairings and correctly activating the reward system, you can harness the natural mechanisms of the brain to facilitate therapeutic change.

Psychodynamic/Psychobiological Approaches

Dynamic therapies have seen a revitalization in recent years (Fig. 19.3). Neuroscience has deepened our understanding about the reality of Freud's conception that the complex interactions of emotion with behavior takes place in the brain, even if there continues to be contention about the content and meaning of the patterns. Freud was a neurologist first. He began his work in quest of understanding the unconscious in terms of the brain. As he said:

Scientifically, I am in a bad way, namely caught up in the "Psychology for Neurologists," which regularly consumes me totally until, actually overworked, I must break off. I have. I have never before experienced such a high degree of preoccupation. And will anything come of it? I hope so, but it is difficult and slow going (Moran 1993, p. 29).

A new area is developing today, known as neuro-psychoanalysis, which interrelates the brain with psychologic functions, and more specifically correlates psychodynamic processes with corresponding brain processes.

The Importance of the Unconscious

We have long known that consciousness is limited. In his famous paper, "The magical number seven, plus or minus two: Some limits on our capacity for processing information," published in 1956, George A Miller showed that working memory is capable of holding only seven plus or minus two bits of information (Miller 1956). Later research clarified and refined Miller's findings, but did not disconfirm that this conscious capacity is still widely believed to be limited to this parameter. By contrast, a great deal of information is processed unconsciously through numerous, non-conscious, automatic pathways found throughout the brain. Freud introduced the idea that much of our functioning is governed by unconscious thoughts, feelings, and wishes, and indeed, neuroscience is finding this to be true. We have a number of unconscious memory systems (Squire and Kandel 2000) in the brain. Fear researchers (LeDoux 1996, 2003) showed that fear responses are also mediated by unconscious brain pathways, bypassing the conscious memory circuits in the hippocampus, such that the stimulus can trigger a fear response that occurs without any conscious control.

One of Freud's principles of therapy was to make the unconscious, conscious. He believed the ego to be the source of organized, rational, reality-directed thinking, which correlates with the frontal lobe executive functions. Psychoanalysis focuses on mastering unconscious impulses by enhancing the strength and functionality of the ego. And research corroborates this. A number of researchers have found greater activation in the prefrontal cortex following successful psychotherapy, including CBT and psychodynamic forms of therapy. (Shore 2012). Thus, Freud's attempts to strengthen and enhance the ego in his patients may well be reconceptualized, using the modern neuroscience model of the brain, as activating and enhancing the executive structures and functions.

Id Impulses

Freud believed that the unconscious mind was driven by primitive instincts and impulses. Neuroscience claims that not only do we have primitive impulses and instincts, we also share many of these with animals (Squire et al. 2003).

Evolutionary theory proposes that many of the same id impulse systems are present in the entire animal kingdom. What sets us apart from the other animals is our larger cortex, not the rest of our brain and nervous system.

Neuroscience uncovered the neural correlates for Freud's conception of libido, by which he meant emotional, physiological, and motivational energy, with the consequent life instinct drives for sex, survival, hunger, and thirst. In fact, the libido may be fueled by the dopamine reward system. We know that the reward pathway becomes involved in addiction and cravings, and that its activation does help perpetuate our life and the continuation of the species. Freud believed that the reservoir of libido was the id, which he identified with the individual, our essence, or what we essentially are (Freud 1960). As libido is extended outwards toward objects, it becomes stabilized. Vitality and life energy interact with the outer world and objects we identify with. We enjoy and love our relationships with others, our work, and our life.

Early Formative Relationships

Early development does not occur in a vacuum. We are always in relationship, from the very beginning of life itself. From the first cell that begins the process of human development, cells divide and form highly complex network of neurons, through interactions at the synapses. In time, the communication broadens to interface with different brain areas, and then after birth, from one brain to another as part of the social matrix of human interactions. Recent research on infant development shows that the infant's nervous system is in harmony with the nervous system of the mother.

Our social interactions, especially with those who are closest to us, have an enduring effect on the development of the brain. Parental imprinting takes place early in life and can continue to influence emotions, thoughts, and behavior ever after. Psychoanalysis addresses problems that may have occurred in early life by working on the original attachment to bring about corrective experiences.

Development of a strong relationship with a parent during critical phases of early development has a lasting effect on the individual's ability to form healthy relationships throughout life. The early attachment between mother and infant stimulates the neuronal connections that shape the brain's development for life. According to one of the founding fathers of this theory, John Bowlby (1907–1990), a healthy relationship with a significant other is a fundamental need that is wired into our very physiology. The theory that libido extends outwards to the world is the basis. This primary relationship is called attachment. Bowlby researched mothers and children and found that when children form a strong attachment with their parent, they are more resilient, confident, and capable of being calm and confident all through life (Johnson 2008).

Attachment goes both ways, between mother and child and back to mother. A mutual regulation model (MRM) explains how this works (Brazelton 1982;

Tronic and Weinberg 1980). The goal is to achieve a state of mutual regulation between mother and child. Together, mother and child interact with reciprocal exchanges of emotion. When the exchanges are well matched, both mother and child feel synchrony, shared delight. Together, they both gain the ability to control themselves appropriately. But when the expectations of one or the other are mismatched, problems may develop. A certain amount of mismatch is normal. Too much mismatching leads to negative affect and later problems (Tronic et al. 1980). From resonance together come the foundations for relationships of the future.

One explanation for the importance of attachment looks at the neuronal level and applies Hebb's rule. Strong interpersonal attachments bring about mutual firing of certain neuronal patterns. Then, because these neurons are typically firing together, they become wired together. These neuronal patterns lay a foundation for the ability to form healthy interpersonal relationships in general. With the hardware in place, people tend to naturally recognize and respond to fulfilling relationships throughout life (Tronic 2007, p. 492). Stability takes hold.

And so, a loving, affectionate bond between mother and infant is essential for an emotionally healthy adult life. Just as Freud would have predicted, evidence is mounting that structural and functional brain abnormalities are associated with child abuse and neglect in the early years (Teicher 2001). Those who have been denied a healthy early attachment often have problems with violence and aggression later in life. These abnormalities can be prevented by affectionate physical contact with the mother or primary caregiver, or later from psychotherapy. The original attachment experiences can be analyzed to free the client for a healthy life.

Love in the Brain

Love is instinctual, as fundamental as sex and food. We are all pre-wired to seek and have close, loving relationships (Johnson 2008). Love engages many regions on both sides of the brain to activate or deactivate (See Fig. 4.4). When people are in love they have increased bilateral activity in the insula and anterior cingulate and in parts of the basal ganglia: the caudate and putamen. They also show reduced activity in the posterior cingulate and in the amygdala as well as the right prefrontal cortex (Bartels and Zeki 2000). Those who are involved in a close, loving relationship feel safe and comfortable in the world, which helps the nervous system to be more able to respond flexibly and realistically to the situations of life.

Our ability to find and maintain loving relationships is influenced by the early primary relationship with our mother or caregiver. One of the core principles of attachment theories is that human beings rely on their primary attachment relationships for our feelings of safety. From the base of a secure attachment, a child will have the confidence to explore other relationships out in the world. Exploration is in balance between relying on the parent and relying on herself. Secure attachment when the parent is available, sensitive, and responsive, encourages the child to develop naturally and confidently. Parents can teach their children how to regulate their emotions even during stressful situations. So, through interpersonal interaction, we learn how to solve conflicts, handle emotions, and cope well with stress.

Learning to self-regulate occurs without direct awareness. The right hemisphere is thought to be dominant during the first 2 years of life (Badenoch 2008), and helps to foster limbic system development (Shore 2012). Many of the functions that are regulated in the right brain such as emotion, body experience, and autonomic processes, take place unconsciously. The quality of attachment develops from the mother's right brain to the child's right brain in an unconscious manner. The emotional processing that arises from early right brain attachment experiences continues to have an unconscious influence on later experiencing.

Unfortunately, not all mother–infant attachments are warm, comforting, and secure. Poor parenting can result in an insecure/avoidant attachment, or an insecure/anxious/ambivalent attachment (Ainsworth et al. 1978). When people form secure attachments, they find the satisfactions of love and enduring relationships (Johnson 2008). But when they have formed avoidant or anxious ways of attaching because of problems in early developmental experiences, difficulties develop later in life. Psychotherapy can correct this, by giving the client a secure attachment experience through the power of a good therapeutic relationship. Psychotherapy research shows that the therapeutic relationship is one of the most important nonspecific factors for effective treatment (Frank and Frank 1991). Therapy can provide an experience of having a healthy attachment, clearing the way for better relationships.

Defenses

When people have had disturbance in their primary relationships, they are at war within. Psychoanalysis has identified certain defenses that people typically engage in to protect themselves from the intrusion of the id impulses into consciousness. Often we see our clients engage in defenses that are detrimental to their mental health, and yet they continue to do so. Why do they feel so compelled? According to V.S. Ramachandran, we may find answers in the specialization of the two hemispheres of the brain. The neuroscience perspective offers an alternative explanation to add another dimension to our psychologic theories.

Brain damaged patients often display classic defense mechanisms. When they suffer damage to one hemisphere, they lose movement on the opposite side of the body. When the right hemisphere is damaged, patents exhibit neglect of the left side of their body, often denying any problem, rationalizing why they aren't moving, and confabulating elaborate reasons.

As we discussed in Chap. 4, the two hemispheres are specialized for different tasks, the left for language and not just for speaking, but also for meaning and syntax of language skills. The right hemisphere does respond to language, but more subtly and globally, through ambiguity and metaphors. Thus, when we work in therapy using stories and multiple meaning of words, we are appealing to the right hemisphere.

Everyday life bombards us with multitudes of inputs and information. When new information comes in, we weave it in, to fit into the cloth of our belief system. If the new information is inconsistent with our narrative, we engage in the Freudian defenses such as repression, denial, or confabulation. Ramachandran suggests that the left hemisphere creates a narrative of our life, screening out what does not fit. The right hemisphere provides some balance. It responds more globally, taking in the larger context. Thus, it acts as devil's advocate to prevent the left hemisphere from going too far (Ramachandran and Blakeslee 1998).

Patients who have right hemisphere damage no longer have the benefit of the right hemisphere's broader perspective to balance out their reactions. And so, when a patient with paralysis in the left arm from a stroke is asked, "Can you move your right arm?" they correctly answer, "Yes." When next asked, "Can you move your left arm?" they also answer, "Yes," even though movement in this arm is impossible for them. They will deny that anything is wrong and adamantly defend that they *can* move their paralyzed arm. Without the right hemisphere to serve its moderating function, the left hemisphere is free to defend the patient's belief system, no matter how far-fetched and out of touch with reality it is. As Ramachandran said,

> Freud's most valuable contribution was his discovery that your conscious mind is simply a façade and that you are completely unaware of 90 percent of what really goes on in your brain…And with regard to psychological defenses, Freud was right on the mark" (Ramachandran, Blakeslee, and Sacks 1999, p. 152).

We have covered a few of the fascinating new interconnections being made between psychoanalysis and the brain. Neuro-psychodynamics has old roots in Freud's neurologic background, but with the new evidence from all that we are learning about the brain, it is just beginning to unfold and evolve. We can find value in revisiting formative relationships, working with non-conscious perceptions, motivations, and instincts, addressing defenses, and fostering healthy rational functioning. In the interplay between the brain and psychodynamics, we see the broader usefulness of correlating brain structures and functions with our psychologic theories.

Meditation

Meditation is a valuable method that can be integrated into most therapy approaches. It includes techniques for following and regulating breathing, directing the mind in varied ways, and coordinating attention and breathing with body positioning. Many of the methods therapists use today are drawn from rich and beautiful ancient traditions of yoga, Daoism, Buddhism, and Zen. Mindfulness and yoga have been most widely adapted to therapy, with a large group of research projects that have shown their efficacy for use in therapy (Simpkins and Simpkins 2011, 2012).

There are different forms of meditation. Some forms focus attention and keep it steady while others broaden attention and keep it flowing. Some emphasize mental methods while others, as in yoga and Chi-Kung, integrate body movement techniques as well. This book offers the different methods for specific applications, to guide in how and when to apply these helpful methods.

The regular practice of meditation brings a number of changes in cognition, emotions, and behaviors. The ongoing flow of thoughts that we all experience in everyday life slows or even stops for a time. Alertness and attention is enhanced as well. People feel emotionally calm and peaceful, and they regulate their emotions better. And by being calmer and more aware, behaviors tend to be less impulsive, more reality based, and more compassionate.

Brain imaging and improvements in EEG technology show that the many positive effects of meditation practice correlate with changes in the brain. Researchers have also provided plausible models for how the brain is altered by meditation. Developing these abilities through the simple practice of meditation can become a helpful way to improve the quality of life for yourself and your clients at many levels of functioning.

Meditation Brings Relaxation

People have known for thousands of years that meditation produces relaxation. In recent years, Western science has shown this claim to be true. A comprehensive meta-analysis (Dillbeck and Orme-Johnson 1987) gathered 31 physiological studies that compared meditation to resting with eyes closed. The study evaluated three key indicators of relaxation and found that meditation provided a deeper state of relaxation than simple eyes-closed rest.

Experiments using EEG showed that meditators produced steady alpha waves associated with relaxed attention whereas beginners only had scattered alpha waves, with frequent interruptions from beta waves, associated with being alert and wakeful. Experienced meditators produced alpha waves, which slowed even more to become theta waves (Johnston 1974).

In the early 1970s, a model was proposed about meditation's relaxing effect on the autonomic nervous system. As you may recall, the sympathetic nervous system prepares us for action and the parasympathetic nervous system facilitates the resting, non-action response. Thus, it made sense to predict that the parasympathetic nervous system would predominate during meditation, with slower heart rate, lower blood pressure, and decreased respiration that most meditators experience (Gellhorn and Kiely 1972). These changes seemed to explain the feelings of calm and peace.

Fig. 19.4 Meditation

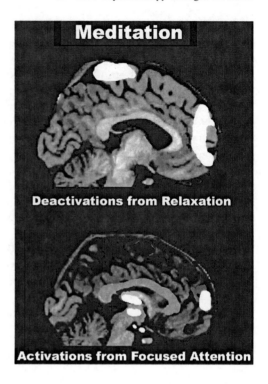

The Dual Effect: Relaxation and Activation

However, meditation is not fully explained by relaxation alone. Meditation has a second effect: It also produces alert attention. This dual effect is found in a number of recent studies that help to explain this interesting dual effect. Not only were the brains of meditating subjects found to be in a more relaxed state, but their brains were also activated at the same time. Newberg et al. (2010) and his group found that meditators have more activity in the frontal areas than non-meditators, specifically in the prefrontal cortex and middle frontal lobe. Other studies found increased gray matter volumes in the frontal lobes (Hotzel et al. 2008; Luders et al. 2009). In general, the increased activity and volume in the frontal lobes and their connection to the limbic system explains better regulation of emotions that come from regular meditation (Fig. 19.4).

Other fMRI studies (Lazar et al. 2005) show activation in the prefrontal cortex for alert attention along with activation in the anterior insula for interoception and inner awareness. Lazar's meditation study showed that meditators stayed highly aroused and focused while also maintaining a low heart rate and slow breathing, qualities of relaxation. These researchers interpreted their results as indicating that meditators have a decoupling of attention from arousal that makes it possible for them to maintain both states simultaneously.

A study by Lutz et al. (2004) add more information about the alertness achieved during meditation. They found that EEG was at the high end of beta waves, known as gamma waves ranging from 26–70 Hz. Furthermore, the gamma waves were highly synchronized across the brain. This finding could account for their heightened powers of perception.

Another result from the prefrontal cortex activation involves activation of the hippocampus, important for attention and memory. The hippocampus becomes activated via the nucleus accumbens, a primary part of the reward pathway. And this hippocampal activation in turn stimulates the amygdala. The fMRI study by Lazar et al. (2005) confirmed an increase in activity in the amygdala and hippocampus during meditation. The hypothalamus–limbic system interconnections alter the autonomic nervous system by increasing parasympathetic activity associated with relaxation, reduction in heart rate and breathing rate, all of which take place during meditation. Truly, meditation is rewarding!

Brain Coherence: More Brain Involved

Meditation practice brings firing of gamma waves. Gamma waves are a pattern of neural activity at a frequency of 25–100 Hz, typically at 40 Hz that fire in synchrony from the thalamus all the way to the frontal areas. Higher gamma wave coherence means that more of the brain is being used and is associated with improved quality of attention. These waves may correlate with the unity of conscious perception according to Crick and Koch (1990). Gamma waves have been observed in meditating subjects in a number of different studies. (Badawi et al. 1984; Lutz et al. 2004; O'Nuallain 2009). Oneness that people feel when meditating seems to be taking place in the brain itself! This gamma activation might explain the overall mental development that comes from meditation.

Hypnosis

Hypnosis is a method for focusing attention inwardly to bring about alteration. People can then feel a deep sense of relaxation and inner calm, though they may have been unable to at will, before. They can also experience modifications in many qualities of experience, such as increase or decrease in sensations, recall of old memories, and alterations in experiences of time. All of these variations can be valuable for helping clients to recognize that change is possible, and that things can be different. After an experience of contrast, the therapist can encourage a more positive adjustment.

Hypnosis, like meditation, brings about a number of characteristic changes in the brain. Neuroscience research on the effects of hypnosis reveals that alterations do indeed take place, bringing many to believe that hypnosis is an altered state

Fig. 19.5 Hypnosis

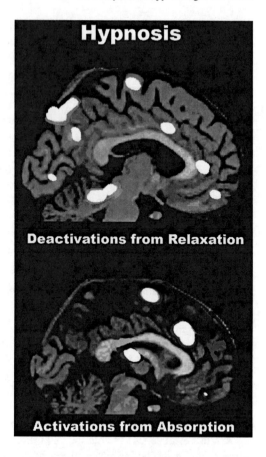

(Jamieson 2007). There are others who argue that hypnosis is better understood not as a unique state, but rather as highly focused attention and absorption or perhaps a deeply enacted role. No matter what theory of hypnosis you subscribe to, the benefits of hypnosis can be helpful at every stage of therapy (Simpkins and Simpkins 2004, 2005, 2010).

Recent Definitions from Neuroscience

One definition of hypnosis that has emerged from recent neuroscience research is that it involves focused attention, then absorption, relaxation, an experience of automaticity and changes in self-agency (Rainville and Price 2003).

A dual effect of absorption and relaxation can be seen in the brain. Rainville et al. (1999, 2002) in a number of studies found a patterned relationship during hypnosis between a network of areas (the anterior cingulate gyrus, thalamus, and brainstem) which differed from subjects who were alert and awake. The thalamus and brainstem areas had lower firing rates, which correlated with relaxation, while higher rates of activation occurred in the anterior cingulate gyrus, correlated with absorption. These two opposite effects remained stable while the subject was under hypnosis. The brainstem, thalamus, and anterior cingulate gyrus all work together in this coordinated pattern of rCBF, reflecting an interrelationship within a network (Fig. 19.5).

Dissociation and Decreased Brain Coherence

Hypnosis has a unique quality of producing automatic behaviors and experiences. For example, people can have a hypnotic phenomenon known as hand levitation in which their hand feels like it is rising up by itself. Or they might have a spontaneous image or memory occur without any deliberate effort to bring it about. These experiences can be explained by the phenomena of dissociation.

Hypnosis has a long history of being characterized as a form of dissociation. Pierre Janet (1859–1947) first proposed the conception of dissociation in his dissertation in 1889. His idea was based on association theory. A coherent complex of thoughts, feeling, and memories could be separated off from the primary consciousness. Under hypnosis, Janet believed these dissociations could be recalled and re-experienced.

Many researchers since have evolved the idea of hypnosis as involving dissociation. For example, Ernest Hilgard (1904–2001), who had a hypnosis research laboratory at Stanford University, developed a modern interpretation he called Neo-dissociation (1977). He pointed out that the unity of consciousness is an illusion. Attention is always partial and shifting among many subsystems such as habits, attitudes, interests, and specialized abilities. Once a subsystem has been activated, it can operate somewhat autonomously, explaining how habits can occur without paying attention to them. In hypnosis, the executive function is split between hypnotist and the hypnotized person.

With the new fMRI and PET research, we see some evidence for a functional dissociation occurring with hypnosis, just as these visionary figures in the history of hypnosis would have predicted. This ability of people under hypnosis to dissociate can be useful for therapy for many problems.

A study that attempted to address how these automatic responses occur found a decrease in functional connectivity (EEG gamma wave coherence) across the hemispheres. The hypnotized individual is characterized by a functional dissociation between conflict monitoring in the anterior cingulate and cognitive control in the prefrontal cortex (Egner et al. 2005). The study used the Stroop effect to test the effects on performance from decoupling between conflict monitoring (from the anterior cingulate) and cognitive control processes (in the PFC). In the Stroop test (1935) two types of processing that compete with each other when paired together are measured. A list of color words, red, blue, green, etc., are shown. The no-conflict situation is when the word red is colored red. For the conflict situation, the word red is given a different color, such as green or yellow, when presented. Subjects are asked to state the color of the letters and disregard the word's meaning. This leads to a delayed reaction time. Decades of Stroop research have shown that people tend to recognize word meanings before they process colors. This is because language is processed automatically and quickly before color.

However, suggestion can alter this language primacy effect. When highly hypnotizable subjects were told that the letters were gibberish, the subjects read the colors equally quickly, even in the conflict situation. In addition, fMRI data showed that the activation in the conflict monitoring areas of the ACC was reduced. The results call into question the assumption that language must always be prior, and cannot be altered. There are times when clients need to break out of typical patterns in order to make the changes they need, and methods like hypnosis and suggestion may be the best choice.

Hypnotic Trance and Suggestion

Studies attempting to understand the hypnotic state have found evidence to support the idea that hypnotic phenomena are "real" in the sense that they differ from simulators and fakers. There are clear neural activation changes that differ from people who are simply simulating trance (Oakley 2008).

The next question is whether suggestions without hypnosis are equally effective as suggestions following induction of hypnotic trance. One study found that there was no difference in removing the Stroop effect between suggestions without trance and suggestions in trance (Raz et al. 2002). And another study comparing suggestion and trance with no-hypnosis and suggestion for pain found no difference (Millinget al. 2005). However, these two studies did not include neuroimaging, which might shed more light on these findings. Another study did perform neuroimaging when comparing hypnosis with suggestion, and suggestion with no formal hypnotic trance for pain in fibromyalgia. These subjects reported significantly more control over their pain when using hypnosis plus suggestion than suggestion alone (Derbyshire et al. 2008). Their brain activations reflected this difference, with stronger effects in the thalamus and insula, as well as in the prefrontal and sensory cortices. Thus, it appears that hypnotic trance might enhance the positive effects of suggestion. These issues are still unresolved, and so as a practitioner, you may find that using suggestion alone suits one situation and client whereas using suggestion along with hypnotic trance works better with another client. We recommend being familiar with both approaches, so that you can use what is best for your client. Trance has many benefits of relaxation, calm, and giving the client a tool of self-hypnosis that they can perform between sessions. It also offers an experience of an alteration in consciousness, something different. For some people, an alteration of consciousness may be just what they need to allow them to recognize that they could possibly change, as mentioned earlier.

Other Brain Activations

Pain has a psychologic component. People are able to alter their experience of pain using mental techniques. One of the well-known effects of hypnosis is pain control. But how does this change in the experience of pain come about? Rainville measured brain alterations under hypnosis when subject's left hand was immersed in painfully hot water. He found increased activation in the occipital regions, because subjects had spontaneously produced visual imagery (Rainville et al. 1999). On the other hand, the subjects had a decrease in activation of parts of their parietal lobes, where sensory input is processed (Rainville et al. 1997). Hypnosis brings a general inattention to external stimuli and body-sense in general. And when working with pain reduction, these hypnotic capacities can be enhanced.

These are but a few of the many alterations that are possible using hypnosis. Therapists who incorporate hypnosis into their practice will find many ways to utilize these alterations to help clients change.

Conclusion

As we learn more about the effects of psychotherapy on the brain, we encourage you to integrate these findings to direct your techniques to target the brain and nervous system. As we hope this chapter has illustrated, neuroscience provides evidence for psychologic theories. Each informs the other. They fit together. From the dialog between them, we get a rationale for better theories and effective treatments.

Chapter 20
Relieving Anxiety

Anxiety is a clear illustration of how intimately the mind, brain, and body are interconnected. Anyone who has ever been anxious knows that anxiety is an embodied experience, felt very definitely in body sensations. And yet, anxiety has strong psychological components that also drive the experience. By understanding these elements and how they work together, you can better help your clients to overcome their problems with anxiety. You can utilize the contributing elements from brain, mind, and body to gain new variations of techniques for accessing the problem and tailoring your treatments to the needs of each individual.

Anxiety is categorized into different types. The distinctions are helpful because different treatments may be used. In addition, different brain areas are involved. Generalized anxiety disorder (GAD) involves a broad, general feeling of anxiety that can inhibit people from doing things and going places. Social anxiety disorder (SAD) is felt in social situations with other people. Sometimes it is specific, such as feeling uncomfortable doing public speaking. People with social anxiety feel as if others are judging them negatively. Panic disorder is often experienced as having a severe physical crisis, such as a heart attack, when there is no real physical danger. These individuals feel an intense panicky feeling, which usually lasts from 1–10 min. Specific phobias are fears of one thing, such as a fear of dogs, heights, spiders, elevators, or open spaces (Agoraphobia). Often these fears are initiated by a traumatic event, but not always. Post-traumatic stress disorder (PTSD) may occur after someone has gone through a very traumatic experience such as rape, war, or torture. Not all people who undergo such experiences have an anxiety reaction, but for those who do, it can be very debilitating. Obsessive compulsive disorder (OCD) is also categorized as an anxiety problem. Sufferers have intrusive thoughts that bring about feelings of anxiety, which leads them to engage in behavioral rituals to try to reduce the anxiety.

C. A. Simpkins and A. M. Simpkins, *Neuroscience for Clinicians*,
DOI: 10.1007/978-1-4614-4842-6_20,
© Springer Science+Business Media New York 2013

An Overview of How Anxiety Changes the Nervous System

Anxiety involves a number of brain systems that have a strong influence on the body and the mind. The autonomic nervous system gets involved through overactivation of the stress pathway. The limbic system also plays a role. And imbalances of neurotransmitters are key as well. We describe each of these systems in the next sections.

The Role of Fear and Stress in Anxiety Disorders

Fear is a natural response built into the mind–brain–body fight or flight system to respond to danger and threat, as we described in Chap. 9. When the sense of danger is sustained over time, it becomes stress. Anxiety, fear, and PTSD all have a strong stress component that sustains the reaction. When people suffer from a long-standing or severe trauma, such as from abuse, illnesses, war, natural disasters, or even from mental disorders, the HPA pathway remains continuously activated. Eventually, the mind–brain–body system forms a balance at this higher level, giving the experience of feeling chronically stressed, anxious, or fearful.

How the Neurotransmitters Balance Changes

The transition from a normal fear response to a stress response brings an alteration in the neurotransmitter patterns. The neurotransmitters that play a key role in anxiety are glutamate, gamma-aminobutyric acid (GABA), serotonin, norepinephrine, and dopamine. As a quick review, glutamate, the excitatory neurotransmitter, and GABA the inhibitory one, provide excitation and inhibition throughout the brain. Serotonin regulates moods, norepinephrine enhances alertness, and dopamine is involved in reward. People with anxiety disorders have changes in serotonin, norepinephrine, GABA, corticotropin-releasing hormone (CRH), and cholecystokinin (Rush et al. 1998). Changes are compounded by the interaction since alteration in one neurotransmitter system invariably elicits changes in another.

Drug therapies for anxiety disorders help to rebalance the system, by reducing the overexcitation. But drugs are not the only way to alter the balance of neurotransmitters. Meditation, hypnosis, and psychotherapy can also be used to calm the system, thereby creating a healthier balance. All of the anxiety patterns, even an entrenched anxiety reaction, can be changed. The methods presented in this chapter help to shift the balance back to a calmer center. They can also be used in conjunction with drug therapy.

Neuroplasticity in the Stress System: You can Alter the Brain's Stress Reaction

People who have been through an actual trauma, as well as those who just regularly feel threatened, whether real or imagined, are under increased stress. They have learned to have an exaggerated fear response, mediated by the amygdala. In essence, the brain cannot shut off the fear reaction. Like keeping a car revving, the nervous system remains balanced in a heightened state of arousal.

Because of neuroplasticity, some changes occur in the brain when the system stays on high alert. Decreased anterior cingulate activity, a crucial area for conflict resolution and decision making, perpetuates the problem by making it harder to resolve conflicts or formulate good decisions. Stress has been shown to inhibit the growth of new neurons in the hippocampus (Gould et al. 1999) and a generalized inhibitory effect of neurogenesis. Adults who suffer from extensive stress have a smaller hippocampus than those who are not under stress (Gage et al. 2008). A smaller hippocampus makes learning more difficult.

The good news is that these processes are dynamic. Although the brain patterns become entrenched, they are reversible. Neuroplasticity works both ways, so with the right experiences, an underactive, smaller hippocampus can be activated to grow. For example, cab drivers tend to be continually engaged in spatial orientation tasks, a function of the hippocampus. They had larger hippocampi than non-cab drivers of the same age (Maguire et al. 2000). This research indicates that with learning experiences that foster hippocampus activity lead to an increase in its size. Neurogenesis can occur in the hippocampus at any age. Even elderly patients who were dying from cancer showed growth in their hippocampi (Gage et al. 1998).

Most people are not born with an overactivated stress response. And, in general, the body wants to return to a normal balance. The fight or flight pathway is always changing in response to situations, and so a process moving toward a lower activation can be started at any point.

Limbic System Involvement in Anxiety

In addition to the stress reaction, the limbic system also plays a role. Here is how the process gets triggered: The thalamus acts as the gateway for signals received from the senses and then sent on to other parts of the brain for processing. The signal is relayed to the hypothalamus, which is the coordinator of internal functions, to put the system on high alert. Recall that the amygdala registers the quality and intensity of emotions. When a particular sensory input seems threatening, the hypothalamus receives a signal that there is something to fear. This triggers the HPA pathway to activate. When the threat signal persists, the nervous system remains on high alert, leading to a stress response. The hippocampus, where short-term memory is processed for long-term storage, monitors and adds to the signal

Fig. 20.1 Anxiety
activations in the Limbic
system

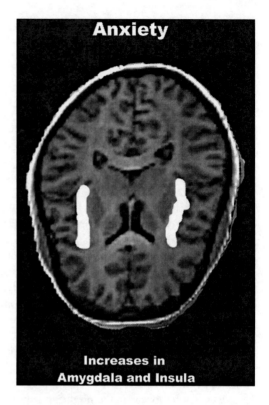

from its experience. And the basal ganglia, involved in motivation and movement, act to coordinate emotions with movements. Finally, the cortex itself is activated, accompanied by more thoughts and ruminations about the anxious or fearful situation (Fig. 20.1).

The Short Path and the Long Path for Emotions and Stress

There are two paths for emotional and stress reactions. One follows a short path and is immediate and often unconscious, a bottom-up brain process. The other is a long one, engaging conscious awareness and taking more time as a top-town process.

Here is how the two paths function in fear and anxiety. In the short path, sensory information enters the lateral amygdala directly from the thalamus. The information passes to the major output nucleus of the amygdala known as the central nucleus of the amygdala. The central nucleus projects, in turn, to multiple brain systems involved in the physiological and behavioral experience of fear. The reaction acts quickly, moving to different regions of the hypothalamus. This activates the sympathetic nervous system and induces the release of stress hormones activating the HPA stress pathway that lead to the feeling of fear and anxiety.

The long path is an alternative way to experience a fearful stimulus. Instead of traveling directly from the thalamus to the amygdala, the stimulus is mediated by the sensory cortex and hippocampus, thereby activating further processing through conscious awareness. When, for example, you see a poisonous snake on the ground right at your feet, you feel an immediate stab of fear and anxiety. Your nervous system has gone on high alert through the short path. But imagine if you look more closely to discern what kind of snake it is, and you realize from your knowledge about snakes that it is just a harmless garter snake. You have engaged a long path of processing that involves your thinking prefrontal cortex and explicit memory systems. This is a model for how psychotherapy can slow down the short path, by teaching strategies to activate the prefrontal cortex and regulate the arousal level.

Using the Short and Long Path in Treatment

Although the stress response may come about automatically, it can be influenced in varying ways. Cognitive methods are effective in switching processing from the short to the long path. The long-path change is that thoughts and memories, correctly processed in the prefrontal cortex, help to calm the system down when no real threat is present. And mindfulness techniques can be effective for transforming processing, from a less adaptive to a more adaptive use of the short path. Even if there is a real threat, and the client needs to be on alert, meditation's dual effect of alertness and relaxation can help the client handle the situation with attention and clarity. The immediate, reflex reaction becomes a calm one.

Our thoughts, emotions, and sensations are always functioning together in a highly interactive network. We can calm an overaroused nervous system through the long brain pathway by altering cognitions, applying reason correctly, challenging troubling assumptions and beliefs that add to distress and discomfort. But sometimes a direct, short-path response is adaptive and appropriate. We can intervene directly through the short path by changing breathing, relaxing muscles, or altering sensations. From either path, long or short, a more comfortable balance can be found. The generalized, overall rebalancing of the brain's stress reaction can be a powerful force for alleviating anxiety and fears.

Calming the System

Slow, relaxed breathing soothes the sympathetic nervous system, activating the parasympathetic nervous system. This shift leads to responses that lessen stress by lowering cortisol levels, relaxing muscles, and lowering blood pressure. Here is one exercise to help calm the system, but you can also use any of the breathing exercises for calming that are given in this book.

Sit on a chair or on a pillow on the floor, whichever is most comfortable. Place your hands on your knees. As you breathe in, arch your back and head as you move slightly forward. You will feel a gentle stretch of your whole spine all the way up through your neck. As you exhale, gently round your back in the opposite direction, tucking your head slightly. Breathe fully but keep the breaths soft, slow, and flowing as you gently arch and round with each breath. Repeat this pattern up to five times. Then return to a centered position and breathe comfortably, meditating on the breath as you allow relaxation and calm to spread through your whole body.

Returning to Balance

People who feel anxiety have an imbalance in their nervous system, with overactivation in some areas and underactivations in others. These physiological imbalances can be addressed through a short-path experience. You do not consciously try to be balanced; instead, you discover it. As the nervous system readjusts, the accompanying thought patterns tend to ease, making them easier to address and work with in therapy.

Begin with sitting. Sit cross-legged on the floor with or without a supportive pillow, or sit in an upright chair. Allow your spine to straighten as you gently lift your head up, keeping eyes facing straight ahead. You will feel a slight lengthening of the spine and neck. Close your eyes and turn your attention to your body. Now, gently rock forward and back, keeping your head and spine relatively straight. Feel how your muscles relax as you sway, noting what muscles become tighter. When you move through the center you will notice a point where muscles relax, breathing is easier. This is your balance point. Stay there for a moment and allow breathing to relax even more. Now try swaying side to side to side, noting the change in muscle tone as you move further away from center on either side. Once again, take note of that middle point where muscles seem most comfortable and breathing is easiest. Here is your balanced center. Stay in this position for several minutes, enjoying the effortless feeling of sitting, aligned with gravity and in tune with the greater whole.

Specific Fears

People who have phobias tend to have hyperactivity in the limbic and paralimbic areas. For example, subjects with social phobias were measured for cerebral blood flow (rCBF) in response to a public speaking task. Before treatment, they showed broad bilateral changes, with increased activation in the amygdala, hippocampus, and neighboring cortical areas of the limbic system that get defensively involved when the person is threatened. Following 8 weeks of successful CBT, or comparison drug treatment, the threat activations of the limbic system decreased significantly (Furmark et al. 2002). Studies of people with specific phobias showed similar results (See Fig. 4.5). One project compared people with spider phobias to normal subjects, by having all subjects watch a film about spiders. The phobic subjects reported fear as they watched. They had a significant activation of the parahippocampal gyrus, the right dorsolateral prefrontal cortex, and the visual associative cortex, bilaterally. Controls did not show these activations. The researchers proposed that the activation in the hippocampus was likely due to an automatic triggering of a fearful memory of spiders. Their activation in the prefrontal cortex was related to an attempt to cognitively regulate the fear response (Paquette et al. 2003). This dual action of an automatic fear response plus higher cortical activation is a typical pattern found in many phobias.

Embracing Fear to Overcome It

Steve was a hard-working medical doctor in his early thirties who worked long hours. He found relief from his long hours at the hospital by taking up the hobby of rock climbing. As his climbing skills improved, he made trips with an advanced group who invited him to participate on an extended climb of Mount Everest. He arranged to take time off his work well in advance and felt excited about the trip. But as the time for the climb grew closer, he began to feel anxiety. He made up his mind to go on the hike no matter what, and so he sought hypnosis to overcome his fear.

We taught him how to enter the hypnotic state and be able to relax at will. He also worked with systematic desensitization during sessions in our office, imagining progressively more challenging climbs while cultivating relaxation in his muscles.

He explored his early memories to discover the origins of the fear. We learned that as a child he had been overweight, inactive, and awkward, with many fears. He outgrew this "awkward stage" during his senior high school year, and had remained fit and confident as an adult. But the awkwardness returned later. We chose to use desensitization by visualizing coping better, followed by relearning better attitudes.

During desensitization visualizations, whenever the fear washed over him, he lost touch with his surroundings and would become that awkward child again, sometimes tripping over branches or even getting too close to the edge. His irrational fear was compounded with his healthy rationally based fear response that was an accurate signal, telling him he was in real danger. He worked with sensory awareness to train his attention to keep focused, even in the midst of strong emotions. He also accepted that the feeling of fear could be a positive warning sign to help him correct faulty technique.

We combined our treatment of systematic desensitization with exposure therapy and awareness skills. He went on some local hikes that became progressively more challenging. As his awareness skills improved, he found that he was enjoying the beauty around him and the feeling of satisfaction within. Whenever he became tense, he would stop, sit down, and do a little self-hypnosis. He took the hiking trip to Tibet and found it one of the most fulfilling and enjoyable experiences of his life.

By developing prefrontal cortex attention skills, Tom learned how to self-regulate his responses. He learned to take command of his fear response and regulate his emotions in the situation. But he also learned to accept healthy fear. He changed several reactions, which then allowed him to overcome his phobia. We helped him to alter his false-alarm fear response that overactivated his limbic areas, by using hypnosis, desensitization, and gradual exposure to situations fraught with the potential for him to experience irrational fear. In sessions between, he worked through the deeper issues that had set it all in motion. As he learned to accept himself, he transitioned from the awkward child of his youth to the successful lawyer he had become. He allowed balance he gained during his climbs into his life, so that he could continue to achieve professionally while also enjoying his life.

PTSD

PTSD involves three general systems, all interacting together (Green and Ostrander 2009). The syndrome begins when the individual is involved in, or witnesses, a traumatic event. First, the memory system (the hippocampus and parahippocampus), encodes the event into memory from all the visual, auditory, olfactory, and somatosensory inputs that the brain receives during the initial traumatic experience. The amygdala responds in close interaction with this memory encoding system to register the emotional qualities of the experience. An associative link forms between the neutral stimuli that took place during the traumatic event and the emotional memory, thereby forging some of the triggers for later anxious responses. Animal research shows the important role the amygdala plays in fear conditioning, and similarly in PTSD sufferers, fear conditioning is involved in the fear response that keeps being expressed long after the traumatic event (Fig. 20.2).

The third area involved is the ventromedial prefrontal cortex, which engages the anterior cingulate gyrus and parts of the orbitofrontal cortex. These areas contribute the conscious element, where people evaluate the event as dangerous and disturbing, thereby triggering a fear response.

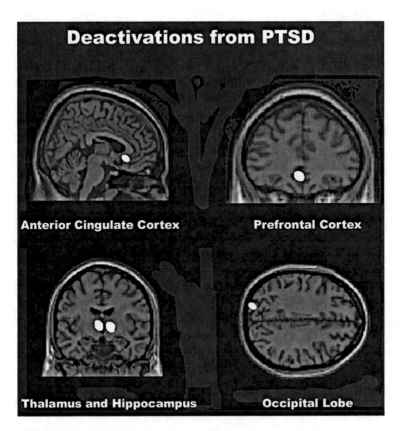

Fig. 20.2 PTSD deactivations

The Role of the Hippocampus in PTSD

Research has found that people suffering from PTSD have a smaller hippocampus, which makes them less capable of drawing on memories to evaluate the nature of the stressor. But if the correct learning experiences are given, the hippocampus can increase in size (McEwen and Magarinos 1997). For example, cab drivers tend to be continually engaged in spatial orientation tasks, also a function of the hippocampus. Researchers found that they had larger posterior hippocampi, as compared to non-cab drivers of the same age (Maguire et al. 2000). So, the hippocampus may grow as well as shrink, depending upon stimuli. Correct learning experiences are likely to stimulate its growth. Given the research that meditation reduces anxiety combined with the capacity of the hippocampus to grow, meditation is likely to help the impaired hippocampus to regenerate (Piver 2008).

Working with PTSD

The three areas involved in PTSD engage both a short- and long-path activation. Therefore, PTSD can be addressed by working with both pathways. For short path interventions, begin by teaching relaxation and calming as in the exercises given in this and other chapters of the book. You can also help by engendering a sense of security from within; this helps open a positive alternative short-path experience.

Accepting Support in Sitting

When people have been through trauma, they feel a sense of threat coming from the environment and ignore any support that might actually be available. This exercise is an example of how to help the anxious person discover a sense of security within, during something very basic to everyday life, by literally attuning to sitting and taking strength from the support of the ground or chair.

This exercise can have a powerful impact if performed sitting on a solid chair. Sit comfortably. We spend a lot of our life sitting in chairs, but rarely do we recognize the support we gain from the chairs in our life. Place your hands, palms down, on the ground, floor, or chair at your sides. Feel the mass of the ground beneath your palms, the strength of the floor or the strength of the chair that holds you. Notice how your body is supported. Are you pushing down on as you sit or holding yourself away? Allow your body to take the support from the ground (or the floor or the chair) by letting your sitting muscles relax. Breathe in and then as you exhale, imagine that you breathe out tension and allow yourself to be comfortably supported as you sit. The chair can provide support for you, and you can allow yourself to accept that support, feeling a sense of comfort in the solidity beneath you.

From Chair Support to Mindful Self-Support

As you are able to take support from the chair, you can extend that experience as you begin to attune to your own sitting. Let your attention turn to just sitting. Notice everything about your sitting. Let your attention roam around your body as you notice your posture. Allow your sitting to be just as it is. Notice your body position, how far it is from your seat to your head, how far apart your shoulders are. Sense your hands as they rest on your lap or on the chair sides. Breathe comfortably and simply allow yourself to be aware of your sitting. Do not try to change it—simply allow yourself to be here now, just sitting. This is your ability, to sit and let be. Often when people just sit in this way, they begin to feel a certain kind of familiarity and comfort. You can always return to your aware sitting as an inner resource, always available to you, no matter where you are, whenever you are sitting. This exercise can also be performed with standing.

The methods included here offer tools that work on several levels. Clients learn how to deliberately focus attention and build skills they need to shift attention away from the trauma. In addition, those who have undergone a traumatic experience have faced some of life's most difficult challenges: tragedy and loss, in life and death situations. Having endured, they have gained recognition of forces acting beyond the self. Therapy can help draw upon positive forces for inner strength to meet life's challenge well.

Long-Path Interventions for PTSD

People suffering from PTSD can benefit from practicing meditations that teach them to direct attention and pinpoint it. In deliberately using attention therapeutically, executive control is strengthened from the frontal areas of the brain, while limbic system short-path reactions are lessened. The mantra meditation that follows helps clients build mental focus while also drawing spiritual strength from their faith.

Mantra Meditation. The use of mantra meditation has been tested with veterans and found to be helpful (Williams et al. 2005). People who are suffering from PTSD need to be able to refocus their attention away from the problem to encourage the nervous system to return to a more comfortable balance. Over time, the allostatic balance shifts from an overactivated state to a calmer balance. Then the anxiety feelings begin to subside as the brain–mind–body system finds a more comfortable balance.

Here is a list of mantras from different spiritual traditions that were used in the Williams et al. (2005) study. Choose one that is meaningful to you to recite. Recite the mantra in its original form rather than reciting the translated words, since the syllables are significant for their vibration of tones in combination, not just for their meaning. Here are some possible choices, but please feel free to choose something else if you have a mantra that you prefer to use.

Buddhist	*Om Mani Padme Hum (Ah-oom Mah-nee Pod-may Hum): An invocation praising the jewel (self) in the lotus of the heart*
Christian	*Kyrie Eleison (Kir-ee-ay Ee-lay-ee-sone): Lord have mercy*
Hindu	*Rama (Rah-mah), Eternal joy within*
Jewish	*Shalom, Peace*
Muslim	*Bismallah Ir-rahman Ir-rahim (Beesemah-lah ir-rah-mun ir-rah-heem), In the name of Allah, the merciful, the compassionate*
Native American	*O Wakan Tanka: Oh, Great Spirit*

Once you have chosen one mantra, sit down in easy pose. Speak the mantra, beginning with 6 min and working up to 30 min each day. Begin with 2 min devoted to each part of the practice and work up to 10 min for each segment. Repeat the practice a second time each day for even stronger results.

Begin by speaking the words aloud slowly, repeatedly, as you relax your chest to allow the words to flow. Focus your attention on the sound of the words, and on the vibration in your chest and voice box as you speak.

Next whisper the words slowly, aloud, keeping your attention focused on the sound and the vibrations.

Now, close your eyes and imagine speaking the mantra for 5 min. Keep your attention focused only on the imagined speaking of the words.

Finally, with attention focused, just sit quietly for several minutes as you allow the calm, centered, quiet experience to be. When you feel ready, open your eyes and stretch gently as you stand up.

OCD

OCD is classified as an anxiety problem partly because it does engage the fear/ stress response with a short-path reaction. Sufferers try to allay their fear response with a long-path ritual action that temporarily eases the anxiety. But the solution often becomes a problem in itself, and so the individual remains stuck. OCD is a good example of how overreliance on the long path does not stop anxiety, nor does it return the nervous system to a healthy balance.

People who suffer from OCD without depression have a distinctive brain pattern. One study of OCD patients found a significant increased blood flow in the right thalamus, left frontotemporal cortex, and bilateral orbitofrontal cortex compared with normal controls (Alptekin 2001). When the orbitofrontal cortex is activated, its signal usually indicates that something is wrong. People with OCD have excessive metabolic activity in this area of the brain, overusing their long path of emotion. They are continually on high alert as if they are in danger, and feel compelled to do something about it. This interferes with their ability to accurately interpret and plan. PET scans reveal that both psychotherapy and drug therapy can reduce the activation in this area, which leads to a more accurate assessment of their situation and a more balanced response to what is happening and what to do about it (Baxter et al. 1992).

Treatments for OCD

Clients with OCD use rituals in an attempt to allay their anxiety, but ultimately these rituals are emotionally unfulfilling; and the anxiety reaction continues. Rituals can be positive and enriching. You can sometimes introduce a ritual that

Fig. 20.3 Tea ceremony

gives the client something healthy to do instead of the symptomatic ritual and at the same time will address the underlying anxiety.

We have often introduced a simple tea ritual, drawn from Zen. It is easy to do and can be performed at home. The ritual helps people to let go of thinking and turn to their senses instead. The result is a natural deactivation of the frontal areas of the brain that are overstimulated in OCD, along with being more in touch with sensory experiencing, to help people be more in touch with their real-life situation.

The tea ceremony, known as Cha-no-yu, is a long-lived tradition that invites centering in a tranquil moment and puts the mind at ease. A sanctuary from anxiety and stress, the tea ceremony brings you into the present moment, with a feeling of calm centered awareness.

Zen tea embodies certain virtues: reverence, harmony, purity, and tranquility. Performing a simple tea ceremony can elicit all these virtues along with relaxation. Like a breath of fresh air, the moment is free of anxiety and fear. Fully immerse yourself in what you are doing, and your thoughts become unified with your action. In this way, you become present. The quality of awareness you develop will stay with you as a resource. By not contemplating the peripheral things, you reduce anxiety associated with them (Fig. 20.3).

Here are the simple instructions for a tea ceremony we have performed with many people over the years. Even a child can do it, yet a few moments immersed in this simple ritual can provide a profound experience. You can perform it alone or with family, friends, or clients.

(1) *What you need*:

    a. *Utensils: A mat, a teapot, a bowl, a spoon, a pot to hold the hot water, and a tea cup for each person.*

    b. *Tea: We perform this ceremony using peppermint tea because of its pleasant aroma and delicious taste. But please use any tea that you like. If you cannot obtain the tea as loose tealeaves, open the teabags and empty several into the bowl.*

(2) *Preheat the water and then bring it to the tea area.*

(3) *Set out your utensils out as pictured and give everyone a teacup, set down in front of them. We often perform tea ceremony sitting in a circle on the floor, but you can also do it sitting or kneeling at a low table.*

(4) *If you are sharing tea with others, begin by explaining briefly about the ceremony. There is no talking once the ceremony begins. Each person will be quietly aware and experience everything deeply. Once the tea is poured, everyone turns their cup before drinking (Instruction 8).*

(5) *Meditate by clearing the mind for several minutes to settle quietly before you begin.*

(6) *Scoop up a spoonful of tealeaves and place them in the pot. Lightly tap the spoon against the teapot three times. Repeat spooning, placing, and tapping several times until all the tea is in the pot. Listen carefully to the sound of the tapping.*

(7) *Pour in the hot water, listening carefully to the sound of the water as it enters the pot and smelling the aroma of the tea as it begins to brew. Make your movements slow and precise without tension. Place the cover on the pot and then rotate the pot three times in a clockwise direction, then three times in a counterclockwise direction.*

(8) *Pick up the teapot and pour each teacup, one by one, and then pour a cup for yourself last.*

(9) *Once all the teacups have been filled, everyone rotates their cup three times clockwise, then three times counterclockwise.*

(10) *Lift the cup slowly, smell the aroma of the tea, look at the color and patterns in the cup, then slowly sip the tea, letting the flavor sit on the tongue for a moment. Keep your mind clear, focused on your actions and sensations. Listen to the sounds, smell the aromas, savor the taste of the tea, and let yourself relax. Refrain from talking. At the end, take a moment for meditation to savor the experience. Paradoxically, as you fill yourself with the experience, you experience emptiness.*

Panic Disorder

Anxiety sometimes involves linking uncomfortable body sensations in the chest to worries about what such physical symptoms usually signify, for example, to worry about having a heart attack, weighing expectations about that possibility, and so on. A real and existing pattern does emerge from associating these components together, even though not originally from an actual heart condition. Even if a medical checkup determines that there is no underlying physical cause, to become free from the anxiety the sufferer still must believe and accept that there is actually no physical cause. The following case illustrates how to deconstruct this pattern of mental construction.

Deconstruction Meditation for Anxiety

Meditation can be applied to relieve anxiety when medical examination has shown that there is nothing wrong. The first step is to contemplate that sensation and cognition can be separated. In other words, a feeling and a thought are not identical. Pure sensation and the thoughts about the sensation can be distinguished, which helps manage the situation. Review the different forms of mindfulness: of body, of emotions, of mind, and of objects of mind. Refer to the mindfulness instructions and practice these meditations when not feeling anxious. As skills build, it becomes possible to draw on them even while feeling tense, especially if already familiar with the meditation.

When practicing mindfulness, make special note of how each sensation is a new one. Start with breathing. Notice that each breath is a new and different breath. Although the last breath resembles the next one, it is not the same breath. Everything is slightly different with each new moment. With the felt understanding of how each breath is unique, consider that each anxiety attack is unique too. Every time the anxiety occurs is a new time, a little different from the last one. Although the experience may seem to be exactly like the last time, it actually is unique and new. People often compress all the anxious experiences together into one dreaded experience. Follow each moment anew, to discover how each moment the anxiety is happening offers a new opportunity to learn to let it go. Relax with each breath as much as possible.

Mentally analyze the anxiety into its component parts. Anxiety can be thought of as a collection of moments during which a frightening interpretation is added to describe an uncomfortable sensation. This process tends to increase the intensity of the sensation, leading to a further interpretation, such as, "It's getting worse." Each successive moment includes an interpretation followed by an even stronger sensation in reaction. First, observe this spiraling process as it unfolds. Then, as soon as possible, observe the separate parts: sensations, interpretations, and emotions.

Try to recognize that the interpretations are separate from and are not the same as the sensations or the emotions.

Now extend the meditation. Notice how the components influence and are inseparably part of each other: worrying interpretations provoke more worry. And the worry brings on more discomfort, in a self-perpetuating cycle. Question the realistic certainty of this reaction with inwardly focused reassurance such as: "I know that my worry may not necessarily be what the sensation will bring; I have checked out my condition and I know that there is no physical problem."

Be patient with the process. Remember that time is a limiting factor. Recall a past anxiety attack when the discomfort passed after a certain amount of time. Remember that even when expecting the feeling would get worse, or that it might be a sign of terrible things, the anxiety sensations eventually passed.

Keep returning to staying in each moment as it comes, rather than anticipating what it might become. Allow breathing to find a natural rhythm appropriate to the moment. Try to carefully follow experiencing mindfully as it happens, including the periods of time that the feeling diminishes. In general, anxiety tends to become easier to handle, and eventually lessens in intensity, when the correct approach is used, and there is no physical cause. But it may take time.

Conclusion

Anxiety is a mind–brain–body pattern that becomes habitual and entrenched. Use several different components during treatment. Cognitive and emotional techniques help to work on the thoughts and feelings that are involved. But you should also work directly with the nervous system, by altering sensations, changing muscle habits, and intervening in the fear–stress pathway. Through multidimensional treatment, your therapeutic interventions will have a stronger, more effective influence.

Chapter 21
Finding a Better Balance for Depression and Bipolar Disorder

The neural circuitry of anxiety and the circuitry of addictions have been fairly well understood thanks to the many years of research on fear conditioning plus understanding of the fear/stress pathway in anxiety and the reward pathway in addictions. The release of dopamine is central. But the neural circuitry of mood disorders such as depression and bipolar disorder are only now beginning to be understood. We have new models of the brain areas involved in mood disorders, thanks to some of the neural imaging capabilities from EEG, TMS, and fMRI. As you become more sophisticated with the nervous system perspective, common sense treatments emerge to counter the disturbed patterns. We provide here some models that have been corroborated by multiple researchers, which offer a way to understand how these patterns influence the problem, along with treatments to overcome it.

The Depressed Brain

In general, when people are feeling depressed, they have decreased activity all over the brain. Areas that are most affected are the frontal lobe, usually more pronounced in the left hemisphere. But the results are mixed. While some studies found decreased blood flow in the anterior prefrontal cortex, others found hyper-metabolism in other prefrontal areas. As we delve more deeply into specific parts of the prefrontal cortex, we can account for the mixed findings. The increases and decreases make sense when we consider the symptoms of depression, increased negative self-appraisal and emotional rumination combined with decreased energy and slowed-down thinking. These levels were found to normalize after successful treatments using cognitive behavioral therapy and interpersonal psychotherapy (Goldapple et al. 2004) (Fig. 21.1).

C. A. Simpkins and A. M. Simpkins, *Neuroscience for Clinicians*, DOI: 10.1007/978-1-4614-4842-6_21, © Springer Science+Business Media New York 2013

Fig. 21.1 The depressed
brain

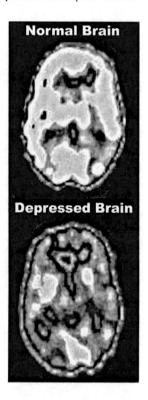

Much research on depression centers on changes in the anterior cingulate cortex (ACC). The ACC is often considered a bridge between attention and emotion (Devinsky et al. 1995), with two main parts. The dorsal part tends to be cognitive, connecting to the dorsolateral prefrontal cortex and the primary motor cortex. This part is active when we are engaged in executive functions and decision-making especially in monitoring conflicts. It connects with other areas of the PFC to enlist effective executive control from other areas to assist in resolving the conflict (Davidson et al. 2002). The rostral part is involved in emotions and connects to the amygdala, nucleus accumbens (where dopamine is made), hypothalamus, and the hippocampus, all key areas in the experiencing of emotions. This network of connections has been found to be involved in how we refer to ourselves internally.

When people are depressed, they show heightened activity in the rostral (emotional) side of their ACC. In addition, they have more connections from the rostral ACC than in people who are not depressed. You have undoubtedly noticed that your depressed clients often engage in repeating loops of rumination and self-doubt. While there is heightened connectivity, depressed people also have reduced activity in key parts of the frontal lobes that help to regulate working memory and executive decision-making. These patterns may explain why people who suffer from major depression have cognitive deficits.

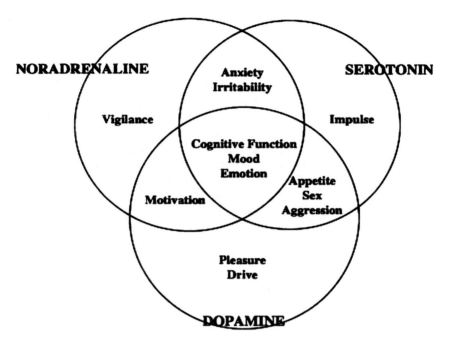

Fig. 21.2 Neurotransmitter interactions

Limbic areas are also affected. The hippocampus volume is consistently found to be smaller in people suffering from major depression (Frodl et al. 2008). Hippocampus dysfunction is also exacerbated by high stress level, leading to poor emotional regulation in depression.

This reduction in hippocampal size may be a result of depression rather than the cause. Reduced size can be altered by treatment. The hippocampus can regenerate from the effects of antidepressants (Chen et al. 2000), and from new learning, positive therapeutic situations, enriched environments (Eriksson et al. 1998), and experiences of novelty (Rossi 2012).

Neurotransmitters and Depression

Depression has been linked to disruption of three neurotransmitters: norepinephrine, serotonin, and dopamine. This diagram in Fig. 21.2 shows how the neurotransmitters are thought to interact.

Norepinephrine is involved in arousal and attention. Norepinephrine travels throughout the entire brain. It is produced in the pons, which is located in the brainstem. Recall that the brainstem regulates our level of alertness and arousal, therefore you would expect that less norepinephrine would correlate with the lower energy experienced during depression. Serotonin modulates mood and sleep. It is released from a part of the brainstem, the raphe nuclei. When depressed, people often have difficulty sleeping and of course suffer with mood problems. Dopamine flows out from the nucleus accumbens and brings feelings of pleasure and reward. Certainly, depressed individuals tend to feel a loss of pleasure in their lives.

Drug therapies alter the balance of neurotransmitters. MAO inhibitors restrain the enzyme monoamine oxidase (MAO). The MAO enzyme breaks down dopamine, norepinephrine, and serotonin at their synapses, so when you inhibit an inhibitor of the neurotransmitter, the result is more neurotransmitter activity. These drugs worked well for depression but have a byproduct: They also inhibit the breakdown of tyramine, which can lead to high blood pressure and risk of stroke. A newer group of antidepressants, the tricyclic antidepressants such as Elavil, Parmelor, and Norpramin were developed. But these drugs still have risks for patients with concurrent cardiovascular disease. More recently, a different approach was found in SSRIs, selective serotonin reuptake inhibitors. SSRIs such as Zoloft, Paxil, and Asima act differently, slowing the reuptake of serotonin at the synapses. Newer versions, SNRIs, which are serotonin-norepinephrine reuptake inhibitors, such as Cymalta, Effexor, and Remeron inhibit both serotonin and norepinephrine. Many people are helped by these medications, but effectiveness varies greatly (Levinthal 2008).

Effects of Treatments

Different types of treatments have diverse effects. For example, psychotherapy makes people happier, whereas drug therapy just stops people from feeling sad. In a comparative study of treatments for depression using either CBT or Paxil, both groups improved equally on the Hamilton Depression Rating Scale (Goldapple et al. 2004). Paxil subjects had increases in their prefrontal cortex along with decreases in the hippocampus and subgenual cingulate. The subgenual cingulate has been labeled the sad cingulate (Mayberg et al. 2000). This finding makes sense, since SSRIs like Paxil do not increase serotonin, they just stop it from decreasing.

Psychotherapy alters the brain chemistry in a different way. In one research study on this, the CBT group had certain accompanying metabolic changes that included increases in the hippocampus and anterior cingulate, two areas that are associated with optimism (Sharot et al. 2007) along with decreases in the dorsal ventral and medial frontal cortex, associated with rumination. CBT does not simply stop negative thinking; it also enhances positive, optimistic thought patterns. When meditation is used as part of psychotherapy, it enhances feelings of wellbeing (Davidson et al. 2003). Hypnosis, which deactivates the ruminating prefrontal cortex, provides moments of relief from depression (Yapko 2006).

Fig. 21.3 The bipolar brain

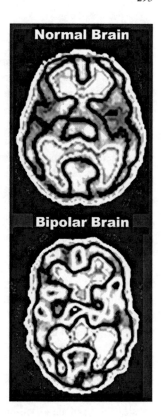

Individual differences can be better addressed by having several options to offer, not just between drug therapy and psychotherapy, but also among the many established forms of psychotherapies, each of which fosters unique brain alterations.

The Bipolar Brain

People who are diagnosed with bipolar disorder are likely to have irregularities in their limbic system. There is no final word on how bipolar disorder changes the limbic system, but trends are emerging. These findings help to explain why the emotional moods of bipolar disorder feel so compelling. Since the late 1999s, studies have found an enlargement of the amygdala (Strakowski et al. 1999). Since bipolar disorder is such a dynamic problem, with strong emotional moods, these results are not surprising. Thus, therapy that normalizes the limbic system is likely to be helpful. Research shows that meditation calms an over-activated limbic system in a number of different ways. This and other therapies that work bottom-up may offer new treatments to help bipolar sufferers find balance (See Simpkins and Simpkins 2012) (Fig. 21.3).

Another finding that seems to be consistent across many studies is that the thinking and control areas of the brain in the temporal and frontal cortices tend to be smaller. Researchers found smaller temporal lobe volumes on both the left and right sides in bipolar disorder patients than in people who were not bipolar (El-Badri et al. 2006). The temporal lobes, being so closely linked to the limbic system, are involved in emotion. Thus, having a smaller volume in these thinking areas of the brain may contribute to the severity and difficulty in controlling moods. The smaller volumes in the prefrontal cortex (Bremmer 2005) helps to explains many of the behaviors associated with bipolar disorder, since these areas are activated when people are performing executive functions like planning, working toward a goal, or making decisions. All of these activities feel difficult to do when people are in the depressed phase. And when they are manic, they often make poor decisions, do not plan well, and are likely to be scattered in their approach to accomplishing their goals. Cognitive and meditative therapies help to address these problems by stimulating key areas in the PFC executive function areas.

Here is another bit of compelling evidence that helps to explain some of the bipolar reactions. You will note in the picture of the limbic system, that the cingulate gyrus, a part of the cerebral cortex, wraps around the limbic system. As described in the section on depression, the cingulate gyrus becomes activated when we want to control our emotions and moods. The cingulate gyrus connects the limbic system structures to the cortex. The cingulate gyrus is involved in self-regulation. Bipolar patients have lower density in their cingulate gyrus as compared to people without bipolar disorder (Benes et al. 2001). In a sense, the brain structures for thought and emotion are not as they should be for optimal functioning in bipolar individuals.

Bipolar disorder has been treated successfully with lithium carbonate and other compounds. Lithium acts on the neurotransmitters serotonin and norepinephrine as well as dopamine to help balance mood, arousal, and emotion.

How Treatments Change the Brain

Therapeutic treatments are known to stimulate helpful brain changes. The brain and mind are an interacting unity of patterned systems in motion. You can utilize these patterns. They reveal dimensions that add to the understanding of clients' problems. Therapists know that effective treatment requires making an intervention to treat the underlying problem, but the real problem may not be the presenting complaint. Sometimes the underlying problem is clearly expressed in the symptom. More often, the problem is complex and hidden. You must apply your expertise to "read" the problem, inferring it through indirect signs communicated in the client's body, behavior, and situation. With the aid of the brain–mind patterns, you can gain a valuable tool to help uncover the patterns involved.

A Comprehensive Three-Component Treatment of Depression: Brain, Mind, and Social

Jerome was an older man who wanted therapy to help him handle the severe discomfort and low energy he was suffering. After a battery of tests, his family doctor and several specialists determined that his physical problems would benefit from psychotherapy, and so recommended it. He still worked part time as an accountant, but was planning to retire in a few more months. His routine was to work in the morning, then come home and lie down on the couch for the rest of the day. We noticed that even though Jerome showed the signs of depression, he still had a charming way of relating.

We taught Jerome some simple relaxation and breathing meditations. We also showed him some of the movement exercises included later in this chapter. He learned non-judgmental awareness and mindfulness meditation. He found these exercises relaxing and said that he truly appreciated being able to relax. He felt better for a few days following each session, but then he felt his physical discomforts and low energy return as the week progressed.

We encouraged him to keep practicing the meditations at home between sessions, but he could only sustain it for a few minutes at a time. It is important to accept what the client can do, and work from there, so we complimented him on trying and suggested he do brief meditations, between one and 5 min in length, multiple times a day. As he became more aware, he had an insight. He had been so sure that it was time to retire, that he had narrowed down his life such an extent that it had lost its meaning, and no longer satisfied him. He felt moments of well-being in meditation which seemed to rekindle his belief that happiness might still be possible for him. He searched back to memories of fulfilling experiences from his past. He recalled that he was happiest when he had worked in sales. He liked interacting with people, which allowed him to express his warm personality. He looked into working part time in sales and found a job at the local university, which hired elderly people to work in the campus bookstore. As soon as his accounting job was finished, he took the bookstore job. He made new friends and found little time for lying on the couch. As he became more active, his discomforts subsided even more.

This case is an example of how treatments are better when they work on several different levels. Jerome learned a valuable tool using meditation to relax and lessen his discomforts. But he also needed more awareness. Being mindful of his everyday experience helped him to cognitively rethink some of his life choices. And as his nervous system balance became more normalized, he had the energy he needed to make real world changes. In his sales job, he interacted with people and involved himself in new relationships. Working on the problem at many different levels, Jerome overcame his depression and opened a new chapter in his life.

You can view depression as a network of interactions among the brain, mind, and interpersonal relationships. Depression affects an individual's brain, which interacts in a network with rational thinking processes and the quality of interpersonal relationships. Thus, you can best treat depression by addressing all three components for the greatest success. The brain is altered by mood problems, but cognitions and relationships have an impact on the brain. Change works both ways, and so wherever you initiate a change, you set a positive process in motion.

Often therapists just treat one aspect, such as when a psychiatrist simply administers drug therapy, or when a therapist only performs a cognitive form of therapy alone. The exercises given here will have a dramatic effect by addressing multiple levels, thereby changing the brain by altering cognitions, working with emotions, and improving relationships. We offer a few possible methods for each of the components of depression. We encourage you to think of other exercises that utilize your own methods and therapeutic approach, as you undoubtedly will.

Rebalancing the Nervous System

Depression involves an imbalance in the brain, with over-activations in some areas and under-activations in others as described earlier in this chapter.

The depressed individual who suffers from an overall lowering of energy may find a return to balance by a 2-fold treatment approach. The overactive pre-frontal cortex that has been engaged in too much rumination and self-criticism must be de-activated and calmed, perhaps using hypnosis or meditation. But at the same time, low affect needs to be stimulated by activating the emotional limbic areas, by applying CBT to engage more thoughtful executive functioning to intervene.

Activating a Sluggish Nervous System: Circling

Movement can alter the balance of the nervous system. Perform these slow, regular movements to energize the nervous system when it is chronically low, as in those suffering from depression. These exercises are easy to do and circulate energy naturally. Plus, they feel good! (Fig. 21.4).

Begin by simply standing and noticing the sensations in your body. After a minute or so, step out with one leg so that you are standing with your legs comfortably apart and knees slightly bent. Extend your arm out and scoop down and across the front of your body, then circle your arm around overhead and then come around again. Repeat this slow, circling motion, allowing your whole body to sway with the motion.. Synchronize comfortable breathing with the movements by inhaling as you extend your arm down in front, and then exhaling as you raise your arm high overhead. We tell children who are practicing this exercise to imagine that they are scooping up water and then dropping it on their heads! Perform this circling movement for several minutes. Then stop and perform with the other arm. Remember to allow your whole body to move together, in harmony with your breathing. After several minutes on both sides, stop, stand upright, and

Fig. 21.4 Circling

sense your body. You may notice that you have tingling or sense of an increase in energy all through your body.

Using the Ideomotor Link to Calm an Over-Activated Nervous System

The mind and body are connected through the ideomotor link, first elucidated by William James. When we hold an idea in mind, it is expressed in the body. For example, vividly imagine sucking on a tart lemon. You probably will find your mouth watering automatically. This direct link between mind and body travels a short path in the brain. You can use the ideomotor link to calm an over-activated ruminating cortex while activating the lower brain areas for alertness and energy, automatically and effortlessly.

Think about a time when you felt good. Perhaps you were with friends, on vacation out in nature or visiting a fascinating city filled with museums, shops, and architecture. Or maybe you were enjoying peaceful time off at home with few responsibilities and the time to engage in your favorite hobby or sport. Vividly imagine being there in this situation. Recall what you saw, thought, and felt, as if you are there now. Keep this idea firmly in mind for several minutes as you imaginatively move around in this situation, being there and experiencing it now again. Do not try to deliberately change how you are feeling, simply allow your natural response to occur as it did when you were enjoying this wonderful experience. Through the ideomotor effect, your brain and body will respond spontaneously with a more comfortable balance.

Cognitive Change: Shifting from Inner Rumination to Mindful Awareness

Maria was a housewife. She worked for her husband part time as a secretary. She came to therapy because she felt depressed. She resented her family, who she felt were giving her a hard time. She decided to try therapy rather than drug therapy, to see if she could make herself feel better naturally. As she described her grievances with us, we could see that she was also looking for a sympathetic ally against her family. We suspected that she had something to learn.

We taught her how to practice mindfulness meditation, to notice her experiencing without judging it good or bad. She learned to accept each moment as it was, simply noticing. She did it while she folded laundry, cooked the dinner, and walked the dog. As she became aware, she noticed her feelings of resentment toward her family. She heard herself saying that they were making her life difficult. We encouraged her to continue to notice her feelings without judging them good or bad, justified or not. As she continued to observe in this way, she began to see that she was behaving in a resentful way toward her family. She saw herself snap at her husband and yell at her children. She looked deeper and could perceive that her family was not really provoking her outbursts. Gradually she began to recognize their kindness and understanding, which she had not noticed before. She realized that her depressed mood was influencing how she experienced her family and herself. Through her daily meditations, she began to accept what she felt as she felt it, and paradoxically she found herself feeling less disturbed and more in touch. She felt pleasure in the simple things she had been ignoring, like a bird in the window, or a flower blooming in the garden. Her mood began to lift and her interactions with her husband and children changed. She felt her love for them, something she had not felt for a long time. By the time her therapy was over, she reported that she enjoyed her time with her family more than ever, and spent her time planning fun things to do together, rather than ruminating about all the bad times.

Sometimes people get out of touch with their real-life situation, and instead project fantasy resentments on those around them. The nervous system becomes habitually over-activated, leading to irritable and nervous reactions. Their interpersonal relationships suffer. If you guide depressed clients into more accurate experiencing, they can become aware of what is actually happening without making disturbing judgments. Then they will rediscover the love that they may have been missing in their relationships.

Cognitive therapy can help people to sort out the issues, using the brain in a top-down manner. But change can also be initiated both top-down and bottom-up, by using meditation. Mindfulness can begin a process that alters the cognitive process itself. By shifting attention away from judgmental ruminations, mindfulness brings people back to real-life experiencing with more neutral, natural, alert attention. Cognitions then are tied to the experience directly, bypassing some of the biased thoughts and feelings that perpetuate depression.

Developing Non-Judgmental Acceptance

Mindful awareness should be non-judgmental. Like a researcher who is gathering data, set aside judgment. Do not jump to conclusions or use the new information gained to form biased opinions. Suspend judgment until more data is gathered. Trust the process of awareness and cultivate an open mind.

If you notice a trait that you do not like, take note of it. This may be a quality to change, but do so without self-criticism. There is an important difference between simply observing something that may need changing and evaluating whether it is good or bad. Non-judgmental observation often brings a lessening of defenses and a more open attitude. So while performing the mindfulness exercises, try to observe without making evaluative judgments: just become aware in the situation, without taking a position.

To become aware of what is there, keep observations clear and descriptive. Learn to accept each experience, without making comparisons or criticisms. Then, the finer qualities of experience can be fully appreciated, just as they are.

To apply this non-judgmental attitude to mindfulness, survey yourself from head to toe and recognize all the different parts. Traditionally, Buddhists counted thirty-two body parts. Describe each part. Notice, for example, the hair, its color, texture, style; the eyes, the eyebrows, etc. But stay factual. For example, observe that the hair is, for example, long, dark brown, and curly. Try not to use evaluative terms, such as unattractive or attractive, too long, too short, or perhaps, not straight enough. What happens?

You may have to guide your clients who want to reject their appearance or believe they have serious flaws toward a more neutral attitude. The great Zen Buddhist teacher Lin Chi often told his disciples that nothing is missing (Watson 1968). Problems occur when people step away from what is actually there. Everyone is fully equipped to be mindful. And if they are willing to truly just notice without adding anything extra, they will find that their negative appraisals will dissipate.

Body Mindfulness in Everyday Routines

Warm up to mindfulness with a generalized sense of body awareness. When going about daily activities, take a moment to notice body sensations. Start by noticing the body positions and movements at times while sitting, standing, lying down, and walking. People often pay little attention to such fundamentals. But body sensations are part of every activity, an important and valuable inroad to attuning. So, when first awakening in the morning, begin by taking a moment to experience lying in bed. Then pay attention to the process of sitting up, stepping onto the floor, and slowly standing up. Take note of various body positions whenever possible through out the day.

Delve a little deeper and pay attention to the body sensations of being in a certain position. For example, when sitting on a chair somewhere at home or at work, stop to pay attention to posture. Notice the quality of sitting: Is it straight, leaning, or slumping into the chair? Do you take support from the chair or push down on the seat? Where are the feet? Pay close attention to all the details.

Turn your attention mindfully to your body at various moments throughout your day. Notice, without judging, just what you are experiencing in your body moment-by-moment as you open a door, lift your fork as you are eating, or hear the sound of your own voice as you speak. Do this regularly, from time to time, for a minute or two at a time. Gradually you will develop a tendency to be aware without ruminating about it.

Mindfulness of Feelings

Emotions are an important component of living, and so mindfulness must include attention to feelings. This gives a strategy for dealing with emotions in a way that will overcome suffering from uncomfortable feelings and maximize fulfillment from positive ones. Mindfulness is key with a clear method for applying it.

Feelings can be evaluated as pleasant, unpleasant, or neutral. People tend to cling to pleasant feelings and reject unpleasant ones. And this clinging and rejecting sets in motion a secondary reaction that interferes with awareness and causes suffering. Instead, pay attention to the feelings themselves. Do not add evaluations. Then the secondary reaction that follows can be dropped or modified. The process leads to more comfort in reactions.

Identify Feelings

Mindfulness of feelings begins with first identifying the feeling being experienced. Studies show that altering the labeling and appraisal of emotions lowers the activation of the amygdala, the area responsible for emotional experiencing. This meditation develops the skill of affect labeling by identifying feelings, using this technique.

To start the process, sit down for a moment and close your eyes. Turn attention inwards. Try to put a name to the emotion or mood. Then match the description with what you felt. If it is not quite right, modify the label until you feel satisfied that the description fits the feeling.

While doing so, try to be like an impartial judge. When presiding over a case, the judge does not become angry. Instead, she tries to calmly attend to the facts and come to the best decision. Objectively observe all your feelings, even the ones that could be labeled unpleasant.

Mindful Attention to Feelings

When out of touch, emotions seem to take on a power of their own, pushing and pulling in many directions. Awareness puts people in the center of their experience, giving greater ability to understand and manage feelings maturely.

Begin by noticing pleasant feelings. While in the midst of a pleasant experience, stop and take a moment to observe. Feel the accompanying sensations. You may want to sit down and close the eyes. With practice, it gets easier to sense a feeling without needing to stop.

Next, try noticing neutral feelings. Work gradually toward becoming aware of the unpleasant ones. For example, practice being aware of mild annoyance before trying to become aware of strong anger. Build awareness skills gradually, and they will be ready when needed. Mindfulness responds to practice.

Pay close attention to the feeling. Notice that the feeling is not just one quality, but rather is a collection of different sensations. For example, when annoyed there may also be accompanying sensations such as butterflies in the stomach, quickening of the breathing, and heat in the face. Get to know the feeling and all its accompanying aspects. In time, you will be able to discern that even the strongest feeling is a combination of sensations. Notice as many aspects as possible, without judgment. With careful, objective attention that deconstructs the feeling into sensations, the strength of the emotion behind it, as seen in all its parts, tends to lessen.

Stay with the feeling, moment-by-moment. Often the feeling begins to change under close observation. Notice how each moment is different. The annoyance felt at first may be altered now. Observe the differences. Embrace the feelings and accept them as just what they are, this feeling now, and that feeling later, an ongoing transformation.

Mindfulness of Mind

Mindfulness of mind involves observing mental activity itself. Cognitive psychology is a modern discipline devoted to understanding the nature of mental processes. These methods of mindfulness of mind are compatible with the findings of cognitive psychology about how the mind works.

Mental processes takes on many different forms, filling our minds with one thought after another. At times, we think with clarity while other times our thoughts are confused. Sometimes the mind is filled with emotion and other times it is completely unemotional. But, if you step back and look at the broader picture, you notice that all of these different states of mind are actually mental processes.

Mindfulness of the mind begins by first recognizing thoughts while thinking them. Sit quietly and close the eyes. Notice each thought as it occurs. Follow the flow of each thought. Be like someone sitting on the bank of a river watching leaves and twigs flow down a river. Do not jump into the river but stay back on shore observing. Keep observing and letting each thought drift past. If you find yourself drifting downstream with a thought, climb back on shore and resume the observing as soon as possible.

Recognize Mental Constructions

Next, try to understand the cognitive process itself. People do not usually experience the world directly. Rather they recognize it through a combination of sensory experiencing and brain processing. The term recognition comes from the Greek, gnosis. Knowing involves a two-part sense of re-cognizing. We now know that primary cognitive processing is complex and multifaceted. The brain disorder of visual agnosia can serve as an example of how perception involves the combination of the visual system and cortical action. Visual agnosia causes blindness, even when there is no injury to the eyes. The brain plays a vital role in vision. Seeing takes place through the eyes, not just with the eyes. Similarly, other areas of the brain are part of hearing, smelling, tasting, or touching.

Thus, when viewing a flower, the flower is perceived through the brain's participation with the information received from the senses. Perception involves meanings of flowers, both personal and shared with the culture, with sensory experience of the flower.

Experience is constructed, and mindful awareness of perceptual processing can clarify how this occurs. The sense of permanence comes from these mental formations—the concepts, the meanings, the abstractions that endure in the mind.

Mindfulness of the Objects of Mind

Typically thinking is intentional. We think about something. Similarly, when feeling an emotion, it is directed toward something or someone. Similarly, sensory experiences are of an object or a person. Consciousness is filled with an object of mind. The Chinese word for perception is made up of two ideograms: sign and mind.

When we try to experience the world outside of consciousness, we discover that we cannot easily do so. How can we step outside consciousness and know the world? The objects we think are outside of ourselves in the world are actually in relationship to us, connected as the objects of mind. Through the mind, experience of the world comes to be. Mindfulness gets one in touch with this constructive process, and meditation helps develop the vision to see beyond what appears.

Mindful in the Moment

Now bring all the ways of being mindful together to the present moment. Scan the body to raise body awareness. To become mindful of emotions, observe the feelings. Notice mental activity: thoughts and perceptions. Pay close attention to the objects of the perceptions. Perform these four qualities of mindfulness does not need to take too long. Once centered in the moment's experiencing, let all this go and just be present.

Notice how experience transforms, moment-by-moment. Whenever possible, turn the attention to experiencing. Get in touch mindfully as often as possible. In time, mindfulness will become habitual and natural. In balance with oneself and the surroundings, accept the flow of life as it comes and act in harmony with what is needed. Stay with each moment of experience anew.

Mindfulness Meditation in Action

Depression often involves focus turned inwards with rumination. Meditation can help to turn the focus outward, to help overcome a depressed mood. By turning attention toward simple activities, you can help clients alter the brain pathways. Meditation on daily chores is a good way to begin, as Maria did. You shift attention away from the ventromedial cortex toward external stimuli, thereby enhancing awareness of sensory input. The nervous system will return to a natural balance, aware and in touch. Any issues that arise during the mindfulness meditations can be discussed and worked on using the usual cognitive and dynamic methods.

Enhancing Relationships with Shared Meditation

The quality of relationships can be enhanced using meditation. This series is written for use with a couple, but it can also be applied to a family or with any significant other.

Meditating as a Couple or Family

Couples and families who meditate together will find a bond begins to form from the quiet seat of the shared experience. *Sit together on a couch or on the floor and quiet the thoughts for a few minutes. Then, sense your body position, wherever you*

are seated. Feel the sensations of sitting on the chair. Pay attention to the rela-
tionship with the seat. Note how you push down as the chair pushes back with an
equal and opposite force to keep you supported. Let awareness range out further to
notice your partner. Can you hear his (her) breathing? Can you sense how the
temperature might be warmer on the side from your partner? Allow yourself to
sense the sharing in this moment, to be present with your partner here and now.
Recognize how both of you are participating together in this shared experience,
and enjoy a moment of sharing. Allow your feelings of good will, perhaps even
love, to extend outward to your partner.

Extending Love and Compassion

As you develop your awareness of this shared moment with your partner, imagi-
natively move out to other family members, housemates, etc. Feel the presence of
the others with feelings of love and compassion for them. Next, range out further to
people in the neighborhood. Sense their being, and feel the connection to them.
Search for feelings of compassion for others by caring about their concerns and
their struggles. Intend good will and kindness toward them, just as in the natural
instinct to care about a child in distress. Keep extending outward, with feelings of
good will to the city, the state, the country, and the world. Become aware of the
many ways of interdependence. As you allow your own feelings of good will to
develop, you will feel your mood lift.

Conclusion

Depression and bipolar disorder can be stubborn problems. But you can make your
treatments more effective by incorporating both bottom-up and top-down methods.
You can alter the brain balance along with the cognitive disturbances while also
treating the symptoms of low or high energy, adding body techniques. Be gentle
and gradual, without pushing clients too hard or too fast. By accepting what they
are capable of, you will help to build their confidence and positive expectancy, so
important for overcoming depression.

Chapter 22
Recovering from Addiction

Brain-based theories of addiction generally follow the idea that a neurological modification occurs from drug use. Dependency is not a matter of loose morals, bad choices, or ethnic background. Substance abuse is a direct result of modifications in hardwiring of the brain (Kauer and Malenka 2007). As a form of learning and memory, addiction involves changes in particular synapses of the reward pathway that rewire around the substance through the process of LTP. Behaviors, emotions, and cognitions become biased by the feedback received from the brain's signals. And so, these alterations in brain structures and functions account for the difficulties people have in giving up their substance.

Psychotherapists are well suited to deal with learning and memory. Therefore, these neurological changes can be addressed therapeutically, and in fact, our interventions have a powerful effect. If you know how addiction alters the brain, treatment strategies to counter these effects become clear. You will know how to devise techniques to effectively counter these changes. And as the brain begins to rewire back to normal, substance abusers will find it easier to give up their drug and stay free of relapses.

A Universal, Non-Specific Mind–Brain Change in Addiction: A Compromised Reward Pathway

Each drug of abuse has specific effects on the mind–brain system. However, there is one brain change, common to all forms of abuse: the reward pathway of the brain is altered. In fact, when someone is addicted to a substance, such as alcohol, cocaine, or heroin, the reward system of the brain rewires to include that drug (Figs. 22.1 and 22.2).

Recall that the reward pathway is a natural, built-in response of the nervous system that helps to reinforce healthy, life-promoting behaviors. The brain reward

C. A. Simpkins and A. M. Simpkins, *Neuroscience for Clinicians*,
DOI: 10.1007/978-1-4614-4842-6_22,
© Springer Science+Business Media New York 2013

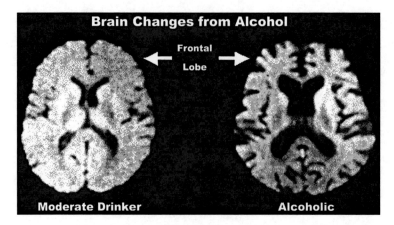

Fig. 22.1 Brain changes for alcohol

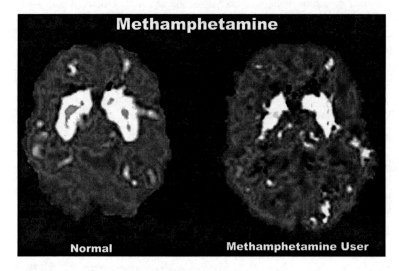

Fig. 22.2 Brain changes from methamphetamine

pathway becomes activated with the release of dopamine (DA), beginning in the midbrain, projecting to the forebrain, and then going back to the midbrain (See Fig. 9.5). Normally, when the cortex receives and processes a rewarding sensory stimulus, such as from eating a favorite food, it sends a signal activating a part of the midbrain, the VTA. The VTA then releases DA to the nucleus accumbens (NA), the septum, the amygdala, and the prefrontal cortex. The NA activates the motor functions while the prefrontal cortex focuses the attention. DA transmits the sensation of pleasure, the reward for eating your favorite food. The endocrine and the autonomic nervous systems interact via the hypothalamus and the pituitary to modulate the reward pathway. We described the normal function of the reward pathway in Chap. 9. This pathway is responsible for driving the normal feelings of pleasure that we get when we engage in behaviors that are necessary for survival, such as eating, drinking, and sex.

When a person takes a drug such as alcohol, cocaine, heroin, or any other of the substances commonly abused, DA circulates through the reward pathway into the prefrontal cortex and into the limbic system. The individual feels pleasure. And so, drug dependence begins with positive reinforcement from the pleasure associated with using the substance. The hippocampus remembers the contexts of the drug reward so that a learned association is made. The midbrain, forebrain, and neurotransmitter systems activate the positive sensation from taking the drug. The drug dramatically alters the action of the synapses, stimulating and wiring the reward pathway in a way that links the drug effect all the way down to the neurons. In a sense, the reward pathway becomes a drug reward pathway.

There is a certain individual level of how much pleasure and vitality each person has, much like Jung's concept of libido and Daoism's idea of Qi. The body tends to self-regulate (Squire et al. 2003). But when an addict abuses a drug, he or she expends all the pleasure at once. The drug inhibits the release of GABA, which is like taking the foot off the brakes, thereby releasing more DA. From the over-expenditure of DA, the reward pathway becomes compromised.

The brain adapts to the continued use of the drug with sensitization. The presence of the drug becomes part of the body's attempt to normalize toward homeostasis. A new allostatic balance forms, different from the normal homeostatic balance, including the drug as part of the pathway. In a sense, the reward pathway seeks stability through change (Koob and Le Moal 2001). More quantities of the drug are required for a similar effect as tolerance builds (Squire et al. 2003). The problem with the allostatic balance is that it is a rigid, fixed system. For the drug addict, there is very little natural response to the everyday pleasures of life. All of the addict's feelings of pleasure are now tied in with drug use. Clients find life flat and colorless without their substance. This loss of interest in life itself is well explained by the compromise of their reward pathway.

Withdrawal feelings that occur when the user has been away from the drug give an opposite experience including negative affect, anxiety, and strong physical discomfort. The marked and long-lasting change in DA and serotonin levels is opposite to the effect when the drug is taken, leaving the addict feeling depleted. Anxiety and stress are common elements of dependence and withdrawal. The homeostatic balance has been altered by the challenge to body, mind, and brain. This reaction is highly stressful and activates a stress response, as was described in Chap. 9.

Treatments for Substance Abuse:
Restoring Normal Functioning

We have seen how the neurological components of substance abuse involve a compromise of the reward pathway in the brain. The organism's natural tendency to enjoy healthy pleasures and avoid harmful ones is no longer functioning properly. Simply using direct, cognitive methods may fall short when the brain balance is giving false signals. Instead, encourage the brain back to healthy functioning through the pathways, and foster more balanced brain functioning. Then the mind, brain, and body can work in harmony making it more likely to avoid relapse and stay on the road to recovery. The exercises that follow offer ways to help shift the reward pathway back to normal functioning.

Activating the Unconscious to Balance the Reward System

Hypnosis and meditation are useful resources for regulating and coping with the process. Relaxing and calming is a helpful starting place. Developing calm deliberately can help to lower activation levels in the fear–stress pathway. Clients will feel some relief of tensions while also gaining the confidence needed to begin a withdrawal process. Use meditation and/or hypnosis to lower the overall tension level. Shift in cycles between relaxing for a brief period, then working on the underlying problems that fuel drug use. We provide hypnotic methods first, followed by meditation methods.

You can use the client's automatic tendencies to help find the way into relaxation naturally. There are inclinations to favor one way of perceiving over another. Everyone uses all of these ways at times, but we tend to rely on one more than others. If you get to know your preferred mode of perceiving, you can then use it to help you relax more naturally.

Close your eyes and think about when you last entered your house. How do you remember your arrival? Do you see a picture of yourself walking through the door? If so, you probably tend to orient visually.

Do you remember the physical sensations of walking to the entrance and opening the door? Then, your tendency is to be sensory in your orientation.

Or maybe you recall the sound of the door opening and closing? If this is more vivid, your preferred perceptual mode is likely to be auditory.

Another possibility is that you might tie your experience to a thought you were having as you opened the door, such as, "I was really looking forward to getting home," or "I was thinking about what I did last night." If so, you may tend to orient cognitively.

Some people recall what they were feeling, such as "I was feeling glad to be home." If you recall feelings most strongly, you tend to orient emotionally to things. Whatever way, you tend to recall, use associated perceptual modes to initiate your relaxation.

Inducing Relaxation Unconsciously

Begin by choosing your most natural way to imagine something. We offer a simple way to induce relaxation using each of the perceptual modes. Use the one that comes naturally to you. At another time, you might want to experiment with others.

Body Sensations: Sit comfortably in a chair with your feet flat on the floor. Place the palms of your hands on your knees, resting gently. Close your eyes and notice the sensations in one of your hands. Ask yourself; does my hand feel light or heavy? Wait for a sensation to happen that will give you your answer. Do you immediately feel like your hand is distinctively light or heavy? Or do you feel a mild sensation? You might feel something else, such as a sensation of warmth, coolness, tingling, or other possible sensations.

Imagery: Sit or lie down comfortably. Imagine a place where you felt very relaxed and comfortable. Perhaps it is a beautiful place where you vacationed, the beach, the woods, a favorite restaurant, a museum, or even somewhere in your own home. Picture this calm scene. Notice as many details as you can: the objects, colors, textures, lighting, etc. If possible, imagine yourself there. Let yourself relax for a few moments as you enjoy the image.

Sound: Close your eyes and suggest to yourself that you can hear a song playing, perhaps some music that you especially like. Begin by remembering the song and then allow it to continue of itself. Can you hear the melody and enjoy listening to it? As you listen, let yourself become more relaxed.

Scent: Imagine your favorite perfume, a fragrant flower, and the smell of pine or fresh-salty sea air. With each breath in, increase the fragrance. Draw on past-experience if helpful.

Feeling: Recall a time when you felt completely calm and happy. Perhaps you were soaking in the tub or taking a relaxed walk with a friend. Let yourself remember the feeling as vividly as possible.

Thought: Think about your idea of hypnosis. What do you think it would be like? Would you feel comfortable? Do you think your muscles would be relaxed? Do you think your thoughts might slow down or center on relaxing ideas? Think about hypnosis and let yourself experience what it might be like for you.

For all perceptual modes: If something different from the suggested sensation happens, accept it as your own unique responsiveness. You will find many ways to express yourself, and all of these responses can be useful. A few people may feel nothing. If this is your response, try to pay close attention to how this nothing feels and let it develop. You can learn to build on whatever your unconscious mind presents to you, even if the experience seems minimal.

Deepening the Experience with Hypnosis

Once you begin to feel some response, even if it is subtle, you can guide toward deepening the experience using hypnosis. Here is an example of a typical narrative to guide your client into a hypnotic experience of absorption and inwardly focused attention.

As you sit quietly and allow yourself to focus, you can become very relaxed. The process engages a subtle communication between your conscious and unconscious processes. Consciously you want something, for example, I want to be relaxed. But the unconscious does not always respond to demands or wishes. Rather, you can offer what you would like to bring about. Then wait, allowing your unconscious the freedom to discover the best way to make it happen. You do not really know just how your unconscious would like to relax, but you can be confident that it will find a way.

Hypnotic Calm through a Memory

Can you recall a memory of a time when you were feeling completely at ease? Perhaps you will recall through your preferred perceptual mode. But do not try to limit how you remember, just ask your unconscious to come up with a memory and then sit back, and wait. You may be surprised with the memory your unconscious provides. Be aware of even the subtlest part of a memory. You may remember vividly or only vaguely, in pieces. Be patient. As you wait, your breathing rate can become slower, more relaxed, as your muscles settle and your thoughts slow down. Enjoy the memory and let yourself relax very deeply as you think about and even re-experience the memory.

Apply a 4-Step Method for Changing Negative Self-Suggestions

Self-suggestion can contribute to the perpetuation of an addiction, as users tell themselves repeatedly that their drug use brings them pleasure. Many such self-suggestions are inaccurate misconceptions. You cannot stop a negative or harmful self-suggestion by simply forcing yourself to stop it. New Year's resolutions are a good example of how consciously directing change rarely works. We provide here a better method to elicit change that combines the conscious and unconscious together. The exercise uses an example of a typical negative self-suggestion that substance abusers often tell themselves: drug use is relaxing, invigorating, and/or somehow pleasurable. Other typical misconceptions might be that substances are "cool," or fit in with others who also use. Each person undoubtedly has his or her own self-suggestions to put through this 4-step method. You can modify these questions to fit any kind of addiction.

Observe, Question, Imagine, Allow

Observe: *What are you telling yourself about your substance? Perhaps you are saying, "I only feel good when I'm using." Or perhaps, "Using is cool, and so when I use, I'm cool." Or even, "I want to quit, but I can't."*

Question: *These statements are your beliefs and take on the form of a self-suggestion that often works against you. When you tell yourself, I only feel good when I am using, you are suggesting that nothing else brings enjoyment. So, question what you might be telling yourself. For example, if you believe that drug use is pleasurable, turn it into a question: Does using this drug really bring as much pleasure as I think? Does it always bring pleasure? What about after the drug wears off? And do the expenses and the risks really bring pleasure? Or is there also a great deal of suffering that goes along with substance abuse? Gather information about its real, physical effects. Substances, such as cigarettes, alcohol, heroin, and cocaine actually harm the body and brain. Do you really want this?*

And if you believe you cannot quit, question this belief. Point out to yourself that you are making a prediction about a future, asserting a certainty about something that is open. Here, the meditative, open mind attitude can help.

Imagine: *Contemplate the situation as it really is: the source of pleasure is within you, not the drug. Vividly imagine having enjoyment without the substance. Picture yourself in a situation when you found pleasure, either one from long ago before you used the substance, or a time recently when you felt enjoyment. Perhaps you were with friends, relaxing at home, on a vacation, or out in nature. Picture this pleasurable experience without your substance as vividly as possible.*

Or, contemplate the fact that the future is open. Vividly imagine yourself succeeding in overcoming your addiction.

Allow: *Now just allow your response as you keep imagining that enjoyable experience. You will feel a comfortable, pleasurable feeling begin to arise, and perhaps it might even intensify and spread. Allow yourself to feel the satisfactions that come from succeeding in your efforts. As you gain freedom from your negative self-suggestions, you may find yourself being more capable of gaining freedom from dependence on something outside of yourself to elicit your feelings of happiness and motivation.*

Trusting Intelligent Non-conscious Brain Pathways for Change

We tend to rely on our conscious, rational mind to solve problems, but when dealing with psychological issues, rational solutions often do not work. Non-conscious brain pathways can be intelligent, such as the ability to respond to the beauty of nature or the natural ability to forget.

We had a middle-aged client who wanted to stop smoking. She prided herself on being a hard-driving professional who could get things done. But lately she was suffering from forgetfulness that was getting in her way. Sometimes she would forget to fill out an important form or take care of a vital detail. And then, feeling annoyed with herself, she would smoke even more. We worked with her using some of the exercises in this chapter, to tap into her natural abilities to relax spontaneously when she was in a beautiful place, and to forget what she did not need to remember. She came back the next week smiling. She told us, "I had the most amazing experience. All week, I kept forgetting to smoke. I ended up smoking only two or three cigarettes the first three days, and then just stopped altogether. And the fascinating thing was that I stopped being so forgetful at work!" So, she had unconsciously solved her problem, forgetting what was better not to remember, to smoke, and remembering what she should at work!

The brain has all the necessary wiring for a healthy life. The fear–stress system protects us from danger and the reward system guides us to do things that will perpetuate our life. Substance abuse takes people away from their natural healthy responses, as we have described, and so, the natural instincts become distorted. As your client's nervous system begins to return to normal functioning, you can help her regain trust in her inner regulatory systems. There is a built-in wisdom. We know when we are tired and know when we have had enough sleep. We know when we are hungry and know when we have eaten enough. If we listen appropriately, we will regulate our sleeping and eating correctly. This exercise can guide clients back to the natural inner wisdom, to sense what is needed. In so doing, a sense of inner confidence can develop, that ability to live a life that is in harmony with the inner needs of the organism.

Sit quietly for several minutes until you feel comfortable. If you have been using hypnosis, go into a comfortable trance. Then let your thoughts drift around these ideas: your body knows how to sleep when tired and eat when hungry. You can trust the unconscious wisdom of your mind–brain–body system to give you cues. Allow yourself to listen to the inner signals, and respond to them with positive, healthy action. As you attune within, you will find a new reservoir of confidence to help you. The discovery of your own untapped inner strength becomes the source of change. When you feel ready, come out of hypnosis, relaxed and refreshed.

Associations and Imaginings of the Future

You can help your clients to activate a more positive future, one that is filled with interesting experiences and new possibilities. It is important to activate talents and potentials for future involvements without drug abuse.

Joey was tough. He had been a gang member who was always ready for a fight. He strongly asserted himself by instilling fear in others around him. He used heroin to stay "cool." But he had found a girl who he loved, and so he decided to give up heroin and start a new life. Joey began hypnotherapy and worked well in trance. He enjoyed the feeling of true calm, something he had never really experienced before. He also began to feel that his life did not really accomplish anything. We saw talent in him, and in trance, he began to realize that he was not using his strength very well. We worked with him to recognize his abilities and to face the guilt he felt for all the pain and suffering he had caused to others. He withdrew from heroin successfully. But when he went back to his life after time in rehab, he found his girlfriend still using and all his friends engaged in the same life of crime. He had visualized a more positive, honest life for himself in trance. Utilizing the strength he had exhibited in creating fear in those around him, he directed it into getting honest work. Years later, he returned to tell us that he had married a girl who knew nothing of drugs and crime and was happily working as a manager of a big company.

Imagine how you would be without your habit. What a positive sense of accomplishment you will have achieved! You may wonder why you did not change it sooner. What thoughts and images do you have as you project yourself into the future? How did you do it? What are you like? Your thoughts and associations flow from the source of change, but the source is not conscious. Let your unconscious help you in positive ways. Then relax and rest calmly for several minutes. Repeat this experience often over the next several days. You might want to emphasize suggestion of one aspect over another at each session. Some will find results are immediate, while others may need several sessions.

You can use brain pathways in better ways, incorporating remembering and forgetting, imagination, and creative responses. You can be calm, when needed, which initiates a shift in the nervous system toward a healthier, less activated balance, free of the substance, and more responsive to the natural pleasures of life.

The Power of the Breath

Breathing can have a powerful influence over the nervous system in ways that are helpful for overcoming addictions. One of the most widely practiced meditation traditions is mindful breathing. Because the breath is so highly interconnected with the heart and the nervous system, it becomes a direct way to bring a return to balance.

Sit down and close your eyes. Allow yourself to breathe naturally and softly through your nose. Observe a breath, beginning with bringing the air in through your nose. Feel the air in your nasal passages and then follow the sensation as the air travels down into your lungs. Concentrate on the movement in your chest and diaphragm as the air moves into these areas. Follow the air moving out as you begin to exhale. Note how your diaphragm pushes down as the air travels up and out through your nose. Feel the sensation of air leaving through your nose.

Treat each breath as a unique experience to be noticed fully and enjoyed for the first time. Approach the next breath as a completely new experience. Follow the air as it moves in and out, with fresh interest. Each moment is distinct, different from the past moment, and unique, open to new potentials. And so, every breath is worthy of your full attention.

After following each breath in this way for several minutes, people often experience calm and ease. This sense of inner peace can become a resource.

Opening Intuition Through Mindful Breathing

Mindful attention to breathing at key moments in the day can become a helpful therapeutic tool. Attention to breathing puts people directly in touch with what is happening within. By turning attention to breathing at various times during the day, people will become aware of different qualities in their breathing and in so doing, be able to recognize deeper feelings that may be occurring, expressed subtly. For example, when hurrying to accomplish tasks, the breathing rate quickens, with shorter breaths. By stopping to notice, other feelings such as impatience might emerge.

When you are feeling tense or anxious, a moment of mindful breathing might help you to get back in touch. After a few minutes of close attention, a previously unnoticed feeling of disturbance might begin to surface. Emotional responses are usually accompanied by certain patterns of breathing, deep or shallow. First, experience your breathing patterns just as they are. Then, accept them by quietly noticing. You may enjoy an easing of some or all of your discomfort.

Always start from where you are, without trying to alter anything. With patient, mindful breathing, changes will happen. Remember that each moment is new, so even an uncomfortable breathing pattern or feeling will eventually transform. Turn your attention to breathing for a few minutes here and there through the day to gain new intuitive understandings.

Keep thinking on a concrete level of simply noticing the momentary experience as it occurs. If an evaluation is added as to whether this experiencing is good or bad, liked or disliked, etc. notice that too. Your ability to sustain mindfulness will improve over time and in the process, you will gain a sense of confidence.

Disengaging from Craving Mindfully Through Non-Attachment

The ability to be non-attached when needed, drawn from Buddhist meditation, is helpful for controlling addiction. Non-attachment is a skill that can be trained. Everyone has had times when they handled discomfort well, perhaps when putting forth effort while playing a competitive game or when meeting a challenging deadline at work. Facing discomfort just as it is gives the opportunity to gain control. This meditative method of non-attachment helps to move beyond seeking pleasure and avoiding pain, thereby breaking the reinforcement cycle, to provide the brain a chance to disengage from an unhealthy reward pathway response, allowing the nervous system to return to balance.

The next series of exercises offers the therapist some ways to present mindfulness meditation that work directly on the cravings. Individualize the instructions to fit the unique needs of the client.

When a craving feeling occurs, sit down and focus attention on it. Notice the sensations with each passing moment. If you start to label them as negative, notice that. Keep returning to the fullness of the experience without being drawn away from the moment-to-moment experiencing. Let the breathing settle and the muscles relax while sitting quietly. Turn attention to breathing as well. While observing, allow any unnecessary tensions to ease. If the thoughts move away from the moment, gently bring them back. Continue to breathe as comfortably as possible while observing. Usually the craving feeling will alter in various ways. Notice the changes as they occur. After a time, the sensations may even ease down or diminish.

Differentiating the Features of Cravings

Closely observing the general craving experience mindfully, you may notice that craving is not just one overwhelming experience: it has various contributing parts. If this has not been noticed, try to distinguish the parts. Differentiate the ongoing sensations. Are sensations felt in one part of the body as well as in another? What is the quality of breathing? Are breaths long or short? Are there emotions occurring right now? If so, notice what they are. Refine the quality of each feeling, such as, "Now I feel sadness, and I also feel fatigue." And what are the types of thoughts? Maintain that meditative awareness, calm and steady.

Extend mindfulness into other settings. Notice the places where the craving is stimulated, people who are associated with it, and personal behaviors and thoughts that accompany it. This will require stopping to observe mindfully at many times through the day and evening. Then, when some of these parts have been distinguished, how might they be interconnected in the momentary experiencing? For example, do thoughts turn to another person, and then the craving increases? Or perhaps a memory of a place where drugs were used occurs, and then a craving feeling follows. Notice any cues that might signal and elicit your craving response.

Make Small Changes: Harm Reduction

As you begin to gain some control over the feelings of cravings, you can minimize the harm that comes from your addiction in small steps. Start the change process from where you are. Moderate by using less harmful practices such as participating in needle exchange programs, drinking less quantities, or drinking in a more controlled manner.

At first people sometimes feel unwilling or incapable of stopping the harmful behavior. But you can gradually work toward that by enhancing your coping skills. Practice these methods regularly. You will help build confidence and mental flexibility. Just learning these skills can bring an experience of mastery. As you begin to feel a source of confidence from within, you form a foundation for better choices in the future.

Maintain Non-attachment

Use the skills developed from practicing non-attachment in the earlier exercise with feelings of pleasure or pain that might emerge. Remember, "my craving" is just a group of sensations, not representative of all that you are. Pleasures and pains come and go, so seek the middle way between, calm and centered in meditation.

Choosing Renunciation

As skills in awareness and detachment develop, people will inevitably gain more insight into what they are doing and how the involvement in the substance actually affects their life in adverse ways. They may begin to feel motivated to give up the harmful substance or behavior that is having a destructive effect. The dependency may seem entrenched. But remind them that the brain pathways have some plasticity. With better habits, the reward pathway can normalize.

Keep in mind that when withdrawing from substance abuse, the client usually requires medical care to monitor the process. You can guide your client in choosing a hospital or medical facility that will ensure a safe and healthy withdrawal.

When ready to take the steps, forgo taking the substance or indulging in the behaviors that lead you there. Notice the tensions and other feelings mindfully.

Mindfulness, relaxation methods, and non-attachment all help you to face the discomfort and accept it is as it occurs. Then you can calm down despite these discomforts.

Whatever you have been before, might be tomorrow, or are even feeling right now can all be part of the mindful moment, filled with potential. Notice any doubts or worries, but do not let them distract you from what is occurring in this moment now. For example, if you are thinking a thought such as, "I might not every get over this problem." or "I've always had troubles," consider it just part of the flow in this moment when fully aware. Such thoughts are accepted as part of the mindful flow.

Working Through

Often the sense of self is distorted by substance abuse. Substance abusers think of themselves as users. But defining the self as a user is really a cognitive construction, built up from the social interactions, behavioral habits, beliefs, and practices over time. Through meditation, hypnosis, and cognitive therapy, people can rediscover their deeper nature: their talents, and their personality. As the nervous system returns to its more open and adaptive nature, a more unbiased and flexible sense of self returns.

Meditation on Your Dynamic, Ever-Changing Being

Begin with a familiar mindful meditation on breathing to notice how each breath is new, a different breath. Sense how the surrounding circumstances change with each passing moment. Perhaps there is a breeze now, and then it stops. Or maybe there is a sound of a passing car, then silence. Gradually shift the focus of meditation to include the sense of yourself right now. Notice how you might be thinking about yourself in this moment while sitting on this seat now. Pay close attention to whatever is noticed. Perform this meditation again at a later time or on a different day. You might have new thoughts or feelings about yourself. Become aware of all the ways that each moment is a unique combination of thoughts, feelings, and sensations, including your personal sense of self and the others in your life who you care about. The next moment is open, yet to be discovered. With your mind–brain–and body in harmony now, you can step into the future hopefully anew.

Nurturing the Positive: Seek Healthy Rewards

The brain, with its quality of plasticity, can readjust back to a healthy balance. Beneficial motivations are part of the whole but are often overlooked as the addiction syndrome takes over conscious awareness. People usually claim that they began drug use in search of happiness, social acceptance, or perhaps, spiritual fulfillment. There is nothing inherently harmful about wanting to be happy, or accepted, or to have spiritual fulfillment, but using drugs to achieve such objectives ultimately does not bring those goals. By shifting away from substance abuse, positive qualities that may have been overlooked can be noticed, and then developed and fostered.

What do you enjoy doing? Often people who are deeply immersed in substance use lose touch with their interests and talents. What did you like to do as a child? Some of the spontaneous interests you had long ago may be the seeds of unrealized potentials. Did you collect things? Were you drawn to sports? Did you love to draw and paint? Were you always outdoors in the garden? Begin in a small way. Plant some seeds to grow, play basketball at a recreation center, or draw a picture. Allow yourself to enjoy the experience of learning as you go. Get books or instruction from others if that is helpful. Explore your talents and let them develop. You may be surprised to find out that you have abilities you did not know were there.

Conclusion

Keep the brain in mind when you work with substance abuse. The reward pathway is central in addiction. Therapeutic interventions that help to activate normal reward processes will help to rebalance the system. Building the resources for withdrawal is helpful, and then reinforcing normal reward processes will not only help to maintain a healthy balance, but will also support clients in finding happiness and rewards in everyday life.

References

Adams, F. (1886). The genuine works of Hippocrates. (Trans.). London: Sydenham Society.

Adolphs, R. (1999). The human amygdala and emotion. *The Neuroscientist, 5*(2), 125–137.

Aftanas, L., & Golosheykin, S. (2005). Impact of regular meditation practice on EEG activity at rest and during evoked negative emotions. *International Journal of Neuroscience, 115*(5), 893–909.

Ainsworth, M. D. S., Blehat, J. C., Waters, E., & Wall, S. (1978). *Patterns of attachment: a psychological study of the strange situation.* Hillsdale: Erlbaum.

Albert, D. Z. (1992). *Quantum mechanics and experience.* Cambridge: Harvard University Press.

Alexander, A., & French, T. M. (1946). *Psychoanalytic therapy: principles and application.* New York: Ronald Press.

Alptekin, K. (2001). Tc-99 m HMPAO brain perfusion SPECT in drug-free obsessive-compulsive patients without depression. *Psychiatry Research: Neuroimaging, 107*(1), 51–56.

Altman, J., & Das, G. D. (1967). Autoradiographic and histological evidence of postnatal hippocampal neurogenesis in rats. *Journal of Comparative Neurology, 124,* 319–335.

Anand, B. K., & Brobeck, J. R. (1951). Hypothalamic control of food intake in rats and cats. *Yale Journal of Biology and Medicine, 24,* 123–140.

Apkarian, A. V., Thomas, P. S., Krauss, B. R., & Szeverenyi, N. M. (2001). Prefrontal cortical hyperactivity in patients with sympathetically mediated chronic pain. *Neuroscience Letters, 311,* 193–197.

Archer, G. S., Piedrahita, J., Nevill, C. H., Walker, S., & Friend, T. H. (2003). Variation in behavior, cloned pigs. *Applied Animal Behaviour Science, 81*(4), 321–331.

Armstrong, D. M. (1968). *A materialist theory of the mind.* London: Routledge & Kegan Paul.

Atkinson, R.C. & Shiffrin, R. M. (1968). Human memory: A proposed system and its control processes. In K. W. Spence & J. T. Spence (Eds.), *The psychology of learning and motivation. Vol. 2. Advances in research and theory.* New York: Academic.

APA. (2007). Mirror, mirror in the brain: Mirror neurons, self-understanding and autism research. Society for Neuroscience, November 7.

Aserinsky, E., & Kleitman, N. (1953). Regularly occurring periods of eye mobility, and concomitant phenomena during sleep. *Science, 118,* 273–274.

Badawi, K., Wallace, R. K., Orme-Johnson, D., & Rouzere, A. M. (1984). Electrophysiologic characteristics of reparatory suspension periods occurring during the practice of the transcendental meditation program. *Psychosomatic Medicine, 46*(3), 267–276.

Baddeley, A. (1990). *Human memory: Theory and practice.* Needham Heights: Allyn & Bacon.

Badenoch, B. (2008). *Being a brain-wise therapist.* New York: W. W. Norton & Company.

Badgaiyan, R. D., & Posner, M. I. (1997). Time course of cortical activation in implicit and explicit recall. *Journal of Neuroscience, 17*, 4904–4913.

Baliki, M., Geha, P. Y., Apkarian, A. V., & Chialvo, D. R. (2008). Beyond feeling chronic pain hurts the brain, disrupting the default-mode network dynamics. *Journal of Neuro Science, 28*(6), 1398–1403.

Bandura, A. (1977). *Social learning theory.* New York: General Learning Press.

Banks, S. J., Eddy, K. T., Angstadt, M., Nathan, P. J., & Phan, K. L. (2007). Amygdala-frontal connectivity during emotion regulation. *Social Cognitive Affect Neuroscience, 2*(4), 303–312.

Bartels, A., & Zeki, S. (2000). The neural basis of romantic love. *NeuroReport, 11*(17), 3829–3834.

Battino, R., & South, T. L. (2005). *Ericksonian approaches: A comprehensive manual.* Norwalk: Crown House Publishing.

Bavelier, D., Tomann, A., Hutton, C., Mitchell, T., Liu, G., Corina, D., et al. (2000). Visual attention to the periphery is enhanced in congenitally deaf individuals. *Journal of Neuroscience, 20*(17), 1–6.

Bayes, T. (1763). An essay towards solving a problem in the doctrine of chances. By the late Rev. Mr. Bayes, communicated by Mr. Price, in a letter to John Canton, M. A. and F. R. S. *Philosophical Transactions of the Royal Society of London.* 53. 370–418.

Baxter, L. R., Schwartz, J. M., & Bergman, K. S. (1992). Toward a neuroanatomy of obsessive compulsive disorder. *Archives of General Psychiatry, 49*, 681–689.

Bear, M. F., Connors, B. W., & Paradiso, M. A. (1996). *Neuroscience: Exploring the brain.* Baltimore: Williams & Wilkins.

Bagley, S. (2007). *Train your mind, change your brain: How a new science reveals our extraordinary potential to transform ourselves.* New York: Ballantine Books.

Bender, E. (2004). Brain data reveal why psychotherapy works. *Psychiatric News, 39*(9), 34.

Benes, F. M., Vincent, S. L., & Todtenkopf, M. (2001). The density of pyramidal and nonpyramidal neurons in anterior cingulate cortex of schizophrenic and bipolar subjects. *Biological Psychiatry, 50*(6), 395–406.

Berger-Sweeney, J., & Hobmann, C. F. (1997). Behavioral consequences of abnormal cortical development: Insights into developmental disabilities. *Behavioral Brain Research, 86*, 121–142.

Bernier, R., & Dawson, G. (2009). The role of mirror neuron dysfunction in autism. In J. Pineda (Ed.), *Mirror neuron systems: The role of mirroring processes in social cognition* (pp. 261–286). New York: Humana.

Berry, T. (1992). *Religions of India: Hinduism, yoga, Buddhism.* New York: Columbia University Press.

Bhatia, M., Kumar, A., Kumar, N., Pandey, R. M., & Kochupilla, V. (2003). Electrophysiologic evaluation of Sudarshan Kriya: an EEG, BAER and P300 study. *Indian Jouranal of Pharmacology, 47*, 157–163.

Blanchard, C., Blanchard, R., Fellous, J. M., Guimarlaes, F. S., Irwin, W., LeDoux, J. E., et al. (2001). The brain decade in debate: III. *Neurobiology of Emotion Brazilian Journal of Medical and Biological Research, 34*, 283–293.

Bliss, G., & Lemo, T. (1973). Long-lasting potentiation of synaptic transmission in the dentate area of the anaesthetized rabbit following stimulation of the preforant path. *J Physiol (Lond), 232*, 331–356.

Blumer, H. (1969). *Symbolic interactionism: Perspective and method.* New Jersey: Englewood Cliffs.

Bogen, J. E., & Bogen, G. M. (1983). Hemispheric specialization and cerebral duality. *BBS, 6*, 517–520.

Bogen, J. E., & Vogel, P. J. (1962). Cerebral commissurotomy in man: Preliminary case report. *Bulletin of the Los Angeles Neurological Society, 27*, 169–172.

Boshuisen, M. S., Ter Horst, F. J., Paans, A. M., & Reinders, A. A. (2002). rCBF differences between panic disorder patients and control subjects during anticipatory anxietyand rest. *Biological Psychiatry, 52,* 126–135.

Bowers, K.S. (1984). On being unconsciously influenced and informed. In Bowers, K. S. & Meichenbaum, D. S. (Eds.), *The unconscious reconsidered.* New York: Wiley. 227–272. http://www.brainexplorer.org/neurological_control/Neurological_Neurotransmitters.shtml.

Bowlby, J. (1983). *Attachment.* New York: Basic Books.

Bowlby, J. (2005). *A secure base: Parent-child attachment and healthy human development.* London: Routledge. (Original work published 1988).

Boyke, J., Driemeyer, J., Gaser, C., Buchel, C., & May, A. (2008). Training-induced brain structure changes in the elderly. *Journal of Neuroscience, 28*(28), 7031–7033.

Brazelton, T. B. (1982). Joint regulation of neonate-parent behavior. In E. Tronick (Ed.), *Social interchanges in infancy: Affect, cognitive, and communication.* Baltimore: University Park Press.

Breedlove, S. M., Rosenzweig, M. R., & Watson, N. V. (2007). *Biological psychology: An introduction to behavioral, cognitive, and clinical neuroscience.* Sunderland: Sinauer Associates, Inc.

Bremmer, J. D. (2005). *Brain imaging handbook.* New York: W. W. Norton & Company.

Briones, T.L., Klintsova, A.Y., Greenough, W.T. (2004). Stability of synaptic plasticity in the adult rat visual cortex induced by complex environment exposure. *Brain Research, 20,* 1018, 1, 130–135.

Broadbent, D. E. (1958). *Perception and communication.* London: Pergamon Press.

Brodmann, K. (1909/1994). *Localisation in the cerebral cortex.* (trans: Laurence, G.). London: Smith-Gordon.

Brooks, D. C., & Bizzi, E. (1963). Brain stem electrical activity during deep sleep. *Archives Italiennes de Biologie, 101,* 648–665.

Brown, S. L. (1994). Serotonin and aggression. *Journal of Offender Rehabilitation, 21*(34), 27–39.

Brown, T. G., & Sherrington, C. S. (1912). The role of reflex response and limb reflexes of the mammal and its exceptions. *Journal of Physiology, 44,* 125–130.

Buccino, G., Binkofski, F., Fink, G. R., Fadiga, L., Fogassi, L., Gallese, R. J., et al. (2001). Action observation activates premotor and parietal areas in a somatotopic manner: An fMRI study. *European Journal of Neuroscience, 13,* 400–401.

Buccino, G., Lui, F., Canessa, N., Patteri, I., Lagravinese, G., & Benuzzi, F. (2004). Neural circuits underlying imitation learning of hand actions: An event-related fMRI study. *Neuron, 42,* 323–334.

Bush, G., Luu, P., & Posner, M. I. (2000). Cognitive and emotional influences in anterior cingulated cortex. *Trends in Cognitive Sciences, 4*(5), 215–222.

Butt, A. M., & Bay, V. (2012). 10th European Meeting on Glial Cells in Health and Disease. Prague Conference Center, Prague, Czech Republic. *Future Neurology, 7*(1), 27–28.

Byrne, J. H., & Berry, W. O. (1989). *Neural models of plasticity.* San Diego: Academic.

Cabeza, R., & Nyberg, L. (2000). Imaging cognition II: An empirical review of 275 PET and fMRI studies. *Journal of Cognitive Neuroscience, 12,* 1–47.

Cacioppo, J. T., & Berntson, G. G. (2007). Affective distinctiveness: illusory or real? *Cognition and Emotion, 21*(6), 1347–1359.

Calvert, G. A., & Campbell, R. (2003). Reading speech from still and moving faces. The neural substrates of visible speech. *Journal of Cognitive Neuroscience, 15,* 50–70.

Campos, J. J., Thein, S., & Owen, D. A. (2003). A Darwinian legacy to understanding human infancy: Emotional expressions as behavior regulators. *Ann. N.Y. Academy of Science, 1000,* 110–134.

Cannon, W. B. (1927). The James-Lange theory of emotion: A critical examination and an alternative theory. *American Journal of Psychology, 39,* 106–124.

Cannon, W. B., & Uridil, J. E. (1921). Studies on the conditions of activity in endocrine glands. VIII. Some effects on the denervated heart of stimulating the nerves of the liver. *American Journal of Physiology, 58*, 353–354.

Carr, D. B., & Sesack, S. R. (2000). Projections from the rat prefrontal cortex to the ventral tegmental area: Target specificity in the synaptic associations with mesoaccumbens and mesocortical neurons. *Journal of Neuro Science, 20*, 3864–3873.

Carr, L., Iacoboni, M., Dubeau, M. C., Mazziotta, J. C., & Lenzi, G. L. (2003). Neural mechanisms of empathy in humans: A relay from neural systems for imitation to limbic areas. *Proceedings of the National Academy of Sciences, U. S. A., 100*, 5497–5502.

Carter, C. S. (1999). The contribution of the anterior cingulated cortex to executive processes in cognition. *Reviews in the Neurosciences, 10*, 49–57.

Cernuschi-Frias, B., Garcia, R. A., & Zanutto, S. (1997). A neural network model of memory under stress. IEEE Transactions on systems, man, and cypernetics—Part B. *Cybernetics, 27*, 2, April, pp. 278–284.

Chatrian, G. E. (1976). The mu rhythms. In A. Remond (Ed.), *Handbook of electroencephalography* (pp. 104–114). Amsterdam: Elsevier.

Chekhov, M. (1953). *To the actor on the technique of acting*. New York: Harper & Row.

Chen, G., Rajkowska, G., Du, F., Seraji-Bozorgzad, N., & Manji, H. K. (2000). Enhancement of hippocampal neurogenesis by lithium. *Journal of Neurochemistry, 75*, 1729–1734.

Cherry, E. C. (1953). Some experiments on the recognition of speech, with one and with two ears. *The Journal of the Acoustical Society of America, 25*(5), 975–979.

Churchland, P. M. (1995). *The engine of reason, the seat of the soul: A philosophical journey into the brain*. Cambridge: The MIT Press.

Churchland, P. M. (1988). *Matter and consciousness revised*. Cambridge: The MIT Press.

Churchland, P. S. (1986). *Neurophilosophy: Toward a united science of the mind-brain*. Cambridge: The MIT Press.

Cimino, G. (1999). Reticular theory versus neuron theory in the work of Camillo Golgi. *Rivista Internazionale Di Storia Della Scienza, 36*(2), 431–472.

Clark, A. (1997). *Being there: Putting brain, body, and world together again*. Cambridge: MIT Press.

Cleare, A. J., & Bond, A. J. (1997). Does central serotonergic function correlate inversely with aggression: A study using D-fenfluramine in healthy subjects. *Psychiatry Research, 69*, 89–95.

Cleary, T. (1993). *The flower ornament sutra: A translation of the Avatamsaka sutra*. Boston: Shamabhala.

Cloninger, R., Svrakic, M., & Przybeck, T. R. (1993). A psychobiological model of temperament and character. *Archives of General Psychiatry, 50*, 975–990.

Cochin, S., Barthelemy, C., Lejeune, B., Roux, S., & Martineau, J. (1998). Perception of motion and qEEG activity in human adults. *Electroencephalography and Clinical Neurophysiology, 107*, 287–295.

Cochin, S., Barthelemy, C., Roux, S., & Martineau, J. (1999). Observation and execution of movement: similarities demonstrated by quantified electroencephalography. *European Journal of Neuroscience, 11*, 1839–1842.

Cohen-Seat, G., Gastaut, H., Faure, J., & Heuyer, G. (1954). Etudes expérimentales de l'activité nerveuse pendant la protection cinematographique. *Revue Internationale de Filmologie, 5*, 7–64.

Cohen-Tannoudji, M., Babinet, C., & Wassef, M. (1994). Early determination of a mouse somatosensory cortex marker. *Nature, 368*, 460–463.

Colcombe, S. J., Erickson, K. I., Rax, N., Webb, A. G., Cohen, N. J., & McAuley, E. (2003). Aerobic fitness reduces brain tissue loss in aging humans. *Journals of Gerontology. Series A, Biological Sciences and Medical Sciences, 58*(2), 176–180.

Conze, E. (1995). *A short history of Buddhism*. Oxford: Oneworld Publications.

Conze, E. (1951). *Buddhism: Its essence and development*. New York: Philosophical Library.

Cook, F. H. (1977). *Hua-yen Buddhism: The jewel net of Indra.* Pennsylvania: The Pennsylvania State University Press University Park.

Corballis, P. M. (2003). Visuospatial processing and the right-hemisphere interpreter. *Brain and Cognition, 53,* 171–176.

Cosmides, L. & Tooby, J. (2000). Introduction: Evolution in M. S. Gazzaniga (Ed.), *The new cognitive neurosciences,* Second Edition, Cambridge: A Bradford Book, The MIT Press, pp. 1163–1166.

Cozolino, L. (2006). *The neuroscience of human relationships: Attachment and the developing social brain.* New York: W. W. Norton & Company.

Craig, A. D. (2003). Interoception: The sense of the physiological condition of the body. *Current Opinion in Neurobiology, 13,* 500–505.

Craik, F. L. M., & Lockhart, R. S. (1972). Levels of processing: A framework for memory research. *Journal of Verbal Learning and Verbal Behavior, 11,* 671–684.

Crick, F., & Koch, C. (1990). Towards a neurobiological theory of consciousness. *Seminars in the Neurosciences, 2,* 263–275.

Crivellato, E., & Ribatti, D. (2007). Soul, mind, brain: Greek philosophy and the birth of neuroscience. *Brain Research Bulletin, 71,* 327–336.

Crossin, K. L., & Krushel, L. A. (2000). Cellular signaling by neural cell adhesion molecules of the immunoglobulin family. *Developmental Dynamics, 216,* 260–279.

Crosson, B. (1992). *Subcortical functions in language and memory.* New York: Guilford Press.

Cuijpers, P., van Straten, A., Andersson, G., & van Oppen, P. (2008). Psychotherapy for depression in adults: A meta-analysis of comparative outcome studies. *Journal of Consulting and Clinical Psychology, 76*(6), 909–922.

Damasio, A. R. (2010). *Self comes to mind: Constructing the conscious brain.* New York: Pantheon Books.

Damasio, A. R., Grabowski, T. J., Bechara, A., Damasio, H., Ponto, L. L., Parvizi, J., et al. (2000). Subcortical and cortical brain activity during the feeling of self-generated emotions. *Nature Neuroscience, 3*(10), 1049–1056.

Damasio, A. R. (1994). *Descartes' error: Emotion, reason and the human brain.* New York: G. P. Putnam's Sons.

Damasio, A. R., Tranel, D., & Damasio, H. (1990). Individuals with sociopathic behavior caused by frontal damage fail to respond automatically to social simuli. *Behavioral Brain Research, 41,* 81–84.

Dappreto, G., Davies, M., Pfeifer, J., Scott, A., Sigman, M., & Bookheimer, S. (2006). Understanding emotions in others: Mirror neuron dysfunction in children with autism spectrum disorders. *Nature Neuroscience, 9,* 28–30.

Darwin, C. (1859/1999). *On the origin of species by means of natural selection: Or the preservation of favored races in the struggle for life.* New York: Bantam.

Darwin, C. (1872). *Expression of the emotions.* London: John Murray.

Das, G. D., & Altman, J. (1970). Postnatal neurogenesis in the caudate nucleus and nucleus accumbens septi in the rat. *Brain Research, 21,* 122–127.

Davidson, R. J., Kabat-Zinn, J., Schumacher, J., Rosenkranz, M., Miller, D., Santorelli, S. F., et al. (2003). Alterations in brain and immune function produced by mindfulness meditation. *Psychosomatic Medicine, 65,* 564–570.

Davidson, R. J., Pizzagalli, D., Nitschke, J. B., & Putnam, K. (2002). Depression: Perspectives from affective neuroscience. *Annual Review of Psychology, 53,* 535–573.

Davis, M., & Whalen, P. J. (2001). The amygdala: Vigilance and emotion. *Molecular Psychiatry, 6*(1), 13–35.

Debner, J. A., & Jacoby, L. L. (1994). Unconscious perception: Attention, awareness, and control. *Journal of Experimental Psychology: Learning Memory and Cognition, 20*(2), 304–317.

Derbyshire, S. W. G., Whalley, M. G., & Oakley, D. A. (2008). Fibromyalgia pain and its modulation by hypnotic and non-hypnotic suggestion: An fMRI analysis. *European Journal of Pain, 13*, 542–550.

Desikachar, T. K. V. (1955). *The heart of yoga.* Rochester: Inner Traditions International.

Descartes, R. (1984). *Meditations on first philosophy.* (J. Cottingham, trans.). Cambridge: Cambridge University Press.

de Lange, F. P., Koers, A., Kalkman, J. S., Bleijenberg, G., Hagoort, P., van der Meer, J. W. M., et al. (2008). Increases in prefrontal cortical volume following cognitive behavioral therapy in patients with chronic fatigue syndrome. *Brain, 13*(8), 2172–2180.

Descartes, R. (1968). *Discourse on method and the meditations.* New York: Penguin Books.

Deuschl, G. (2009). Neurostimulation for Parkinson's disease. *JAMA, 301*(1), 104–105.

Deutsch, J. A., & Deutsch, D. (1963). Attention: Some theoretical considerations. *Psychological Review, 70*, 80–90.

Devinsky, O., Morrell, M. J., & Vogt, B. A. (1995). Contributions of anterior cingulate cortex to behavior. *Brain, 118*, 279–306.

Diamond, M. C., Scheibel, A. B., Murphy, G. M, Jr, & Harvey, T. (1985). On the brain of a scientist: Albert Einstein. *Experimental Neurology, 88*, 198–204.

Dillbeck, M. C., & Orme-Johnson, D. W. (1987). Physiological differences between transcendental meditation and rest. *American Psychologist, 42*(9), 879–881.

Dobson, K. S., Hollon, S. D., Dimidjian, S., Schmaling, K. B., Kohlenberg, R. J., Gallop, R. J., et al. (2008). Randomized trial of behavioral activation, cognitive therapy, and antidepressant medication in the prevention of relapse and recurrence in major depression. *Journal of Consulting and Clinical Psychology, 76*(3), 468–477.

Drevets, W. C., & Raichle, M. E. (1998). Riciprocal suppression of regional cerebral blood flow during emotional versus higher cognitive processes: implications for interactions between emotion and cognition. *Cognition Emotion, 12*, 353–385.

Dudek, F. E., & Traub, R. D. (1989). Local synaptic and electrical interaction in hippocampus: Experimental data and computer simulations. In J. H. Byrne & W. O. Berry (Eds.), *Neural Models of plasticity: Experimental and theoretical approaches* (pp. 378–402). San Diego: Academic.

Duyvendak, J. J. L. (1992). *Tao Te Ching, The book of the way and its virtue.* Boston: Charles E. Tuttle, Co. Inc.

Ecker, B. (2008). Unlocking the emotional brain: Finding the neural key to transformation. *Psychotherapy Networker, 32*, 5, 42–47, 60.

Egner, T., Jamieson, G., & Gruzelier, J. (2005). Hypnosis decouples cognitive control from conflict monitoring processes of the frontal lobe. *NeuroImage, 27*, 969–978.

Ekman, P. (1992). An argument for basic emotions. *Cognition and Emotion, 6*, 169–200.

Ekman, P. (2003). *Emotions revealed: Recognizing faces and feelings to improve communication and emotional life.* New York: Times Books.

El-Badri, S. M., Cousins, D. A., Parker, S., Ashton, H. C., Ferrier, I. N., & Moore, P. B. (2006). Magnetic resonance imaging abnormalities in young euthymic patients with bipolar affective disorder. *British Journal of Psychiatry, 189*, 81–82.

Elias, A. N., Guich, S., & Wilson, A. F. (2000). Ketosis with enhanced GABAergic tone promotes physiological changes in transcendental meditation. *Medical Hypotheses, 54*, 660–662.

Epel, E. S., Blackburn, E. H., Lin, J., Dhabhar, F. S., Adler, N. E., & Morrow, J. D. (2004). Accelerated telomere shortening in response to life stress. *Proceedings of the National academy of Sciences of the United States of America, 101*(49), 17312–17315.

Erickson, M. H., & Rossi, E. L. (2006). *The Neuroscience editions: Healing in hypnosis; Life reframing in hypnosis; Mind-body communication in hypnosis; Creative choice in hypnosis.* Phoenix: Milton H. Erickson Foundation Press.

Erickson, M. H. (1964). Initial experiments investigating the nature of hypnosis. *American Journal of Clinical Hypnosis, 7*, 254–257.

Erik, S., Mikschi, A., Stier, S., Claramidaro, A., Gapp, V., Weber, B., et al. (2010). Emotion regulation in major depression. *Journal of Neuroscience, 30*(47), 15725–15734.

Eriksson, P. S., Perfilieva, E., Bjork-Eriksson, T., Alborn, A., & Nordborg, C. (1998). Neurogenesis in the adult human hippocampus. *Nature Medicine, 4,* 1313–1317.

Esiri, M. (1994). Dementia and normal aging: Neuropathology. In F. A. Huppert, C. Brayne, & D. W. O'Connor (Eds.), *Dementia and normal aging* (pp. 385–436). Cambridge: Cambridge University Press.

Evans, B. (2005). *Space shuttle Columbia: Her missions and crews.* New York: Springer.

Evans, J. (2009). Biological Psychology, Psychology 106. University of California, San Diego, January 5–March 7.

Fadiga, L., Fogassi, L., Pavesi, G., & Rizzolatti, G. (1995). Motor facilitation during action observation: A magnetic stimulation study. *Journal of Neurophysiology, 73,* 2608–2611.

Fadiga, L., Craighero, L., Buccino, G., & Rizzolatti, G. (2002). Short communication: Speech listening specifically modulates the excitability of tongue muscles: A TMS study. *European Journal of Neuroscience, 15,* 399–402.

Fancher, R. E. (1979). *Pioneers of psychology.* New York: W. W. Norton & Co.

Feldman, R. S., Meyer, J. S., & Quenzer, L. F. (1997). *Principles of neuropsychopharmacology.* Sunderland: Sinauer Associates, Inc., Publishers.

Ferdinand, R. D. & White, S. A. (2000). Social control of brains: From behavior to genes. M. S. Gazzaniga (Ed.) *The new cognitive neurosciences,* Second Edition, Cambridge: A Bradford Book, The MIT Press. pp. 1193–1208.

Fisher, H. E., Aron, A., & Brown, L. L. (2006). Tomantic love: a mammalian brain system for mate choice. *Philosophical Transactions of the Royal Society B, 361,* 2173–2186.

Fitzgerald, P. B., Fountain, S., & Daskalakis, Z. J. (2006). A comprehensive review of the effects of rTMS on motor cortical excitability and inhibition. *Clinical Neurophysiology, 117*(12), 2584–2596.

Foa, E. B., Hearst-Ikeda, D. E., & Perry, K. J. (1995). Evaluation of a brief cognitive-behavioral program for the prevention of chronic PTSD in recent assault victims. *Journal of Consulting and Clinical Psychology, 63*(6), 948–955.

Foder, J. A. (1983). *The modularity of mind.* Cambridge: MIT Press.

Foder, J. A., & Pylyshyn, Z. (1988). Connectionism and cognitive architecture: A critical analysis. *Cognition, 28,* 3–71.

Fogassi, L., Ferrari, P. F., Gesierich, B., Rozzi, S., Chersi, F., & Rizzolatti, G. (2005). Parietal lobe: From action organization to intentional understanding. *Science, 308,* 662–666.

Francis, D. D., Szegda, K., Campbell, G., Martin, W. D., & Insel, T. R. (2003). Epigenetic sources of behavioral differences in mice. *Nature Neuroscience, 6*(5), 445–446.

Frank, J. D., & Frank, J. (1991). *Persuasion and healing.* Baltimore: Johns Hopkins University Press.

Frank, J. D., Hoehn-Saric, R., Imber, S., Liberman, B., & Stone, A. (1978). *Effective ingredients of successful psychotherapy.* New York: Brunner/Mazel.

Freud, S. (1999). *The interpretation of dreams: Original 1899 version.* Oxford: Oxford University Press.

Freud, S. (1953). *The complete psychological works of Sigmund Freud.* (trans: Strachey, J.). Toronto: Hogarth Press.

Freud, S. (1960). *The Ego and the Id.* New York: W. W. Norton & Company.

Fritsch, G.T., Hitzig, E, (1870; 1960). On the electrical excitability of the cerebrum In: Von Bonin G (trans.) *Some Papers on the Cerebral Cortex.* Springfield IL, Charles C. Thomas. pp. 3–21.

Frodl, A., Schaub, A., Banac, S., Charypar, M., Jager, M., & Kummier, P. (2008). Reduced hippocampal volumes correlates with executive dysfunctioning in major depression. *Journal of Psychiatry Neurosis, 31*(5), 316–323.

Furmark, T., Tillfors, M., Marteinsdottir, I., Fischer, H., Pissiota, A., Langstrom, B., et al. (2002). Common changes in cerebral blood flow in patients with social phobia treated with citalopram or cognitive-behavioral therapy. *Archives of General Psychiatry, 59*, 425–433.

Gabrieli, J. D. E., Corkin, S., Mickel, S. F., & Growdon, J. H. (1993). Intact acquisition and long-term retention of mirror-tracing skill in Alzheimer's disease and in global amnesia. *Behavioral Neuroscience, 107*(6), 899–910.

Gage, F. H., Eriksson, P. S., Perfilieva, E., & Bjork-Eriksson, T. (1998). Neurogenesis in the adult human hippocampus. *Nature Medicine, 4*(11), 1313–1317.

Gage, F. H., Kempermann, G., & Hongjun, S. (2008). Adult neurogenesis: A prologue. *Adult Neurogenesis, 52*, 1–5.

Gall, F.J. & Hufeland, C.W. (2010). *Some account of Dr. Gall's new theory of physiognomy, founded upon the anatomy and physiology of the brain, and the form of the skull: With the critical strictures of C.W. Hufeland, M.D.* Charleston, South Carolina: Nabu Press.

Gallese, V., Fadiga, L., Fogassi, L., & Rizzolatti, G. (1996). Action recognition in the premotor cortex. *Brain, 119*, 598–609.

Gallese, V., Keysers, C., & Rizzolatti, G. (2004). A unifying view of the basis of social cognition. *Trends in Cognitive Science, 8*(9), 397–401.

Gallese, V. (2009). Mirror neurons and the neural exploitation hypothesis: From embodied simulation to social cognition. In J. A. Pineda (Ed.), *Mirror neuron systems: The role of mirroring processes in social cognition* (pp. 163–184). New York: Humana Press.

Ganguly, K., Kiss, L., & Poo, M. M. (2000). Enhancement of presynaptic neuronal excitability by correlated presynaptic and cerebellar interneurones. *Journal of Physiological, 527*, 33–48. (London).

Garcia, R., & Cacho, J. (2006). Psosopagnosia: Is it a single or multiple entity? *International Congress of Neuropsychology, 19*(13), 42.

Gardner, M. (1974). Mathematical games: The combinatorial basis of the "I Ching", the Chinese book of divination and wisdom. *Scientific American, 230*(1), 108–113.

Gardner, H. (1976). *The shattered mind.* New York: Vintage Books.

Gascoigne, S. (1997). *The Chinese way to health: A self-help guide to traditional Chinese medicine.* Boston: Tuttle Publishing.

Gastaut, H. J., & Bert, J. (1954). EEG changes during cinematographic presentation. *Electroencephalography and Clinical Neurophysiology, 6*, 433–444.

Gauthier, L. V., Taub, E., Perkins, C., Ortmann, M., Mark, V. W., & Uswatte, G. (2008). Remodeling the brain: Plastic structural brain changes produced by different motor therapies after stroke. *Stroke, 39*, 1520.

Gazzaniga, M. S. (Ed.). (2000). *The new cognitive neurosciences* (2nd ed.)., A Bradford Book Cambridge: The MIT Press.

Gazzaniga, M. S. (1973). The split brain in man. In R. Ornstein (Ed.), *The nature of human consciousness.* San Francisco: W. H. Freeman.

Gellhorn, E., & Kiely, W. F. (1972). Mystical states of consciousness: Neurophysiological and clinical aspects. *Journal of Nervous and Mental Disease, 154*, 399–405.

Gershon, A. A., Dannon, P. N., & Grunhaus, L. (2003). Transcranial magnetic stimulation in the treatment of depression. *American Journal of Psychiatry, 160*, 835–845.

Giles, L. (1959). *Taoist teachings translated from the book of Lieh-Tzu.* London: John Murray.

Goldapple, K., Segal, Z., Garson, C., Lau, M., Bieling, P., Kennedy, S., et al. (2004). Modulation of cortical-limbic pathways in major depression: Treatment-specific effects of cognitive behavior therapy. *Archives of General Psychiatry, 61*(1), 34–41.

Goldman-Rakic, P. S., Isseroff, A., Schwartz, M. L., & Bugbee, N. M. (1982). The neurobiology of cognitive development. In P. H. Mussen (Ed.), *Infancy and developmental psychobiology (Vol II* (pp. 281–344). New York: Wiley.

Golomb, J., Kluger, A., de Leon, M. J., Ferris, S., Convit, A., Mittelman, M., et al. (1994). Hippocampal formation size in normal human aging: A correlate of delayed secondary memory performance. *Learning and Memory, 1*, 45–54.

Gould, E., Beylin, A., Tanapat, P., Reeves, A., & Shors, T. J. (1999). Learning enhances adult neurogenesis in the hippocampal formation. *Nature Neuroscience, 2,* 260–265.

Green, E. E., Green, A. M., & Walters, A. D. (1970). Voluntary control of internal states: Psychological and physiological. *Journal of Transpersonal Psychology, 1*(1), 1–26.

Green, R. L., & Ostrander, R. L. (2009). *Neuroanatomy for students of behavioral disorders.* New York: W. W. Norton & Company.

Greenough, W. T., Black, J. E., & Wallace, C. S. (1987). Experience and brain development. *Child Development, 58,* 539–559.

Gregory, P. S. (1988). *Predatory dinosaurs of the world.* New York: Simon & Schuster.

Grevert, P., Albert, L. H., & Goldstein, A. (1983). Partial antagonism of placebo analgesia by naloxone. *Pain, 16,* 129–143.

Grigg, Ray. (1995). *The new Lao Tzu: A contemporary Tao Te Ching.* Boston: Charles E. Tuttle, Inc.

Grutzendler, J., Kasthuri, N., & Gan, W. B. (2002). Long-term dendritic spine-stability in the adult cortex. *Nature, 420,* 812–816.

Guillery, R. W. (2007). Relating the neuron doctrine to the cell theory. Should contemporary knowledge change our view of the neuron doctrine? *Brain Research Reviews, 55,* 411–421.

Halfhill, T. R. (2008). *What is the electric brain?* http://www.halfhill.com/ebrain.html.

Hamilton, A.F., de C., Brindley, R.M. & Frith, U. (2007). Imitation and action understanding in autistic spectrum disorders: How valid is the hypothesis of a deficit in the mirror neuron system? Neuropsycholologia, *45,* 1859–68.

Hankey, A. (2006). Studies of advanced stages of meditation in the Tibetan Buddhist and Vedic traditions. I: A comparison of general changes. *Evidence-based Complementary and Alternative Medicine.* 3, 4, 513–521.

Hannibal, J. Hindersson, P., Knudsen, S.M., Georg, B., & Fahrenkrug, J. (2001). The photopigment melanopsin is exclusively present in pituitary adenylate cyclase-activating polypeptide-containing retinal ganglion cells of the retinohypothalamic tract. *Journal of Neuroscience, 21,* RC191: 1–7.

Harlow, H.F. (1964). Early social deprivation and later behavior in the monkey. In: Abrams, H. H. Gurner & J.E.P. Tomal, (Eds.), *Unfinished tasks in the behavioral sciences* (pp. 154–173). Baltimore: Williams & Wilkins.

Harman, G. (1999). The intrinsic quality of experience. In W. G. Lycan (Ed.), *Mind and cognition* (pp. 14–29). Oxford: Blackwell.

Harris, A.J. (1947, 1974). Harris test of lateral dominance: Manual for administration and interpretation, Psychological Corporation, New York.

Haug, H. (1985). Are neurons of the human cerebral cortex really lost during aging? A morphometric examination. In J. Tarber & W. H. Gispen (Eds.), *Senile dementia of Alzheimer type* (pp. 150–163). New York: Springer.

Hayward, J. W., & Francisco, J. V. (1992). *Gentle bridges: Conversations with the Dalai Lama on the sciences of mind.* Boston: Shambhala Publications.

Hebb, D.O. (1949). *The organization of behaviour.* New York: Wiley. [Cited on pp. 18-21].

Heidegger, M. (1962). *Being and time.* San Francisco: Harper San Francisco.

Hendelman, W. J. (2000). *Atlas of functional neuroanatomy.* Boca Raton: CRC.

Hendry, S. H., Hsiao, S. S., Brown, M. C. (2003). Fundamentals of sensory systems. In L. R. Squire, F.E. Bloom, S. D. McConnell, J. L Roberts. N.C. Spitzer, M. J. Zigmond (2003). Fundamental Neuroscience, Second Edition. (pp. 577–589). New York: Academic.

Henricks, R. C. (1989). *Lao-tzu Te Tao Ching.* New York: Ballantine Books.

Hetherington, A. W., & Ranson, S. W. (1940). Hypothalamic lesions and adiposity in the rat. *The Anatomical Record, 78*(2), 149–172.

Hess, E. H. (1973). *Imprinting: Early experience and the developmental psychobiology of attachment.* New York: Van Nostrand-Reinhold.

Hilgard, E. R. (1977). *Divided consciousness: Multiple controls in human thought and action.* New York: Wiley.

Hirai, T. (1974). *Psychophysiology of Zen*. Tokyo: Igaku Shoin Ltd.

Hoe, P.R., Trapp, B.D., De Vellis, J., Claudio, L. & Colman, D.R. (2003). Cellular components of nervous tissue. In L. R. Squire, F.E. Bloom, S. D. McConnell, J. L. Roberts, N.C. Spitzer, Zigmond, M. J. (Eds). (2003). *Fundamental Neuroscience, Second Edition*. New York: Academic, pp. 49–75.

Horn, N. R., Dolan, M., Elliott, R., Deakin, J. F. W., & Woodruff, P Wr. (2003). Response inhibition and impulsivity: An fMRI study. *Neuropsychologia, 41*(14), 1956–1966.

Hotzel, B. K., Ott, U., Gard, T., Hempel, H., Weygandt, M., & Morgen, K. (2008). Investigation of mindfulness meditation morphometry. *Social Cognition and Affective Neuroscience, 3*, 55–61.

Howard, R.J., Ffytche, D.H., Barnes, J., McKeery, D. Ha, Y., Woodruff, P. W., Bullmore, E. T., Simmons, A., Williams, S.C.R., David, A. S., & Brammer, M. (1998). The functional anatomy of imagining and perceiving color. Neuro Report, 9, 1019–1023.

Hubel, D. H., & Wiesel, T. N. (1962). Receptive fields, binocular interaction and functional architecture in the cat's visual cortex. *The Journal of Physiology, 160*, 106–154.

Hugdahl, K. (1996). Cognitive influences on human autonomic nervous system function. *Current Opinion in Neurobiology, 6*, 252–258.

Husserl, E. (1900, 1970). Logical investigations, Vol I. & II. New York: Humanities Press.

Hutchins, E. (1995). *Cognition in the wild*. Cambridge: MIT Press.

Iacoboni, M., Molnar-Szakacs, I., Gallese, V., Buccino, G., Mazziotta, J.C., & Rizzolatti, G. (2005). Grasping the intentions of others with one's own mirror neuron system. *Plos Biology,* www.plosbiology.org. 3, 3, 0001–0007.

Jacobson, E. (1929). *Progressive relaxation*. Chicago: University of Chicago Press.

Jain, N., Florence, S. L., Qi, H. X., & Kaas, J. H. (2000). Growth of new brain stem connections in adult monkeys with massive sensory loss. *Proceedings of the National academy of Sciences of the United States of America, 97*, 5546–5550.

James, W. (1896). *The principles of psychology* (Vol. I). New York: Henry Holt and Company.

Jamieson, G. A. (Ed.). (2007). *Hypnosis and conscious states: The cognitive neuroscience perspective*. Oxford: Oxford University Press.

Jenkins, W. M., Merzenich, M. M., Ochs, M. T., Allard, T., & Guic-Roble, E. (1990). Functional reorganization of primary somatosensory cortex in adult owl monkeys after behaviorally controlled tactile stimulation. *Journal of Neurophysiology, 63*, 82–104.

Joffe, R., Segal, Z., & Singer, W. (1996). Change in thyroid hormone levels following response to cognitive therapy for major depression. *American Journal of Psychiatry, 153*(3), 411–413.

Johansson, B. (1991). Neuropsychological assessment in the oldest-old. *International Psychogeriatrics*. 3 (Suppl), 51–60.

Johnson, S. (2008). *Hold me tight*. New York: Little Brown and Company.

Johnston, W. (1974). *Silent music: The science of meditation*. New York: Harper & Row.

Kaas, J.H. & Preuss, T. M. (2003). Human brain evolution. In L. Squire, F. E. Bloom, S. K. McConnell, J. L. Roberts, N. C. Spitzer, & M. J. Zigmond. (Eds.), *Fundamental neuroscience* (pp. 1147–1166). New York: Academic.

Kabat-Zinn, J. (2003). *Coming to our sense: healing ourselves and the world through mindfulness*. New York: Hyperion Press.

Kalat, J. W. (2007). *Biological psychology*. Belmont: Thomson Wadsworth.

Kandel, E., Schwartz, J., & Jessell, T. (2000). *Principles of neural science*. New York: McGraw-Hill Medical.

Kanwisher, N., McDermott, J., & Chun, M. M. (1997). The fusiform face area: a module in human extrastriate cortex specialized for face perception. *Journal of Neuro Science, 17*(11), 4302–4311.

Kant, E. (2005). *Critique of judgment*. In J. H. Bernard, (ed.) New York: Barnes and Noble, Inc.

Kaufman, A. S., Reynolds, C. R., & McLearn, J. E. (1989). Age and WAIS-R intelligence in a sample of adults in the 20–74 year age range: A cross sectional analysis with educational level controlled. *Intelligence, 13*, 235–253.

Kauer, J. A., & Malenka, R. C. (2007). Synaptic plasticity and addiction. *Nature Reviews Neuroscience, 8*(11), 844–868.

Kerr, D. S., Huggertt, A. M., & Abraham, W. C. (1994). Modulation of hippocampal long-term potentiation and long-term depression by corticosteroid receptor activation. *Psychobiology, 22*, 123–133.

Kesner, R. P., & Williams, J. M. (1995). Memory for magnitude of reinforcement: Dissociation between the amygdala and hippocampus. *Neurobiology of Learning and Memory, 64*, 237–244.

Kimura, D. (1996). *Understanding the human brain* (pp. 136–141). New York: Encyclopedia Britannica, Inc.

Kohler, E., Keysers, C., Umilta, M. A., Fogassi, L., Gallese, V., & Rizzolatti, G. (2002). Hearing sounds, understanding actions: Action representation in mirror neurons. *Science, 297*, 846–848.

Kohr, R. L. (1977). Dimensionality in the meditative experience. *Journal of Transpersonal Psychology, 9*(2), 193–203.

Konishi, M. (1985). Birdsong: From behavior to neuron. *Annual Review of Neuroscience, 8*, 125–170.

Koob, G. F., & Le Moal, M. (2001). Drug addiction, dysregulation of reward, and allostasis. *Neuropsychopharmacology, 24*, 97–129.

Kosslyn, S. M., Thompson, W. L., Costantini-Ferrando, M. F., Alpert, N. M., & Spiegel, D. (2000). Hypnotic visual illusion alters color processing in the brain. *American Journal of Psychiatry, 157*, 1279–1284.

Krawitz, A., Fukunaga, R., & Brown, J. W. (2010). Anterior insula activity predicts the influence of positively framed messages on decision making. *Cognitive, Affective, & Behavioral Neuroscience, 10*(3), 392–405.

Kroger, W. S. (1977). *Clinical and experimental hypnosis*. Philadelphia, Pennsylvania: J.B. Lippincott.

Krose, B., & van der Smagt, P. (1996). *An introduction to neural networks*. Amsterdam: University of Amsterdam.

Kubie, L. S. (1961). *Neurotic distortion of the creative process*. New York: The Noonday Press.

Kuhn, H. G., Dickinson-Anson, H., & Gage, F. H. (1996). Neurogenesis in the dentate gyrus of the adult rat: Age-related decrease of neuronoal progenitor proliferation. *The Journal of Neuroscience, 16*(6), 2027–2033.

Kuyken, W., Byford, S., Taylor, R. S., Watkins, E., Holden, E., White, K., et al. (2008). Mindfulness-based cognitive therapy to prevent relapse in recurrent depression. *Journal of Consulting and Clinical Psychology, 76*(6), 966–978.

Lane, R. D., Reiman, E. M., Ahern, G. L., Schwartz, G. E., & Davidson, R. J. (1997a). Neuroanatomical correlates of happiness, sadness, and disgust. *The American Journal of Psychiatry, 154*(7), 926–933.

Lane, R. D., Reiman, E. M., Bradley, M. M., Lang, P. J., Ahern, G. L., Davidson, R. J., et al. (1997b). Neuroanatomical correlates of pleasant and unpleasant emotion. *Neuropsychologia, 35*(11), 1437–1444.

Lange, F. P., Koers, A. K., Kalkman, J. S., Bleijengerg, G., Hagoort, P., van der Meer, J. W. M., et al. (2008). Increase in prefrontal cortical volume following cognitive behavioural therapy in patients with chronic fatigue syndrome. *Brain, 131*, 2172–2180.

Larson, J., & Lynch, G. (1986). Induction of synaptic potentiation in hippocampus by patterned stimulation involves two events. *Science, 232*, 985–988.

Lazar, S. W., Kerr, C. E., Wasserman, R. H., Gray, J.R. Greve, M., Treadway, T., MCGarvey, M., Quinn, B.T., Dusek, J. A., Benson, H., Rauch, S.L, Moore, C. L.and Fishi, B. 2005. Meditation experience is associated with increased cortical thickness. NeuroReport. *16*, 17, 1893–1897.

Lazarus, R. S. (1991). *Emotion and adaptation*. New York: Oxford University Press.

LeDoux, J. E. (1996). *The emotional brain: The mysterious underpinnings of emotional life*. New York: Touchstone.

LeDoux, J. E. (2000). Emotion circuits in the brain. *American Review of Neuroscience, 23*, 155–161.

LeDoux, J. E. (2003). The emotional brain, fear, and the amygdala. *Cellular and Molecular Neurobiology. 23*, 4, 5, 727–738.

Legge, J. (1962). *The texts of Taoism* (Vol. I). New York: Dover Publications, Inc.

Leidermann, P. (1981). Human mother-infant social bonding: Is there a sensitive phase? In K. Immerlmann, G. W. Barlow, L. Petrinovich, & M. Main (Eds.), *Behavioral development* (pp. 454–468). Cambridge: Cambridge University Press.

Lentz, J. (2011). *Trance altering epiphanies you can create*. Jeffersonville: Healing Words Press.

Leung, Y., & Singhai, A. (2004). An examination of the relationship between Qigong meditation and personality. *Social Behavior and Personality: An International Journal, 32*(4), 313–320.

Leven, S. (1998). A computational perspective on learned helplessness and depression. In R. W. Parks, D. S. Levine, & D. L. Long (Eds.), *Fundamentals of neural network modeling* (pp. 141–163). Cambridge: The MIT Press.

Leven, S. (1988). Memory and Learned helplessness: A triune approach. Presented at the M.I.N.D. Conference on Motivation, Emotion, and Goal Direction in Neural Networks. Dallas, May 23.

Levenson, R. W. (2003). Blood, sweat, and fears: The autonomic architecture of emotion. *Annals of the New York Academy of Sciences, 1000*, 348–366.

Levi-Montalcini, R. (1982). Developmental neurobiology and the natural history of nerve growth factor. *Annual review of Neuroscience, 5*, 341–362.

Levine, D.S. & Aleksandrowicz, A.M.C. (2005). Neural network approaches to personal change in psychotherapy. *Proceedings of International Joint Conference on Neural Networks*, Montreal, Canada, July 31–August 4, 0-7803-9048.

Levine, J. (1945a). Studies in the interrelations of central nervous structures in binocular vision: I. The lack of bilateral transfer of visual discriminative habits acquired monocularly by the pigeon. *Journal of Genetic Psychology, 67*, 105–129.

Levine, J. (1945b). Studies in the interrelations of central nervous structures in binocular vision: II. The condition under which interocular transfer of discriminative habits takes place in the pigeon. *Journal of Genetic Psychology, 67*, 131–142

Levine, J. (1952). Studies in the interrelations of central nervous structures in binocular vision: III. Localization of the memory trace as evidenced by the lack of intra and interocular habit transfer in the pigeon. *Journal of Genetic Psychology, 18*, 19–27.

Levine, S. (1998). Creativity. In K. H. Pribram (Ed.), *Brain and values* (pp. 427–470). Mahwah: Erlbaum.

Levinthal, C. F. (2008). *Drugs, behavior, and modern society*. Boston: Pearson Education, Inc.

Liberman, A. M., & Mattingly, I. G. (1985). The motor theory of speech perception revised. *Cognition, 21*, 1–36.

Linden, D. E. J. (2006). How psychotherapy changes the brain—the contribution of functional neuroimaging. *Molecular Psychiatry, 11*, 528–538.

Lindvall, O., Sawle, G., Widnet, H., & Rothwell, J. C. (1994). Evidence for long-term survival and function of dopaminergic grafts in the progressive Parkinsoon's disease. *Annals of Neurology, 35*, 172–180.

Luders, E., Toga, A. W., Lepore, N., & Gaser, C. (2009). The underlying anatomical correlates of long-term meditation: volumes of gray matter. *Neuroimage, 45*, 672–678.

Lumsden, A., & Kintner, C. (2003). Neural induction and pattern formation. In L. R. Squire, F. E. Bloom, S. D. McConnell, J. L. Roberts, N. C. Spitzer, & M. J. Zigmond (Eds.), *Fundamental neuroscience* (2nd ed., pp. 363–390). New York: Academic.

Luk, C. (1991). *The secrets of Chinese meditation*. York Beach: Samuel Weiser, Inc.

Lupien, S. J., de Leon, M., de Santi, S., Convit., A., Tarshish, C., Nair, N. P. (1998). Cortisol levels during human aging predict hippocampal atrophy and memory deficits. *Nature Neuroscience 1*, 1, 69–73.

Lutz, A., Gretschar, L. L., Rawlings, N., Ricard, M., & Davidson, R. J. (2004). Long-term meditators self-induce high-amplitude gamma synchrony during mental practice. *Neuroscience, 101*(46), 16369–16373.

Lycan, W. G. (1999). *Mind and cognition*. Oxford: Blackwell Publishing.

Maclean, P. D. (1952). Some psychiatric implications of physiological studies on frontotemporal portion of limbic system (visceral brain). *Electroencephalography and Clinical Neurophysiology, 4*(4), 407–418.

MacLean, P. D. (1985). Brain evolution relating to family, play and the separation call. *Archives of General Psychiatry, 42*, 405–417.

MacLean, P. D. (1967). The brain in relation to empathy and medical education. *Journal of Nervous and Mental Disease, 144*, 374–382.

Maguire, E. A., Gadian, N., Ssrude, J. S., Good, C. D., Ashburner, J., Fractowist, R. S., et al. (2000). Navigation related structural changes in the hippocampi of taxi drivers. *PNAS, 97*(8), 4398–4403.

Maren, S. S., & Quirk, G. J. (2004). Neuronal signaling of fear memory. *Nature Reviews Neuroscience, 5*, 844–852.

Martin, K. C., Bartsch, D., Bailey, C. H., & Kandel, E. R. (2000). Molecular mechanisms underlying learning-related long-lasting synaptic plasticity. In M. S. Gazzaniga (Ed.), *The New Cognitive Neurosciences*. Cambridge: The MIT Press.

Martin, K. C. & Zukin, R.S. (2006). RNA trafficking and local protein synthesis in dendrites: an overview. The Journal of Neuroscience, 5, 26, 27, 7131–4.

Massimini, M., Ferrarelli, F., Huber, R., & Esser, S. K. (2005). Breakdown of cortical effective connectivity during sleep. *Science, 309*, 2228–2232.

Matus, A. (2000). Actin-based plasticity in dendritic splines. *Science, 290*, 754–758.

Mayberg, H. S., Brannan, S. K., Tekell, J. L., Silva, J. A., Mahurin, R. K., McGinnis, S., et al. (2000). Regional metabolic effects of fluoxetine in major depression: serial changes and relationship to clinical response. *Biological Psychiatry, 48*, 830–843.

McConnell, S. K. (1992). The genesis of neuronal diversity during development of cerebral cortex. *Seminars in the Neurosciences, 4*, 347–356.

McCulloch, W. S., & Pitts, W. H. (1942). A logical calculus of the ideas immanent in nervous activity. *Bulletin of Mathematical Biology, 5*, 115–133.

McCulloch, W. S. (1965). *Embodiment of mind*. Cambridge: The MIT Press.

McKeon, R. (1947). *The basic works of Aristotle*. New York: Random House.

McEwen, B. S., & Magarinos, A. M. (1997). Stress effects on morphology and function of the hippocampus. *Annals of the New York Academy of Sciences, 821*, 271–284.

Meaney, M. J. (2001). Maternal care, gene expression, and the transmission of individual differences in stress reactivity across generations. *Annual Review of Neuroscience, 24*, 1161–1192.

Medina, J. F., Christopher Repa, J., Mauk, M. D., & LeDoux, J. E. (2002). Parallels between cerebellum and amygdala dependent conditioning. *Nature Reviews Neuroscience, 3*, 122–131.

Mednick, S. A., & Mednick, M. T. (1962). *Examiner's manual: Remote associates test*. Boston: Houghton-Mifflin.

Mesulam, M. M. (2000). *Principles of behavioral and cognitive neurology* (2nd ed.). New York: Oxford University Press.

Metz, A. E., Yau, H.-J., Centeno, M. V., Apkarian, A. V., & Martina, M. (2009). Morphological and functional reorganization of rat medial prefrontal cortex in neuropathic pain. *PNAS, 106*(7), 72423–72428.

Meyer, A. (1950). *The collected papers of Adolph Meyer Neurology* (Vol. I). Baltimore: The Johns Hopkins Press.

Miller, G. A. (1956). The magical number seven, plus or minus two: Some limits on our capacity for processing information. *Psychological Review, 63*(2), 81–97.

Miller, M. B., Van Horn, J. D., Wolford, G. L., Handy, T. C., Valsangkar-amyth, M., Inati, S., et al. (2002). Extensive individual differences in brain activations associated with episodic retrieval are reliable over time. *Journal of Cognitive Neuroscience, 14*(8), 1200–1214.

Milling, L. S., Kirsch, I., Allen, G. J., & Reutenauer, E. L. (2005). The effects of hypnotic and nonhypnotic suggestion on pain. *Annals of Behavioral Medicine, 29*, 116–127.

Milner, D., & Goodale, M. A. (1995). *The visual brain in action*. Oxford: Oxford University Press.

Minsky, M., & Papert, S. (1969). *Perceptrons: An introduction to computational geometry*. Cambridge: MIT Press.

Mirsky, A. F. (1996). Disorders of attention: A neuropsychological perspective. In G. R. Lyon & N. A. Krasnegor (Eds.), *Attention, memory and executive function* (pp. 71–96). Baltimore: Paul H. Brookes.

Mitchell, D. B. (2006). Nonconscious priming after 17 years: Invulnerable implicit memory? *Psychological Science, 17*(11), 925–929.

Mitchell, D. B., & Brown, A. S. (1988). Persistent repetition priming in picture naming and its dissociation from recognition memory. *Journal of Experimental Psychology: Learning Memory and Cognition, 14*, 213–222.

Moldakarimov, S., Bazhenov, M., & Sejnowski, T. J. (2010). Perceptual priming leads to reduction of gamma frequency oscillations. *Proceedings of the National academy of Sciences of the United States of America, 107*(12), 5640–5645.

Moran, F. M. (1993). *Subject and agency in psychoanalysis: Which is to be master?* New York: New York University Press.

Movius, H. L., & Allen, J. J. B. (2005). Cardiac vagal tone, defensivenss, and motivational style. *Biological Psychology, 68*, 147–162.

Moruzzi, G., & Magoun, H. W. (1959). Brain stem reticular formation and activation of EEG. *Electroencephalography and Clinical Neurophysiology, 1*, 455–473.

Muller, J. P. (2003). *Elements of physiology (facsimile)*. London: Thoemmes Continuum.

Murphy, G. (1960). *Human Potentialities*. London: George Allen & Unwin, Ltd.

Naqvi, N. H., Rudrauf, D., Damasio, H., & Bechara, A. (2007). Damage to the insula disrupts addiction to cigarette smoking. *Science, 26*, 315, 5811, 531–534.

Nelson, J. P., McCarley, R. W., & Hobson, J. A. (1983). REM sleep burst neurons, PGO waves, and eye movement information. *Journal of Neurophysiology, 50*(4), 784–797.

New, A. S., Trestman, R. L., Mitropoulou, V., Benishay, D. S., Coccaro, E., Silverman, J., et al. (1997). Serotonergic function and self-injurious behavior in personality disorder patients. *Psychiatry Research, 69*, 17–26.

Newberg, A. B., & Iversen, J. (2003). The neural basis of the complex mental task of meditation: neurotransmitter and neurochemical considerations. *Medical Hypothesese, 61*(2), 282–291.

Newberg, A. B., Wintering, N., Waldman, M. R., Amen, D., Khalsa, D. S., Alavi, A. (2010). Erebral blood flow differences between long-term meditators and non-meditators. Consciousness and Cognition. *Consciousness and Cognition, In Press, Elsevier Inc.*

Newell, A. (1980). Physical symbol systems. *Cognitive Science, 4*, 135–183.

Newell, A., Rosenbloom, P. S., & Laird, J. E. (1989). Symbolic architecture for cognition. In M. L. Posner (Ed.), *Foundations of cognitive science*. Cambridge: MIT Press.

Nicola, K. (1941). *The natural way to draw: A working plan for art study*. Boston: Houghton Mifflin Company.

Nienhauser, W. H. (Ed.). (1994). *Ssu-ma Chien, The grand scribe's records. aVol. 1*. Bloomington: Indiana University Press.

Nobre, A. C., Coull, J. T., Maquet, C. D., Frith, C. D., Vandenberghe, R., & Mesulam, M. M. (2004). Orienting attention to locations in perceptual versus mental representations. *Journal of Cognitive Neuroscience, 16*, 363–373.

Northoff, G., & Bermpohl, F. (2004). Cortical midline structures and the self. *Trends in Cognitive Sciences, 8*(3), 102–107.

Novelly, R. A. (1992). The dept of neuropsychology to the epilepsies. *American Psychologist, 47*(9), 1126–1129.

Oakley, D. A. (2008). Hypnosis, trance and suggestion: Evidence from neuroimaging. In M. R. Nash & A. Barnier (Eds.), *Oxford handbook of hypnosis* (pp. 365–392). Oxford: Oxford University Press.

Oberman, L. M., Ramachandran, V. S., & Pineda, J. A. (2008). Modulation of mu suppression in children with autism spectrum disorders in response to familiar or unfamiliar stimuli: The mirror neuron hypothesis. *Neuropsychologia, 46*, 1558–1565.

Oberman, L. M., & Ramachandran, V. S. (2009). Reflections on the mirror neuron system: Their evolutionary functions beyond motor representation. In J. A. Pineda (Ed.), *Mirror neuron systems* (pp. 39–62). New York: Humana Press.

O'Nuallain, S. (2009). Zero power and selflessness: what meditation and conscious perception have in common. *Cognitive Sciences, 4*(2), 46–64.

Orne, M. T. (1959). The nature of hypnosis: Artifact and essence. *Journal of Abnormal and Social Psychology, 58*, 277–299.

Ownby, R. L. (1998). A computational model of alcohol dependence: Simulation of genetic differences in alcohol preference and of therapeutic strategies. In R. W. Parks, D. S. Levine, & D. L. Long (Eds.), *Fundamentals of neural network modeling* (pp. 123–140). Cambridge: The MIT Press.

Papez, J.W. (1937). A proposed mechanism of emotion. The Journal of Neuropsychiatry and Clinical Neuroscience. 1995 7, 1, 103–112.

Paquette, V., Lévesque, J., Mensour, B., Leroux, J. M., Beaudoin, G., Bourgoulin, P., et al. (2003). Change the mind and you change the brain: effects of cognitive-behavioral therapy in the neural correlates of spider phobia. *Neuroimage, 18*(2), 401–409.

Pauling, L. (1977). Crusading scientist: Transcript of broadcast of NOVA, Angier, J. Executive producer, 417.

Paulus, M. P., Rogalsky, C., Simmons, A., Feinstein, J. S., & Stein, M. B. (2003). Increased activation in the right insula during risk-taking decision making is related to harm avoidance and neuroticism. *NeuroImage, 19*, 1439–1448.

Pellionisz, A., & Llinas, R. (1982). Space-time representations in the brain. The cerebellum: Distributed processor for productive coordination. *Neuroscience, 4*, 325–348.

Petri, H. L., & Mishkin, M. (1994). Behaviorism, cognitivism, and the neuropsychology of memory. *American Scientist, 82*, 30–37.

Pineda, J. A. (Ed.). (2009). *Mirror neuron systems: The role of mirroring processes in social cognition.* New York: Humana Press.

Pineda, J. A., Moore, R., Elfenbeinand, H., & Cox, R. (2009). Hierarchically organized mirroring processes in social cognition: The functional neuroanatomy of empathy. In J. A. Pineda (Ed.), *Mirror neuron systems: The role of mirroring processes in social cognition* (pp. 135–162). New York: Humana.

Pineda, J. (2007) Neuroanatomy and physiology: Cognitive Science 107a lecture, University of California, San Diego. September–December. 10/1/07.

Pineda, J. (2007) Neuroanatomy and physiology: Cognitive Science 107a lecture, University of California, San Diego. September–December. 10/15/07.

Pitman, R. K. (1989). Post-traumatic stress disorder, hormones, and memory. *Biological Psychiatry, 26*, 221–223.

Piver, S. (2008). The surprising self-healing benefits of meditation. *Weil Lifestyle.* http://www.drweil.com/drw/u/ART02791/self-healing.

Place, U. T. (1999). Is consciousness a brain process? In W. G. Lycan (Ed.), *Mind and cognition* (pp. 14–29). Oxford: Blackwell.

Plato, (1997). *Complete works* (trans: Cooper J. M.). (Ed.) Indianapolis: Hackett Publishing Company.

Pons, T. P., Garraghty, P. E., Ommaya, A. K., Kaas, J. H., Taub, E., & Mishkin, M. (1991). Massive cortical reorganization after sensory deafferentation in adult macaques. *Science, 252*(5014), 1857–1860.

Popper, K. (1965). *Conjectures and refutations: The growth of scientific knowledge.* New York: Harper and Row.

Porges, S. W. (2011). *The Polyvagal theory: Neurophysiological foundations of emotions, attachment, communication, and self-regulation.* New York: Norton.

Porges, S. W. (1992). Vagal Tone: A physiologic marker of stress vulnerability. *Pediatrics, 90*(3), 498–504.

Porges, S. W., Doussard-Roosevelt, J. A., & Maiti, A. K. (1994). Vagal tone and the physiological regulation of emotion. *Monograph for the Society for Research in Child Development., 59*, 167–186.

Portas, C. M., Rees, G., Howseman, A. M., et al. (1998). A specific role for the thalamus in mediating the interaction attention and arousal in humans. *Journal of Neuro Science, 18*, 8979–8989.

Posner, M. L. (1978). *Chronometric explorations of mind.* Hillsdale, N. J. Erlbaum.

Posner, M. I., Rothbart, M. K., Sheese, B. F., & Tang, Y. (2007). The anterior cingulate gyrus and the mechanisms of self-regulation. *Cognition Affect Behavior Neuroscience, 7*(4), 391–395.

Posner, M. L., & Peterson, S. E. (1990). The attention system of the human brain. *Annual Review of Neuroscince, 13*, 25–42.

Posner, M. L. & Fan, J. (2004). Attention as an organ system. In J. Pomerantz (ed.), *Neurobiology of perception and communication: From synapse to society the IVth De Lange conference.* Cambridge: Cambridge University Press.

Posner, M. L., & Rothbart, M. K. (1998). Attention, self-regulation, and consciousness. *Philosophical Transactions of the Royal Society Biological Sciences, 29*(1377), 1915–1927.

Prasko, J., Horácek, J., Zálensky, R., Kopecek, M., Novák, t., Pasková, B., Skrdiantová, L, Belohlávek, O, Hörschi, C. (2004). The change of regional brain metabolism (18FDG PET) in panic disorder during the treatment with cognitive behavioral therapy or antidepressants. Neuro Endocrinol Letters *25*, 5, 340–8.

Prescott, J. W. (1979). Alienation of affection. Psychology Today. (December).

Rainville, P., Duncan, C. H., Price, D. D., & Carrier, B. (1997a). Pain affect encoded in human anterior cingulated but not somatosensory cortex. *Science, 277*, 968–971.

Rainwille, P., Hofbauer, R. K., Bushnell, M. C., Duncan, G. H., & Price, D. D. (2002). Hypnosis modulates activity in brain structures involved in the regulation of consciousness. *Journal of Cognitive Neuroscience, 14*(6), 887–901.

Rainville, P., Hofbauer, R. K., Paus, T., Duncan, G. H., Bushnell, M. C., & Price, D. D. (1999). Cerebral mechanisms of hypnotic induction and suggestion. *Journal of Cognitive Neuroscience, 11*, 110–125.

Rainville, P., Duncan, G. H., Price, D. D., Carrier, B., & Bushness, M. C. (1997b). Pain affect encoded in human anterior cingulated but not somatosensory cortex. *Science, 277*, 968–971.

Rainville, P., & Price, D. D. (2003). Hypnosis phenomenology and the neurobiology of consciousness. *International Journal of Clinical and Experimental Hypnosis, 51*, 105–129.

Ramachandran, V. S., Rogers-Ramachandran, D. C., & Stewart, M. (1992). Perceptual correlates of massive cortical reorganization. *Science, 258*(5085), 1159–1160.

Ramachandran, V.S. (1997). Brain damage and mental function: An introduction to human neurophysiology. University of California, San Diego. August 4–September 4).

Ramachandran, V. S., Blakeslee, S., & Sacks, O. (1999). *Phantoms in the brain: Probing the mysteries of the human mind.* New York: Harper Perennial.

Ramachandran, V.S. (2000). Mirror neurons and imitation learning as the driving force behind "the great leap forward" in human evolution. Retrieved August 25, 2006, from http://www.edge.org/documents/archive/edge69.html.

Raskin, S. A., Borod, J. C., & Tweedy, J. (1990). Neuropsychological aspects of Parkinson's disease. *Neuropsychology Review, 1*, 185–221.

Rauch, S. L., Shin, L. M., Whalen, P. J., & Pitman, R. K. (1998). Neuroimaging and the neuroanatomy of post-traumatic stress disorder. *CNS Spectrums, 75*, 30–41.

Raz, A., Shapiro, T., Fan, J., & Posner, M. I. (2002). Hypnotic suggestion and the modulation of Stroop interference. *Archives of General Psychiatry, 59*, 1155–1161.

Restivo, L., Ferrari, F., Passino, E., Sgobio, C., Bock, J., Oostra, B. A., et al. (2005). Enriched environment promotes behavioral and morphological recovery in a mouse model for the fragile X syndrome. *PNAS (Proceedings of the National Academy of Science), 102*(32), 11557–11562.

Reynolds, J. H., Gottlieb, J. P., & Kastner, S. (2003). Attention. In L. R. Squire, F. E. Bloom, S. K. McConnell, J. L. Roberts, N. C. Spitzer, & M. J. Zigmond (Eds.), *Fundamental neuroscience*. Amsterdam: Academic.

Richardson, M., Strange, B. A., & Dolan, R. J. (2004). Encoding of emotional memories depends on amygdala and hippocampus and their interactions. *Nature Neuroscience, 7*, 278–284.

Richter, C. (1927). Animal behavior and internal drives. *The Quarterly Review of Biology, 2*, 307–343.

Rieber, R. W., & Robinson, D. (Eds.). (2001). *Wilhelm Wundt in history*. New York: Kluwer/ Plenum.

Rizzolatti, G., Craighero, L., & Fadiga, L. (2002). The mirror system in humans. In M. I. Stamenov & V. G. (Eds). *Mirror neurons and the evolution of brain and language* (pp. 37-59). Amsterdam: John Benjamins Publishing Company.

Rizzolatti, G., & Arbib, M. A. (1998). Language within our grasp. *Trends Neuroscience, 21*, 181–194.

Roediger, H. L. & McDermott, K. B. (1993). Implicit memory in normal human subjects. *Handbook of Neuropsychology*. Vol 8.In F. Boiler & J. Gralman (Eds.), Elvesner Science Publishers. (pp. 63–122).

Ross, N. W. (1960). *The world of Zen*. New York: Vintage.

Ross, W.D. (1924; 1953). Aristotle, Metaphysics, 2 Vols. Oxford: Clarendon Press.

Rosenblatt, F. (1958). The perceptron: A probabilistic model for information storage and organization in the brain. *Psychological Review, 65*(6), 386–408.

Rossi, E. L., Erickson-Klein, R., & Rossi, K. L. (2006). *The neuroscience edition: The collected papers of Milton H. Erickson M.D.: On hypnosis, psychotherapy and rehabilitation*. Phoenix: Milton H. Erickson Foundation Press.

Rossi, E. L. (2012). *How therapists can facilitate wonder, wisdom, truth, and beauty*. Phoenix: Milton H. Erickson Foundation Press.

Rossi, E. L. (2002). *The psychobiology of gene expression: Neuroscience and neurogenesis in hypnosis and the healing arts*. New York: W. W. Norton & Co.

Rusak, B., & Zucker, I. (1979). Neural regulation of circadian rhythms. *Physiological Reviews, 59*, 449–536.

Rush, A. J., Stewart, R. S., Garver, D. L., & Waller, D. A. (1998). Neurobiological bases for psychiatric disorders. In R. N. Rosenberb & D. E. Pleasure (Eds.), *Comprehensive neurology* (2nd ed., pp. 555–603). New York: Wiley.

Sadato, N., Pascual-Leone, A., Grafman, J., Deiber, M. P., & Dold, G. (1996). Activation of the primary visual cortex by Braille reading in blind subjects. *Nature, 380*, 526–528.

Salenius, S., Schnitzler, A., Salmelin, R., Jousmaki, V., & Hari, R. (1997). Modulation of human cortical rolandic rhythms during natural sensorimotor tasks. *Neuroimage, 5*, 221–228.

Sala, M., Perez, J., Soloff, P., Ucelli di Nemi, S., Caverzase, E., Soares, J. C., et al. (2004). Stress and hippocampal abnormatlities in psychiatric disorders. *European Neuropsychopharmacology, 14*(5), 393–406.

Salthouse, T. A., Babcock, R. L., Skovronek, E., Mitchell, D. R. D., & Palmon, R. (1990). Age and experience effects in spatial visualization. *Developmental Psychology, 26*, 128–136.

Sergent, J., Ohta, S., & MacDonald, B. (1992). Functional neuroanatomy of face and object processing. *A Positron Emission Tomography Study. Brain, 115*(1), 15–36.

Saxena, S., Gorbis, e., O'Neill, J., Baker, S.K., Mandlkern, M.A., Maidment, K.M., Chang, S., Salamon, N., Brody, A.L., Schwartz, J.M. & London, E.D. (2009). Rapid effects of brief intensive cognitive-behavioral therapy on brain glucose metabolism in obsessive-compulsive disorder. *Molecular Psychiatry, 14*, 197–205.

Schacter, D. L., Wig, G. S., & Stevens, W. D. (2007). Reductions in cortical activity during priming. *Current Opinion in Neurobiology, 17*, 171–176.

Schacter, D. L. (1985). Priming of old and new knowledge in amnesic patients and normal subjects. *Annals of the New York Academy of Sciences, 444*(13), 41–53.

Schacter, D. L. (1987). Implicit memory: History and current status. *Journal of Experimental Psychology: Learning Memory and Cognition, 13*, 501–518.

Schachter, S., & Singer, J. (1962). Cognitive, social, and physiological determinants of emotional state. *Psychological Review, 69*, 379–399.

Schmolck, H., Kensinger, E. A., Corkin, S., & Squire, L. R. (2002). Semantic knowledge in patient H.M. and other patients with bilateral medial and lateral temporal lobe lesions. *Hippocampus, 12*, 520–533.

Schneider, A. M., & Tarshis, B. (1986). *An introduction to physiological psychology*. New York: Random House.

Schneider, P., Scherg, M., Dorsch, G., Specht, H. J., Gutschalk, A., & Rupp, A. (2002). Morphology of Heschl's gyrus reflects enhanced activation in the auditory cortex in musicians. *Nature Neuroscience, 5*, 688–694.

Schoenemann, P. T., Sheehan, M. J., & Glotzer, L. D. (2005). Prefrontal white matter volume is disproportionately larger in humans than in other primates. *Nature Neuroscience, 8*, 242–252.

Schore, A. N. (2005). Back to basics: Attachment, affect regulation, and the right brain: Linking developmental neuroscience to developmental neuroscience to pediatrics. *Pediatrics in Review, 26*(6), 204–217.

Schore, A. N. (2003). *Affect regulation and the repair of the self*. New York: Norton.

Schultz, W., Dayan, P., & Montague, P. R. (1997). A neural substrate of prediction and reward. *Science, 275*(5306), 1593–1599.

Schultz, W. (2001). Reward signaling by dopamine neurons. *The Neuroscientist, 7*(4), 293–302.

Scoville, W. B., & Milner, B. (1957). Loss of recent memory after bilateral hippocampal lesions. *Journal of Neurology, Neurosurgery and Psychiatry, 20*, 11–22.

Seligman, M. E. (1975). *Helplessness*. New York: W. H. Freeman.

Serizawa, K. (1989). *Tsubo: Vital points of oriental therapy*. Tokyo: Japan Publications, Inc.

Shannahoff-Khalsa, D. (2006). *Kundalini Yoga meditation: Techniques specific for psychiatric disorders, couples therapy, and personal growth*. New York: W.W. Norton & Company.

Sharot, T., Riccardi, A. M., Raio, C. M., & Phelps, E. A. (2007). Neural mechanisms mediating optimism bias. *Nature, 450*, 102–106.

Shen, K. (2003). Think globally, act locally: Local translation and synapse formation in cultured aplysia neurons. *Neuron, 49*(3), 323–325.

Sherrington, C. S. (1906). *The integrative action of the nervous system*. Cambridge: Cambridge University Press.

Sherwood, C. C., Holloway, R. L., Semendeferi, K., & Hof, P. R. (2005). Is prefrontal white matter enlargement a human evolutionary specialization? *Nature Neuroscience, 8*, 537–538.

Shore, A. (2012). *The science and art of psychotherapy*. New York: W. W. Norton & Company.

Siegel, D. (2007). *The mindful brain*. New York: W.W. Norton & Company.

Simpkins, C. A., & Simpkins, A. M. (2012). *Zen Meditation in psychotherapy: Techniques for clinical practice*. Hoboken: Wiley.

Simpkins, C. A., & Simpkins, A. M. (2011). *Meditation and yoga in psychotherapy: Techniques for clinical practice*. Hoboken: Wiley.

Simpkins, C. A., & Simpkins, A. M. (2010a). *Neuro-hypnosis: Using self-hypnosis to activate the brain for change*. New York: W. W. Norton & Company.

Simpkins, C. A., & Simpkins, A. M. (2010b). *The dao of neuroscience: Eastern and western principles for optimal therapeutic change*. New York: W. W. Norton & Company.

Simpkins, C. A., & Simpkins, A. M. (2009). *Meditation for therapists and their clients.* New York: W. W. Norton & Co.

Simpkins, C. A., & Simpkins, A. M. (2005). *Effective self hypnosis: Pathways to the unconscious.* San Diego: Radiant Dolphin Press.

Simpkins, C. A., & Simpkins, A. M. (2004). *Self-hypnosis for women.* San Diego: Radiant Dolphin Press.

Simpkins, C. A., & Simpkins, A. M. (1997). *Zen around the world: A 2500-year journey from the Buddha to you.* Boston: Charles E. Tuttle Co., Inc.

Simpkins, A., de Callafon, R., & Todorov, E. (2008). Optimal trade-off between exploration and exploitation. In W. A. Seattle, (Ed.), *2008 American Control Conference.* June 11–13.

Simpkins, A. & Todorov, E. (2009). Practical numerical methods for stochastic optimal control of biological systems in continuous time and space. To appear in IEEE April, 2009.

Singer, T., Symour, B., O'Dobery, J., Kaube, H., Dolan, R. J., & Firth, C. D. (2004). Empathy for pain involves the affective but not sensory components of pain. *Science, 303,* 1157–1162.

Skinner, J. E., & Yingling, C. (1977). Central gating mechanisms that regulate event-related potentials and behavior. In J. E. Desmedt (Ed.), *Attention, voluntary contraction and event-related cerebral potentials* (pp. 30–69). New York: Basal.

Snowdon, D. A., Greiner, L. H., Kemper, S. J., Nanayakkara, N., & Mortimer, J. A. (1999). Linguistic ability in early life and longevity: Findings from the Nun Study. In J. M. Robine, B. Forette, C. Franceschi, & M. Allard (Eds.), *The paradoxes of longevity.* Berlin: Springer.

Sofonov, M. G. (1996). Focusing on the knowable: Controller invalidation and learning. In A. S. Morse (Ed.), *Control using logic-based switching* (pp. 224–233). Berlin: Springer.

Soyen, S. (1987). *Zen for Americans.* New York: Dorset Press.

Spiegel, D. (1988). Dissociation and hypnosis in posttraumatic stress disorder. *Journal of Traumatic Stress, 1,* 17–33.

Sohlberg, M. M., Catherine, A., & Mateer, C. A. (1989). *Introduction to cognitive rehabilitation: theory and practice.* New York: Guilford Press.

Spear, L. (2010). *The behavioral neuroscience of adolescence.* New York: W. W. Norton & Company.

Spencer, D. G., & Lal, H. (1983). Effects of anticholinergic drugs on learning and memory. *Drug Development Research, 3,* 489–502.

Sperry, R. W. (1974). Lateral specialization in the surgically separated hemispheres. In F. O. Schmitt & F. G. Worden (Eds.), *Neurosciences: Third study program* (pp. 5–19). Cambridge: MIT Press.

Sperry, R. W. (1961). Cerebral organization and behavior. Science, 133–1749.

Sperry, R. W. (1945). Restoration of vision after crossing of optic nerves and after transplantation of eye. *Journal of Neurophysiology, 8,* 15–28.

Springer, S., & Deutsch, G. (1981). *Left brain, right brain.* San Francisco: W. H. Freeman.

Squire, L. R., Bloom, F. E., McConnell, S. D., Roberts, J. L., Spitzer, N. C., & Zigmond, M. J. (2003). *Fundamental Neuroscience* (2nd ed.). New York: Academic.

Squire, L. R., & Knowlton, B. J. (2000). The medial temporal lobe, the hippocampus, and the memory systems of the brain. In M. S. Gazzaniga (Ed.), *The new cognitive neurosciences* (2nd ed., pp. 765–779). Cambridge: MIT Press.

Squire, L. R., & Kandel, E. R. (2000). *Memory: From mind to molecules.* New York: Henry Holt & Company.

Squire, L. R. (1986). Mechanisms of memory. *Science, 232*(4578), 1612–1619.

Squire, L. R. (1987). *Memory and the brain.* New York: Oxford University Press.

Squires, E. J., Hunkin, N. M., & Parkin, A. J. (1996). Memory notebook training in a case of severe amnesia. *Neuropsychological Rehabilitation, 6,* 55–66.

Stahl, S. M. (2000). *Essential psychopharmacology: Neuroscientific basis and practical applications* (2nd ed.). Cambridge: Cambridge Press.

Stanford, C. B. (1999). *The hunting apes: Meat eating and the origins of human be havior.* Princeton: Princeton University Press.

Stanislavski, C. (1936, 1984). An actor prepares. New York: Theatre Arts Books.

Strafella, A. P., Ko, J. H., & Monchi, O. (2006). Therapeutic application of transcranial magnetic stimulation in Parkinson's disease: the contribution of expectation. *Clinical Neurophysiology, 31*(4), 1666–1672.

Strakowski, S. M., DelBello, M. P., Sax, K. W., Zimmerman, M. E., Shear, P. K., Hawkins, J. M., et al. (1999). Brain magnetic resonance imaging of structural abnormalities in bipolar disorder. *Archives of General Psychiatry, 56*, 254–260.

Strehl, U., Leins, U., Goth, G., Klinger, C., Hinterberger, T., & Birbaumer, N. (2007). Self-regulation of slow cortical potentials: A new treatment for children with attention deficit/hyperactivity disorder. Pediatrics. doi: 10. 1542/peds.2005-2478.

Stengel, R. F. (1994). *Optimal control and estimation*. New York: Dover Publications.

Sternberg, R. J. (1996). *Cognitive psychology*. San Diego: Harcourt Brace College Publishers.

Steward, O., & Levy, W. B. (1982). Preferential localization of polyribosomes under the base of dendritic spines in granule cells of the dentate gyrus. *Journal of Neuro Science, 2*, 284–291.

Suzuki, D. T. (1973). *Zen and Japanese culture*. Princeton: Princeton University Press.

Suzuki, S. (1979). *Zen mind, beginner mind*. New York: Weatherhill.

Swaminathan, N. (2011). *Glia—The other brain cells*. Jan-Feb: Discover Magazine.

Taylor, S. E., Gonzaga, G. C., Klein, L. C., Hu, P., Greendale, G. A., & Seeman, T. E. (2006). Relation of oxytocin to psychological responses and hypothalamic-pituitary-adrenocortical axis and activity in older women. *Psychosomatic Medicine, 68*, 236–245.

Teicher, M. H. (2001). Wounds that time won't heal: The neurobiology of child abuse. *Cerebrum 3,1*, 8–9 & 124.

Thorndike, E. L. (1931, 1977). *Human learning*. Cambridge, Massachusetts: M.I.T. Press Paperback Edition.

Thorndike, E. L. (1918). The nature, purposes, and general methods of measurements of educational products. In G. M. Whipple (Ed.), *Seventeenth yearbook of the national society for the study of education* (Vol. 2, pp. 16–24). Bloomington: Public School Publishing.

Tichener, E. (1909). *Experimental psychology of the thought processes*. New York: Macmillan.

Tomberg, C. (1999). Unconscious attention manifested in non-averaged human brain potentials for optional short-latency cognitive electrogeneses without subsequent P300. *Neuroscience Letters. 263*, 2, 26, 181–184.

Tooby, J. & Cosmides, L. (2000). Mapping the evolved functional organization of mind and brain. In Gazzaniga, (Ed). *The new cognitive neurosciences*, Second Edition. (pp. 1167–1178) Cambridge: A Bradford Book, The MIT Press.

Toropov, B., & Hansen, C. (2002). *The complete idiot's guide to Taoism*. New York: Alpha Books.

Totorov, E. (2006). Optimal control theory. In K. Doya (Ed.), *Bayesian brain*. Cambridge: MIT Press.

Tow, P. M., & Whitty, C. W. (1953). Personality changes after operations on the cingulated gyrus in man. *Journal of Neurology Neurosurgery and Psychiatry, 18*, 159–169.

Trachtenberg, J. T., Chen, B. E., Knott, G. W., & Feng, C. (2002). Long-term in vivo imaging of experience-dependent synaptic plasticity in adult cortex. *Nature, 420*, 788–794.

Tran, P. B., & Miller, R. J. (2003). Chemokine receptors: Signposts to brain development and disease. *Nature Reviews Neuroscience, 4*, 444–453.

Treisman, A. M. (1969). Strategies and models of selective attention. *Psychological Review, 76*, 282–299.

Tronic, E.Z. (2007). *The neurobehavioral and social-emotional development of infants and children*. New York: W. W. Norton & Co.

Tronic, E.Z. & Weinberg, M. (1980). Emotional regulation in infancy: Stability of regulatory behavior. Paper presented at International Conference on Infant Studies.

Tronic, E. Z., Als, H., & Brazelton, T. B. (1980). Monadic phases: A structural descriptive analysis of infant-mother face-to-face interaction. *Merrill-Palmer Quarterly of Behavior and Development, 26*, 3–24.

Tronson, N. C., & Taylor, J. R. (2007). Molecular mechanisms of memory reconsolidation. *Nature Reviews Neuroscience, 8,* 262–275.

Tulving, E. (1983). *Elements of episodic memory.* New York: Oxford University Press.

Tulving, E. (1989). Remembering and knowing the past. *American Scientist, 77,* 361–367.

Umilta, M. A., Kohler, E., Gallese, V., Fogassi, L., Fadiga, L., & Keysers, C. (2001). I know what you are doing: A neurophiysiological study. *Neuron, 31*(1), 155–165.

Unis, A. S., Cook, E. H., Vincent, J. G., Gjerde, D. K., Perry, B. D., Mason, C., et al. (1997). Platelet serotonin measures in adolescents with conduct disorder. *Biological Psychiatry, 42*(7), 553–559.

Vallée, M., Mayo, W., Dellu, F., Le Moal, M., Simon, H., & Maccari, S. (1997). Prenatal stress induces high anxiety and postnatal handling induces low anxiety in adult offsprings: Correlations with corticosterone secretion. *Journal of Neuroscience, 17*(7), 2626–2636.

van der Kolk, B. A. (1997). The psychobiology of post-traumatic stress disorder. *J Clinical Psychiatry, 585,* 16–24.

Vanderwolf, C. H. (1992). The electrocoricogram in relation to physiology and behavior: A new analysis. *Electroencephalography and Clinical Neurophysiology, 82,* 165–175.

Van Essen, D.C. (2007). Cause and effect in cortical folding. Nature Reviews Neuroscience. doi: 10.1038/nrn2008-c1.

Van Gelder, T. (1995). What might cognition be, if not computation? *Journal of Philosophy, 91,* 345–381.

Vesalius, A. (1998–2009). *On the fabric of the human body,* (trans: Richardson, W.F., Carman, J.B.). 5 vols. San Francisco and Novato: Norman Publishing.

Vythilingam, M., Anderson, E. R., Goddard, A., & Woods, S. W. (2000). Temporal lobe volume in panic disorder—A quantitative magnetic resonance imaging study. *Psychiatry Research, 99,* 75–82.

Vogt, B. A. (2005). Pain and emotion interactions in subregions of the cingulated gyrus. *Nature Reviews Neuroscience, 6,* 533–544.

Wager, T. D., Rilling, J. K., Smith, E. E., & Sokolik, A. (2004). Placebo-induced changes in fMRI in the anticipation and experience of pain. *Science, 303,* 1162–1167.

Waley, A. (1958). *The way and its power.* New York: Grove Weidenfeld.

Watson, B. (1968). *The complete works of Chuang Tzu.* New York: Columbia University Press.

Watson, J. B., & Rayner, R. (1929). Conditioned emotional reactions. *Journal of Experimental Psychology, 3,* 1–14.

Watts, A. (1957). *The way of Zen.* New York: Vintage Books.

Weiskopf, N., Veit, R., Erb, M., Mathlak, K., Grodd, W., Goebel, R., et al. (2003). Physiological self-regulation of regional brain activity using real-time functional magnetic resonance imaging (fMRI): Methodology and exemplary data. *Neuroimage, 19*(3), 577–586.

Weitzman, E. D. (1981). Sleep and its disorders. *Annual Review of Neuroscience, 4,* 381–417.

Whitehorn, J. C. (1947). Psychotherapeutic strategy. *Acta Medica Scandinavica Supplementum, 196,* 626–633.

Whitehorn, J. C. (1963). The situation part of diagnosis. *International Journal of Group Psychotherapy, 8*(3), 290–299.

Whitehorn, J. C. (1972). *Personal Interview.* Maryland, September: Baltimore. 15.

Wicker, B., Keysers, C., Plailly, J., Royet, J. P., Gallese, V., & Rizzolatti, G. (2003). Both of us disgusted in *my* insula: The common neural basis of seeing and feeling disgust. *Neuron, 40,* 655–664.

Widrow, B., & Hoff, M. E. (1960) Adaptive switching circuits. In 1960 WESCON Convention Record (pp. 96-104) Dunno [Cited on pp. 23–27].

Wilhelm, R. (1990). *Tao te ching: The book of meaning and life.* London: Arkana.

Williams, M. A., Morris, A. P., McGlone, F., Abbott, D. F., & Mattingley, J. B. (2004). Amygdala responses to fearful and happy facial expressions under conditions of binocular suppression. *Journal of Neuroscience, 24*(12), 2898–2904.

Williams, J., Hadjistavropoulos, T., & Sharpe, D. (2006). A meta-anaylsis of psychological and pharmacological treatments for Body Dysmorphic Disorder. *Behavior Research and Therapy, 44*(1), 99–111.

Williams, A. L., Selwyn, P. A., Liberti, L., Molde, S., Njike, V. Y., McCorkle, R., et al. (2005). Efficacy of frequent mantram repetition on stress, quality of life, and spiritual well-being in veterans: a pilot study. *Journal of Palliative Medicine, 5,* 939–952.

Wilshire, B. (1984). *William James: Essential writings.* Albany: The State University of New York Press.

Wilson, D. H., Reeves, A., Gazzaniga, M., & Culver, C. (1977). Cerebral commissurotomy for control of intractable seizures. *Neurology, 27,* 708–715.

Winkelman, P., Niedenthal, P. A., & Oberman, L. M. (2009). Embodied perspective on emotion-cognition interactions. In J. A. Pineda (Ed.), *Mirror neuron systems: The role of mirroring processes in social cognition* (pp. 235–260). New York: Humana.

Witkiewitz, K., Marlatt, G. A., & Walker, D. (2005). Mindfulness-based relapse prevention for alcohol and substance use disorders. *Journal of Cognitive Psychotherapy, 19*(3), 211–228.

Wohlschlager, A., & Bekkering, H. (2002). Is human imitation based on a mirror-neurone system? Some behavioural evidence. *Experimental Brain Research, 143,* 335–341.

Yapko, M. (2001). *Treating depression with hypnosis: Integrating cognitive-behavioral and strategic approaches.* New York: Brunner/Routledge.

Yapko, M. (2003). *Tranceworks: An introduction to the practice of clinical hypnosis.* New York: Routledge.

Yapko, M. (2006). *Hypnosis and treating depression: Applications in clinical practice* (3rd ed.). New York: Routledge.

Yutang, L. (1942). *The wisdom of China and India.* New York: Random House.

Zanchin, G. (1992). Considerations on "the sacred disease" by Hippocrates. *Journal of the History of the Neurosciences, 1,* 91–95.

Zeelenberg, R., Pecher, D., Shiffrin, R., & Raaijmakers, F. G. W. (2003). Semantic context effects and priming in word association. *Psychonomic Bulletin & Review, 10*(3), 653–660.

Zeig, J. K. (2006). *Confluence: The selected papers of Jeffrey K. Zeig.* Phoenix, AZ: Zeig, Tucker, & Theisen, Inc.

Zeld, D. H., & Pardo, J. V. (1997). Emotion, olfaction, and the human amygdala: Amygdala activation during aversive olfactory stimulation. Proceedings of the National Academy of Sciences, *15*, 94, 8, 4119–4124.

Zehr, J. L., Todd, B. J., Schultz, K. M., McCarthy, M. M., & Sisk, C. L. (2006). Dendrite pruning of the medial amygdala during pubertal development of the male Syrian hamster. *Journal of Neurobiology, 66,* 578–590.

Zigova, T., Snyder, E., & Sanberg, P. R. (2003). *Neural stem cells for brain and spinal cord repair.* Totowa: Humana Press.

Zillmer, E. R., Spiers, M. V., & Culbertson, W. C. (2008). *Principles of neuropsychology.* Belmont: Thomson Wadsworth.

Zola-Morgan, S. M., & Squire, L. R. (1990). The primate hippocampal formation: Evidence for a time-limited role in memory storage. *Science, 250,* 228–290.

Index

C. A. Simpkins and A. M. Simpkins, *Neuroscience for Clinicians*,
DOI: 10.1007/978-1-4614-4842-6,
© Springer Science+Business Media New York 2013